W0037164

Zukunftstechnologien für den multifunktionalen Leichtbau

Series Editor

Open Hybrid LabFactory e.V., *Open Hybrid LabFactory e. V., Wolfsburg, Niedersachsen, Germany*

Ziel der Buchreihe ist es, zentrale Zukunftsthemen und aktuelle Arbeiten aus dem Umfeld des Forschungscampus Open Hybrid LabFactory einer breiten Öffentlichkeit zugänglich zu machen. Es werden neue Denkansätze und Ergebnisse aus der Forschung zu Methoden und Technologien zur Auslegung und großserienfähigen Fertigung hybrider und multifunktionaler Strukturen vorgestellt. Insbesondere gehören neue Produktions- und Simulationsverfahren, aber auch Aspekte der Bauteilfunktionalisierung und Betrachtungen des integrierten Life-Cycle-Engineerings zu den Forschungsschwerpunkten des Forschungscampus und zum inhaltlichen Fokus dieser Buchreihe.

Die Buchreihe umfasst Publikationen aus den Bereichen des Engineerings, der Auslegung, Produktion und Prüfung materialhybrider Strukturen. Die Skalierbarkeit und zukünftige industrielle Großserienfähigkeit der Technologien und Methoden stehen im Vordergrund der Beiträge und sichern langfristige Fortschritte in der Fahrzeugentwicklung. Ebenfalls werden Ergebnisse und Berichte von Forschungsprojekten im Rahmen des durch das Bundesministerium für Bildung und Forschung geförderten Forschungscampus veröffentlicht und Proceedings von Fachtagungen und Konferenzen im Kontext der Open Hybrid LabFactory publiziert.

Die Bände dieser Reihe richten sich an Wissenschaftler aus der Material-, Produktions- und Mobilitätsforschung. Sie spricht Fachexperten der Branchen Technik, Anlangen- und Maschinenbau, Automobil & Fahrzeugbau sowie Werkstoffe & Werkstoffverarbeitung an. Der Leser profitiert von einem konsolidierten Angebot wissenschaftlicher Beiträge zur aktuellen Forschung zu hybriden und multifunktionalen Strukturen.

This book series presents key future topics and current work from the Open Hybrid LabFactory research campus funded by the Federal Ministry of Education and Research (BMBF) to a broad public. Discussing recent approaches and research findings based on methods and technologies for the design and large-scale production of hybrid and multifunctional structures, it highlights new production and simulation processes, as well as aspects of component functionalization and integrated life-cycle engineering.

The book series comprises publications from the fields of engineering, design, production and testing of material hybrid structures. The contributions focus on the scalability and future industrial mass production capability of the technologies and methods to ensure long-term advances in vehicle development. Furthermore, the series publishes reports on and the findings of research projects within the research campus, scientific papers as well as the proceedings of conferences in the context of the Open Hybrid LabFactory.

Intended for scientists and experts from the fields of materials, production and mobility research; technology, plant and mechanical engineering; automotive & vehicle construction; and materials & materials processing, the series showcases current research on hybrid and multifunctional structures.

Klaus Dröder · Thomas Vietor

Editors

Circularity Days 2024

 Springer Vieweg

Editors
Klaus Dröder
TU Braunschweig – Institute of Machine
Tools and Production Technology
TU Braunschweig – Institut für
Werkzeugmaschinen und Fertigungstechnik
Braunschweig, Niedersachsen, Germany

Thomas Vietor
Institut für Konstruktionstechnik (IK)
TU Braunschweig
Braunschweig, Niedersachsen, Germany

ISSN 2524-4787   ISSN 2524-4795 (electronic)
Zukunftstechnologien für den multifunktionalen Leichtbau
ISBN 978-3-658-45888-1   ISBN 978-3-658-45889-8 (eBook)
https://doi.org/10.1007/978-3-658-45889-8

This Springer Vieweg imprint is published by the registered company Springer Fachmedien Wiesbaden GmbH, part of Springer Nature.
The registered company address is: Abraham-Lincoln-Str. 46, 65189 Wiesbaden, Germany

If disposing of this product, please recycle the paper.

Contents

Sustainable Materials & Surfaces

Life-Cycle Engineering for Circular Economy

AI-based Optimisation

Circular Product Design

Life-Cycle Assessment

Innovative and Smart Production

Concept for Direct Recycling of Battery Electrodes and Recovery of Active Material

Florian Denk[1(✉)], Patrick Wiechers[2], Lukas Lödige[3], Sebastian Schabel[1],
Marco Gleiß[2], Philip Scharfer[3], Wilhelm Schabel[3], and Jürgen Fleischer[1]

[1] Karlsruhe Institute of Technology - KIT, wbk Institute of Production Science, Kaiserstraße 12, 76131 Karlsruhe, Germany
{florian.denk,sebastian.schabel,juergen.fleischer}@kit.edu
[2] Karlsruhe Institute of Technology - KIT, Institute of Mechanical Process Engineering and Mechanics (MVM), Kaiserstraße 12, 76131 Karlsruhe, Germany
{patrick.wiechers,marco.gleiss}@kit.edu
[3] Karlsruhe Institute of Technology - KIT, Thin Film Technology (TFT), Kaiserstraße 12, 76131 Karlsruhe, Germany
{lukas.loedige,philip.scharfer,wilhelm.schabel}@kit.edu
https://www.wbk.kit.edu, https://www.mvm.kit.edu,
https://www.tft.kit.edu

Abstract. To achieve the goal of sustainable and competitive production of battery electrodes, the integration of recycling approaches into the product life cycle is becoming increasingly important. In addition to the fixed recycling quotas for end-of-life batteries, which are legally stipulated for the future, production scrap also plays a decisive role.

Production scrap is a result of quality requirements, which are directly correlated to production parameters. As a result, up to 40% of the material used in battery cell production is scrap, which also needs to be recycled to close the value chain. Currently, there are two established processes for recycling the functional materials of lithium ion batteries (LIBs), namely pyrometallurgical and hydrometallurgical recycling. An alternative process, which is not yet established due to the complexity of the physical and chemical relationships, is direct battery recycling. This function-preserving recycling enables a direct return to the LIBs value chain while simultaneously processing the materials in an environmentally friendly and resource-saving manner. Therefore, this paper presents technological approaches in the form of a process chain for the direct recycling of active materials (lithium iron phosphate and graphite) in order to significantly improve the battery sustainability. Anode layers with defined structural properties are treated in the process chain consisting of decoating, dispersing and thickening.

Initial tests have looked at two different decoating methods. The mechanical removal of anode active material from the collector foil using brushes and removal in a high-power ultrasonic bath are demonstrated. Furthermore, the process for the preparation of the recyclate is presented. This involves dispersing the decoated material to break up large agglomerates. Finally, the anode suspension is thickened so that the concentrate has the same solids concentration as a newly produced paste.

Keywords: Direct recycling · Battery electrodes · Scrap · Active material

© The Author(s) 2025
K. Dröder and T. Vietor (Eds.): CD 2024, Proceedings, pp. 3–16, 2025.
https://doi.org/10.1007/978-3-658-45889-8_1

1 Introduction

Closing value chains through recycling and reuse is essential to ensure a stable supply of raw materials while demand for environmentally friendly products grows in order to achieve the goal of a high-performance ecosystem. A sustainable circular economy is the goal of important strategies of the European Union (EU). The EU directive on batteries and waste batteries EU No. 2019/1020 requires recycling rates of 95% for cobalt (Co), copper (Cu) and nickel (Ni) and 70% for lithium (Li) in 2030 [1]. The initiative "Resource-efficient battery cycles - driving electromobility with a circular economy" even aims to achieve a recycling rate of 90% for Co, Cu and Ni and 85% for Li [2]. These high recycling rates are essential, as the market for LIBs with its steadily increasing number of applications in automotive, e-bikes, crafts, DIY, mobile phones and notebooks is likely to lead to a shortage of raw materials. The targeted interaction of product development, production and recycling along complex value chains is essential in order to meet the growing demand for raw materials for battery cell production in a sustainable way. The focus of recycling is particularly on the efficient recovery of the cathode active material and the current collectors. However, the recovery of the anode active material graphite is also becoming increasingly important. Although graphite is significantly cheaper than cathode active materials, there is a strong dependency on China as the world market leader in graphite production [3], which has been made even more apparent by the recent export regulations. Furthermore, production facilities for LIBs are usually not operated at their optimum operating point, as production capacity needs to be increased to meet the growing demand. Production scrap is caused by quality requirements that are directly related to production parameters or material properties.

This can result in an overall scrap in battery cell production of up to 40% [4], which must also be subjected to a recycling step in order to close the value chain. There are currently two established processes for recycling the functional materials of LIBs: pyrometallurgical and hydrometallurgical recycling. In the pyrometallurgical recycling process the lithium ion batterie cells are thermally decomposed at high temperatures, recovering the metallic elements as melt and slag. In addition to the high energy input, the low yield of the process is a disadvantage, as graphite and conductive carbon black are completely lost in the harmful exhaust gas stream. The principle of hydrometallurgical recycling is based on the leaching of lithium and critical elements such as nickel and cobalt in their elemental form, followed by gradual selective precipitation and filtration. However, the high use of chemicals and the technical complexity of the equipment contradict a sustainable and environmentally friendly value chain. An alternative to the processes described, but one that has not yet been established due to the complexity of the physical and chemical relationships, is direct battery recycling. It enables material-friendly reprocessing of the cathode and anode active materials at the particulate level. In the case of function-preserving recycling, a direct return to the value chain of the LIBs is possible, while at the same time processing the materials in an environmentally friendly and resource-saving manner [5]. Due to the high level of complexity, purely mechanical recycling of the black mass has not yet been implemented on a technical scale. Initial tests on the fractionation of a mixture of lithium iron phosphate (LFP) and carbon black show a yield of approximately 97% in the processing of cathode material and confirm the promising concept of direct recycling [6]. The aim of the concept described in this paper is to develop technological approaches in the form of a process chain for the direct recycling of active materials (LFP and graphite) in order to significantly improve the battery ecosystem. The starting point are coated and dried electrode sheets, which are processed in the process chain consisting of decoating, dispersing, centrifuging and filtering. For the proof of function of the process routes under consideration, the focus of this paper is on the recovery of anode material, as shown schematically in Fig. 1.

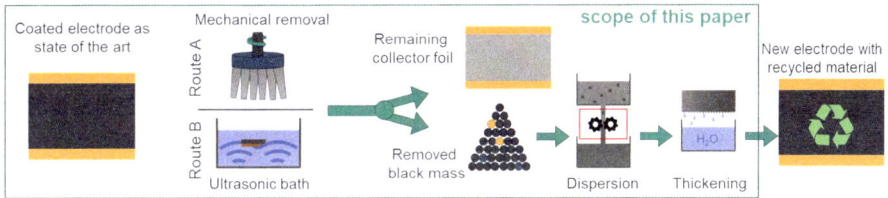

Fig. 1. Schematic flow of the direct recycling process routes with mechanical removal (A) or ultrasonic bath (B).

The electrodes under consideration, typically from the first stages of production, are single-variety electrodes that have not yet been in contact with electrolyte. For this purpose, corresponding electrodes are produced and characterized in a defined manner, which ensures comparable and reproducible results. The focus of this paper is on two process routes for decoating. On the one hand, mechanical removal using brushes is considered (Fig, 1-A), which is intended to achieve high productivity and low solvent consumption in continuous webs. Alternatively, the active material is removed from the collector foil in an ultrasonic bath filled with an electrode-specific solvent (Fig, 1-B). This route promises easy handling and the processability of already shredded or damaged material. Subsequent dispersing and thickening is considered for both process routes.

2 Impact of Material and Production on Recycling

When processing the material in the two process routes under consideration, the delamination behavior of the electrodes is of particular interest for the success of the decoating step. The mechanical integrity of electrodes relies on the cohesion force between the particles and particle adhesion force to the current collector foil. High adhesion force between the active layer and the current-collector foil is a decisive quality criterion in electrode production, and particularly determines the delamination behavior.

In graphite anodes, styrene-butadiene rubber (SBR) usually acts as adhesion agent, while in LFP cathodes, polyvinylidene fluoride (PVDF) is mostly used. The adhesion force is closely related to the selection, amount and local distribution of the binder in the coating [7, 8], but is also significantly influenced by the formation of the network in the dried electrode. This in turn depends on the surface properties, including particle morphology, of the active materials and conductive additive, as well as the processing of the electrode. During processing, these are mainly the drying and the calendering step. Drying conditions substantially affect the adhesion force, with increasing drying rates often leading to a decrease in adhesion [8–11]. This phenomenon is attributed to the migration of the mobile binder within the interparticle pore network, resulting in its accumulation at the top of the electrode and depletion at the current collector [8–14]. In contrast, the structural altering caused by calendering, in which the porous structure of the electrode film is compacted from the initial dry film height to the final porosity, leads to an increase in the adhesion force [8, 15].

Hierarchically structured multilayer electrodes (example of a cathode shown in Fig. 2) with different binder proportions and/or particle morphologies in the individual layers and special pre-coatings with very thin primer layers are effective methods for an even more targeted increase/ adjustment of the adhesion force of electrodes [15, 16]. These approaches hold the potential to establish robust mechanical integrity in the electrodes throughout the battery's operational lifetime, concurrently facilitating the decoating under stress during the recycling process.

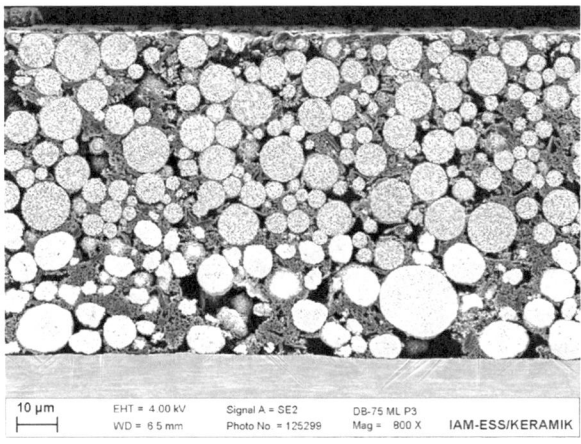

Fig. 2. Two-layer cathode with hierarchic structure of particle morphologies and binder proportions: the bottom layer contains porous and the top layer compact active material particles. To compensate for possible binder migration, the bottom layer is produced with increased binder content. [14]

3   Methods and Results

The aim of the approaches presented is the direct recycling of the active material from battery electrodes and the subsequent use of the recyclate for the coating of new electrodes. The two approaches, removal by brush or an ultrasonic bath, target different forms of scrap. In addition to the form of the starting material, for example shredded or continuous coil, different solvent requirements are expected.

3.1   Electrode Production and Characterization

The water-based anode with the two binders carboxymethyl cellulose (CMC) and SBR with relevant mechanical properties is produced from a slurry of the composition shown in Table 1. The use of the same composition and solids content as in previous works [13] provides an indication of the expected adhesive force. The anode slurry is applied to the copper conductor foils in a continuous slot die coating system. Subsequently, the wet film is dried in a convective dryer with an air inlet temperature of 120 °C at an average heat transfer coefficient of 40 W m^{-2} K^{-1}. In the presented paper only uncalendered anodes were considered.

Table 1 Composition of the anode slurry.

Graphite / wt%	Carbon Black / wt%	CMC / wt%	SBR / wt%	Solids / wt%
93.00	1.40	1.87	3.73	43.00

The electrodes are then characterized by measuring the adhesion force between the substrate and the dried coating using 90° peel tests in an AMETEK LS1 (Lloyd Instruments Ltd., UK) universal testing machine and a 10 N load cell. Sample strips with a width of 30 mm were cut out of the anodes and attached to an adhesive strip with the coated side. The current collector foil was then peeled off the coating at a constant speed of 100 mm min^{-1} at a 90° angle by the testing machine. The films delaminate at the interface between the graphite layer and the substrate with only small amounts of the film sticking to the substrate. For better comparability with literature values the measured pull-off force was divided by the sample width to obtain the line adhesion force. In addition, the dry film thickness is determined. Table 2 shows the measured values.

Table 2 Properties of the uncalendered anode with two-dimensional scanning with 3 cm spacing at 20 dots for the dry film thickness and three repetitions for adhesion force and area weigh.

Dry film thickness / μm	Adhesion force / N·m^{-1}	Area weight / g·m^{-2}
105.1 ± 3.1	10.4 ± 0.1	92 ± 1.9

3.2 Mechanical Delamination of the Active Material

The mechanical removal of the active material using brushes offers various advantages in the context of battery recycling. This process allows mechanical force to be applied to the active material, reducing the use of solvents. In order to reduce an additional thickening step to a minimum, a solvent content lower or equal to the coating process is aimed for. Sword brushes with a rotating belt enable the processing of continuous webs in a similar way to cleaning or deburring systems. The question of how a similar setup can be used for the decoating of battery electrodes was addressed by carrying out manual tests on the removal behavior of various brushes. The aim of these tests was to check the suitability of different brushes and to demonstrate the general feasibility of this approach. The tests were carried out on uncalendered anodes, with wet and dry processing. For wet processing water was used. The test setup consists of a clamping device which holds the electrode in place (see Fig. 3 - (1)). The accessible area is 40 · 160 mm^2, which contains an active material of 0.589 g for the anodes used. Three different state of the art brushes for cleaning and abrasion were used for the tests. In a first screening these have each shown the best delamination results with either dry or wet processing. The materials and dimensions of the filaments are shown in Table 3. A precision scale was used to determine the weight of the electrode piece and the material removed.

Table 3 Material and dimensions of the brush filament with decoating results (three repetitions each).

Brush Nr.	filament diameter	filament length	filament material	processing	material removal (% of coated material)	material collected
B1	0.5 mm	25 mm	polyamide 6	dry	0.532 g (90%)	0.442 g
B2	0.2 mm	25 mm	brass	dry	0.565 g (96%)	0.530 g
B3	0.3 mm	10 mm	polyamide 6	wet	0.579 g (98%)	-

The active material was successfully detached from the collector foil in all of the experiments carried out. The following effects in particular were observed. As can be seen in Fig. 3 - (2), partially wrinkles form in the edge areas of the coating. These wrinkles can initially be attributed to the drying process. Nevertheless, the brushing movement intensifies them and can lead to the film tearing if the applied pressure is too high. In general, good removal was achieved in the tests with all brushes and the recyclate could be collected as powder after dry processing (Fig. 3 - (5)). Especially promising are the high removal rates of 90% (B1 with dry processing) up to 98% (B3 with wet processing) as well as the rates of the removed material recovered. For B1 an average of 83% of the removed material could be collected with a maximum of 94% for one repetition. This fluctuation between the repetitions can be attributed to residues in the brush used and should reach a constant high level for a continuous operation. For the manual tests with wet processing the recovery of the material was only possible partially, as a larger amount of the material remains in the brush. In a continuous process after reaching a saturation of the residues in the brush it is expected to achieve similar rates as shown in the dry processing.

When the tests were carried out manually, a lower contact pressure was observed for B2. Nevertheless, more significant scratches appeared in the film. The optically visible removal result was judged to be comparable for B1 and B2. Average values over three repetitions with each brush for the material removed from the electrode and the material collected are shown in Table 3. Further, a clear gray residue of the active material on the foil and grooves in the direction of processing can be observed. The film is therefore scratched by the treatment. Figure 3 - (3) shows a treated area of the electrode with and without coating. In contrast, hardly any visible residue of the active material remains after wet processing. For this, the brush is immersed in water beforehand in order to achieve only slight moistening. Since it is expected that the moistening would lead to weakening of the mechanical integrity of the electrode, a finer brush is used (B3). This is suitable for gentle cleaning of sensitive surfaces and therefore does not scratch the film. Dry processing is not possible with this brush, this only produces a polishing effect on the surface and no recognizable removal.

Fig. 3 (1) Anode clamped on sample carrier; (2) Waviness in the edge area of the coating; (3) Dry removal with scratches in the conductor foil; (4) Electrode after wet removal; (5) Recyclate after dry removal.

In addition to efficient removal, the avoidance of impurities is essential for the successful recovery of recyclate. Due to the high purity requirements of the battery industry, even small amounts of foreign material are critical. As the tests with dry processing (B1 and B2) partially produced visible scratches in the collector foil, the result was also examined using SEM imaging and EDX analyses. It was found that no copper residues could be detected for either B1 or B3. In the case of B2, up to 1.3 wt% copper was detected. Figure 4 clearly shows the traces of the brass brush (B2) on both the uncoated and coated area. Decoating is therefore possible with all of the brushes considered, but due to the copper residues it cannot be classified as suitable with B2.

Fig. 4. SEM-Analysis of an electrode after processing with B2. Areas can be divided in coated (right half) and brushed (upper half).

3.3   Ultrasonic Delamination

In addition to the option of using mechanical brushes to remove the coatings from the collector foil, there is also the ultrasonic decoating process. While previous work in this field mainly investigated the delamination of the coating in pure solvent, the ultrasonic decoating process presented in the paper is also intended to function as a thickener. After the ultrasonic treatment process, the anode suspension should therefore already have a high solid mass concentration in order to simplify the subsequent filtration step. To investigate the delamination characteristics, defined anode-samples are placed in an aqueous ultrasonic bath, where the coating can be dissolved using ultrasonic waves. Initial tests show that this method can be used to remove the coating without damaging the foil or the coating material. The copper foils emerge from the process step as valuable material, whereas the coating materials are present in particulate form in an aqueous solution. This makes the aqueous suspension directly suitable for further processing in the subsequent filtration step. In addition to the influence of the decoating on the particles of the anode paste, the solid mass concentrations that can be achieved in the ultrasonic bath are of particular interest. This raises the question of the critical solid mass concentration in the ultrasonic bath up to which there is still a high level of decoating efficiency. If the aim is to return the recycled paste directly to lithium-ion battery production, a mass concentration that corresponds to the mass concentration of newly produced anode paste should be aimed for. If high mass concentrations can already be achieved in the ultrasonic decoating process, less water has to be removed in a downstream filtration process at great expense. Figure 5 shows a schematic representation of ultrasonic decoating. On the left-hand side, a coated anode foil is shown before the decoating process. The right part of the illustration shows a picture of the same foil after ultrasonic decoating. The good cleaning result is clearly visible. No conductive carbon black or graphite residues can be seen on the foil after the cleaning process. The film shown was decoated in particle-free, fully demineralized water.

Fig. 5. Schematic process of the decoating in an ultrasonic bath.

Initial tests show that the decoating results of identical, coated anode foils depend significantly on the treatment time, frequency and solid mass concentration in the ultrasonic bath. The decoating efficiency of the copper foils can be determined gravimetrically. For this purpose, coated anode foils were produced which had a reproducibly equal area of copper. The coating mass can be determined by weighing the foils and subtracting the copper foil mass. After the decoating process, drying took place in a drying oven at 50 °C for at least 4 h. The mass of the residual coating can then be determined by reweighing. The decoating efficiency ε is then calculated from Eq. (1) as follows:

$$\varepsilon = \frac{m_{coating\ before\ treatment} - m_{coating\ after\ treatment}}{m_{coating\ before\ treatment}} \tag{1}$$

Assuming that no copper foil comes off or solves, a decoating efficiency of 100% means that the entire mass of the coating was removed during decoating. A decoating efficiency of 0%, on the other hand, means that no coating was removed at all. In the course of decoating tests, an average decoating efficiency of (98.6 ± 0.7) % was achieved across solid mass concentrations from 0% to 21% in the ultrasonic bath at a frequency of 40 kHz, an ultrasonic power of 200 W and a decoating time of 10 min. Good results are also expected for higher solid content which will be continued with the aim of direct reuse in production. Potential is also seen in varying the frequency. These investigations suggest that similarly high degrees of decoating can be achieved at optimized frequencies in significantly less time.

3.4 Determination of the Particle Size Distribution of the Recyclate

In order to be able to recover the anode materials graphite and conductive carbon black in a holistic and function-preserving manner as part of mechanical direct recycling, the particle sizes and properties of both materials must not be affected by the recycling process. Only then the materials can be used directly as starting products for new LIBs. In order to be able to assess the particle quality of graphite and conductive carbon black, a particle size distribution of both materials can be determined after the recycling step. Subsequent comparison with the particle size distributions of the materials before recycling makes it possible to assess the influence of recycling on the particles. As an ideal separation of the graphite and carbon black materials for an analytical determination of the separate particle size distributions is not possible, a method for the qualitative and quantitative analysis of anode pastes has been developed. Using a spatially and time-resolved extinction measurement in the LUMiSizer® measuring device, qualitative and quantitative statements can be made about the carbon black and graphite particles in anode pastes using an optimized analysis method. More detailed information on the exact application of the method, in particular the sample preparation and the analyzer used, is described by Yildiz et al. [17]. Figure 6 shows three particle size distributions created using the measurement method. The blue curve shows the equivalent particle size distribution of an anode paste without graphite, and the red one a particle size distribution of a paste without carbon black. An anode paste containing both substances produces a curve like the black one. The height and position of the saddle point depends on the ratio of conductive carbon black and graphite. Further reference measurements and explanations of the distributions can also be found in Yildiz et al. [17].

The measurement method with the LUMiSizer® as presented here can be used to assess the influence of decoating with brushes, the ultrasonic bath and of the thickening process in the cross-flow filter on the particle size distribution of the anode slurry. It therefore is seen as a useful tool for investigating and evaluatingprocesses for direct recycling of battery electrodes.

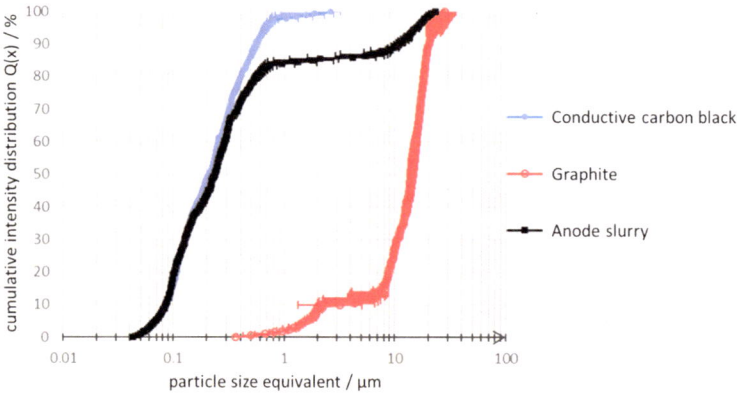

Fig. 6. Equivalent particle size distribution of conductive carbon black (blue), graphite (red) and an anode slurry (black) before processing, determined using the described analytical method. [17]

3.5 Thickening of Anode Slurries in a Cross-Flow Filtration Plant

As already shown, the coating of an anode foil can be recovered to a high degree with brushes or in an ultrasonic bath. After decoating with a solvent, it is possible to thicken the suspension in a cross-flow filter to further reduce the fluid content. This should make the recycled paste directly usable for the production of new lithium-ion batteries. Cross-flow filters are particularly promising for demanding pastes. The filter resistance can be reduced by applying high shear forces, which increases the filter speed compared to other processes. In order to investigate the possibilities for recycling production waste from lithium-ion battery production, a system with a dynamic shear-gap cross-flow filter was set up (see Fig. 7). The cross-flow in the filtration chamber is generated by a rotor driven by a motor. The suspension, on the other hand, is conveyed by a peristaltic pump. This allows the degree of cross-flow and the flow rate to be adjusted independently of each other. The process flowchart is shown in Fig. 7.

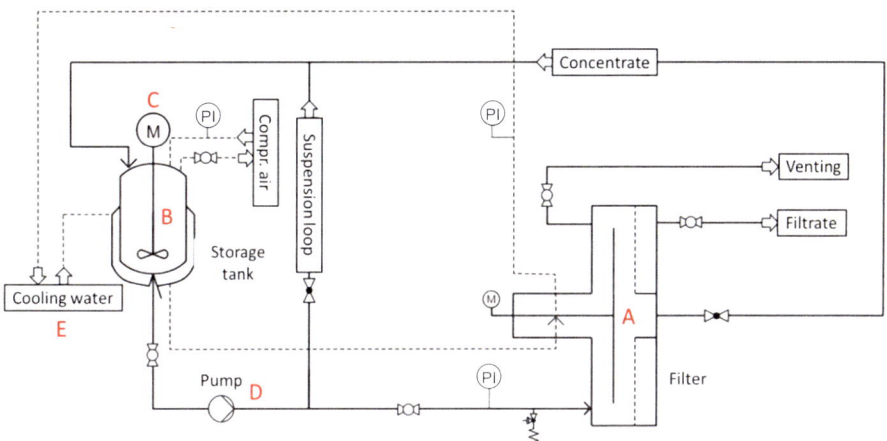

Fig. 7. Filter system with filter chamber and motor (A), storage tank (B), storage tank stirrer (C), peristaltic suspension pump (D) and cooling water periphery (E).

Initial tests have shown that the filtration system can be used to thicken aqueous anode suspensions to the initial solids content of 43 wt%. This corresponds to the mass concentration of anode pastes commonly used in industry as shown in Sect. 3.1. The filter curve of a test at varying rotor speed is shown as an example in Fig. 8. The filtration took place with a 1 μm membrane and an overpressure of 1 bar. The membrane therefore has a pore size through which conductive carbon black can pass. The primary particles and aggregates of conductive carbon black are smaller than the pore size of the membrane. It has been found that the effects of blocking pores and the formation of a covering layer lead to the retention of such particles during filtration. The concentrate was pumped through the filter chamber and then mixed again with the suspension in the receiver tank. An agitator in the receiver tank ensures sufficient mixing. The filtration test was started at a solids mass concentration of 17%. After an initial cloudiness of soot, the filtrate clarified due to the resulting covering layer and the blocking of pores. As expected, the filtrate curve (black) flattens out in the course of thickening. This is due to the filter resistances that build up. Similarly, the flux is reduced from an initial 85 L per hour and square meter of membrane surface $(L \cdot h^{-1} \cdot m^{-2})$ to 32 $L \cdot h^{-1} \cdot m^{-2}$ at a mass concentration of 36%. After this point, the rotor speed was increased during the investigation, which increased the shear forces in the filter chamber.

This leads to a reduction in the filter resistance and the flux increases again. However, this is accompanied by the penetration of carbon black particles, as the increased shear forces reduce the filtering surface layer and can also unblock pores. The test was terminated after a mass concentration of 45% was reached.

The experiment demonstrates the possibility of filtering aqueous black mass. The anode materials graphite and conductive carbon black were successfully filtered. The binder additives CMC and SBR are not the focus of the investigations shown. However, ongoing tests with finer membranes show that it is also possible to retain or pass these substances in the concentrate. Insights from investigations of the filtrates will contribute to the understanding and design of the process. The extent to which the individual process

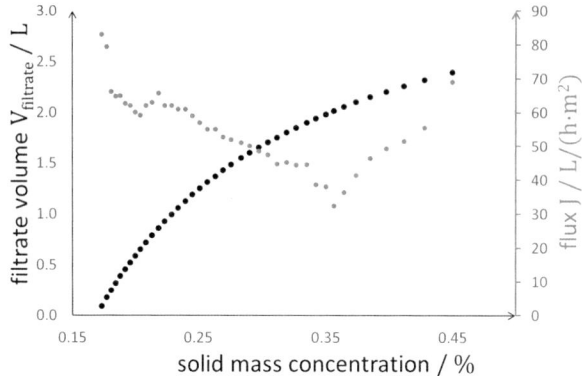

Fig. 8. Filtrate and flux curve over the solid mass concentration of the thickening concentrate.

steps affect the properties of the coating and the battery cell is also the subject of current research.

4   Conclusion and Outlook

It has been successfully demonstrated that electrode material can be recovered from production scrap using suitable brushes or an ultrasonic bath. According to the results of the EDX analysis, both dry operation with a dry polyamide 6 (PA6) brush and wet operation with a fine PA6 brush are possible. Despite the good decoating result, a brush with brass filament is not suitable due to the copper abrasion. In addition, a design with PA6 brushes in wet operation (B3) should be pursued, particularly because of the high removal rates and the low force requirement. The lower force requirement in particular promises more robust operation in the event of minor damage to the electrode, such as more pronounced wrinkles of any kind or cracks in the substrate. A semi-automated set-up will provide a more stable test to be carried out and allow continuous operation with a sword brush to be simulated close to the application. When recovering the material in the ultrasonic bath, it has been shown that a high decoating efficiency can be achieved even with solids content up to 21%. With this process route the high decoating efficiency of (98.6 ± 0.7) % after 10 min is especially promising for scrap parts or residues from the singularization process. Ongoing tests show, that similar high decoating efficiencies can be achieved at optimized frequencies in significantly less time. Due to the achieved mass concentrations of 21% in the ultrasonic bath, an additional thickening step is required to attain a solid content of 43% for coating. Therefor the presented dynamic shear-gap filtration plant provides a suitable and energy efficient method for further thickening the suspension. An important aspect is to consider the effects of the recyclate on cell performance and stability compared to conventional material, as this is the only way to conclusively evaluate the recycling process.

The optimization of the slurry composition and electrode processing with a focus on enhancing recyclability will be subject of future research. In this context, the interdependencies between input material, recycling process, output material and eventually cell quality have to be considered. It is to be investigated to what extent approaches such as hierarchically structured multilayers and thin primer layers can be used to improve the mechanical properties within the electrode coating with respect to delamination in the recycling process in the sense of design for recycling. Furthermore, it is essential to assess the content, distribution and structure of the binder in the recyclate, the evaluation of potential material degradation and the determination of optimum admixture proportions in slurry production. Answering these questions is crucial for the successful integration of sustainable processes into industrial processes. Parallel to the increasing process understanding also the further development of the realization of these concepts on an industrial scale should be further advanced and investigated to work towards rapid implementation.

Acknowledgements. The authors would like to express their appreciation to all research partners and the Ministry of Economic Affairs, Labor and Tourism Baden-Wuerttemberg for supporting the DiRecFM project. This work contributes to the research performed at KIT-BATEC (KIT Battery Technology Center) and at CELEST (Center for Electrochemical Energy Storage Ulm Karlsruhe).

References

1. "Proposal for a Regulation of the European Parliament and of the Council Concerning Batteries and Waste Batteries of the Council Concerning Batteries and Waste Batteries," in *Repealing Directive 2006/66/EC*, 2020.
2. A. Kwade *et al.*, "Ressourcenschonende Batteriekreisläufe – mit Circular Economy die Elektromobilität antreiben," 2020, https://doi.org/10.48669/ceid_2021-1.
3. P. Dolega, J. Betz, and M. Buchert, *Ökologische und sozio-ökonomische Herausforderungen in Batterie-Lieferketten: Graphit und Lithium.* Kurzstudie erstellt im Rahmen des BMBF-Verbundprojektes Fab4Lib.
4. J. Wessel, A. Turetskyy, F. Cerdas, and C. Herrmann, "Integrated Material-Energy-Quality Assessment for Lithium-ion Battery Cell Manufacturing," *Procedia CIRP*, vol. 98, pp. 388–393, 2021, https://doi.org/10.1016/j.procir.2021.01.122.
5. S. Doose, J. K. Mayer, P. Michalowski, and A. Kwade, "Challenges in Ecofriendly Battery Recycling and Closed Material Cycles: A Perspective on Future Lithium Battery Generations," *Metals*, vol. 11, no. 2, p. 291, 2021, https://doi.org/10.3390/met11020291.
6. A. Wolf *et al.*, "Centrifugation based separation of lithium iron phosphate (LFP) and carbon black for lithium-ion battery recycling," *Chemical Engineering and Processing - Process Intensification*, vol. 160, p. 108310, 2021, https://doi.org/10.1016/j.cep.2021.108310.
7. B. Westphal, H. Bockholt, T. Günther, W. Haselrieder, and A. Kwade, "Influence of Convective Drying Parameters on Electrode Performance and Physical Electrode Properties," *ECS Trans.*, vol. 64, no. 22, pp. 57–68, 2015, https://doi.org/10.1149/06422.0057ecst.
8. M. Baunach, S. Jaiser, S. Schmelzle, H. Nirschl, P. Scharfer, and W. Schabel, "Delamination behavior of lithium-ion battery anodes: Influence of drying temperature during electrode processing," *Drying Technology*, vol. 34, no. 4, pp. 462–473, 2016, https://doi.org/10.1080/07373937.2015.1060497.
9. S. Lim, K. H. Ahn, and M. Yamamura, "Latex migration in battery slurries during drying," *Langmuir : the ACS journal of surfaces and colloids*, early access. https://doi.org/10.1021/la4013685.

10. S. Jaiser *et al.*, "Microstructure formation of lithium-ion battery electrodes during drying – An ex-situ study using cryogenic broad ion beam slope-cutting and scanning electron microscopy (Cryo-BIB-SEM)," *Journal of Power Sources*, vol. 345, pp. 97–107, 2017, https://doi.org/10.1016/j.jpowsour.2017.01.117.

11. M. Müller *et al.*, "Investigation of binder distribution in graphite anodes for lithium-ion batteries," *Journal of Power Sources*, vol. 340, pp. 1–5, 2017, https://doi.org/10.1016/j.jpowsour.2016.11.051.

12. B. G. Westphal and A. Kwade, "Critical electrode properties and drying conditions causing component segregation in graphitic anodes for lithium-ion batteries," *Journal of Energy Storage*, vol. 18, pp. 509–517, 2018, https://doi.org/10.1016/j.est.2018.06.009.

13. J. Kumberg *et al.*, "Drying of Lithium-Ion Battery Anodes for Use in High-Energy Cells: Influence of Electrode Thickness on Drying Time, Adhesion, and Crack Formation," *Energy Technol.*, vol. 7, no. 11, p. 1900722, 2019, https://doi.org/10.1002/ente.201900722.

14. J. Klemens *et al.*, "Drying of Compact and Porous NCM Cathode Electrodes in Different Multilayer Architectures: Influence of Layer Configuration and Drying Rate on Electrode Properties," *Energy Tech*, vol. 11, no. 8, 2023, Art. no. 2300267, https://doi.org/10.1002/ente.202300267.

15. R. Diehm *et al.*, "High-Speed Coating of Primer Layer for Li-Ion Battery Electrodes by Using Slot-Die Coating," *Energy Technol.*, vol. 8, no. 9, p. 2000259, 2020, https://doi.org/10.1002/ente.202000259.

16. R. Diehm *et al.*, "In-situ Investigations of Simultaneous Two-Layer Slot Die Coating of Component Graded Anodes for improved High Energy Li-Ion Batteries," *Energy Technol.*, 2020, https://doi.org/10.1002/ente.201901251.

17. T. Yildiz, P. Wiechers, H. Nirschl, and M. Gleiß, "Direct recycling of carbon black and graphite from an aqueous anode slurry of lithium-ion batteries by centrifugal fractionation," *Next Energy*, p. 100082, 2023, https://doi.org/10.1016/j.nxener.2023.100082.

Assessing Automation Opportunities
in End-of-Life Vehicle Disassembly

Severin J. Görgens[1,2]([email]), Sönke Hansen[2], Patrick Schumacher[3], Kolja Meyer[1,2],
and Klaus Dröder[2]

[1] Junior Research Group Urban Flows and Production, Technische Universität Braunschweig,
Langer Kamp 19b, 38106 Braunschweig, Germany
`s.goergens@tu-braunschweig.de`
[2] Institute of Machine Tools and Production Technology, Technische Universität Braunschweig,
Langer Kamp 19b, 38106 Braunschweig, Germany
[3] Institute of Automotive Management and Industrial Production, Technische Universität
Braunschweig, Mühlenpfordtstraße 23, 38106 Braunschweig, Germany

Abstract. The lack of circularity in the automotive industry leads to the loss
of valuable and important raw materials. Resulting economic dependencies in
procurement and environmental burdens could be reduced by implementing a
strategy to systematically retrieve secondary raw materials and components for
vehicle production. In this context, in 2023 the EU proposed a new regulation for
handling End-of-Life Vehicles (ELVs) to support circular economy practices in
the automotive industry. Semi-automated processes offer efficiency and scalability
for the handling of ELVs. This results in the need to analyze the feasibility of
automation steps in order to make decisions for process development.

For this purpose, a systematic approach to identify automation potentials for
automotive disassembly and to embed associated processes into a larger economic
assessment is developed. In order to achieve this, the complexity and suitability
of the automated processes is evaluated, based on relevant automation criteria and
theoretical approaches from literature. In combination with the economic assessment of the alternative manual route, a comparison of both variants is performed
that delivers insights into the conditions under which automation is possible and
sensible as well as the circumstances under which manual process control is more
appropriate. The application of our evaluation methodology allows for a detailed
assessment of the consequences of establishing automated disassembly and thus
helps in the development of component-specific disassembly process decisions.
The presented case study exemplarily demonstrates the methodological approach
for the evaluation of economic effects of automation technologies. It is pointed
out that the quantity of processes to be performed as well as the period under
consideration have an effect on the economic differentiation between manual and
automated disassembly.

Keywords: Automation potential · Automated disassembly · Economic
evaluation · End of Life Vehicles (ELV) · ELV treatment

K. Dröder and T. Vietor (Eds.): CD 2024, Proceedings, pp. 17–31, 2025.
https://doi.org/10.1007/978-3-658-45889-8_2

1 Introduction

The nine End of Life (EoL) strategies outlined in the DIN standardization roadmap aim to increase material reclamation and reduce the environmental impact of products at their EoL stage [1]. Strategies R3 (reuse) to R7 (repurpose), focusing on an extension of the period of use of a product, require a certain degree of non-destructive dismantling in order to be applied successfully. Automation is crucial when anticipating high volumes of product returns with similar designs, as it facilitates scale effects [2]. The utilization of robots can enhance the productivity and affect the number and skillset of workers required. This is dependent on the nature of the specific process in question [3]. Automation in the disassembly of EoL products is thus far only applied to a selected few of companies that use automation to treat their own products. This is at least partially due to the fact that recycling companies lack information on the EoL product in question [4]. It often seems unclear to what extent EoL processing is actually economically, ecologically, and technically feasible to be automated as this depends on a large variety of factors [4–8]. One specific factor distinguishing the automation of disassembly from assembly processes is the uncertainty of the condition of the component and its connections [7], which hinders the implementation. It would be beneficial for these companies to find ways to quantify both, the economic and the technological feasibility of EoL automation and compare it with manual disassembly in order to allow conclusions to be drawn about optimal process design. The result might often be a mixture of automated and manual disassembly, depending on the specific task.

Overview of literature on automated EoL processing
Various studies examine the implementation and appropriateness of automation in manufacturing processes [9]. They identify financial and non-financial information to be necessary to provide an informed decision support [10]. The latter considers the design of the components, such as their weight and physical dimensions [11], whereas the economic criteria can for example include the number of parts to be processed in a given time frame. Numerous EoL automation studies focus on "Design for Disassembly," integrating automated disassembly requirements into product development. Studies such as [12] and [13] utilize quantification methods to calculate design-dependent disassembly times. However, this forward-looking approach does not address immediate challenges posed by products returning in the short term without disassembly-friendly designs. This applies particularly for usage scenarios that can span several decades, as is common in the automotive industry. Herrmann et al. applied a criteria catalogue that covers the evaluation of the technical possibilities of automated process operation and the necessity to carry out these processes automatically for the case of automotive lithium-ion battery disassembly. The catalogue covers various criteria to assess the suitability and usefulness of automating individual processes and respects the specific circumstances of disassembly (in contrast to assembly). However, this approach only includes a rough estimation of the equipment cost, thereby subordinately treating the financial evaluation [5].

Literature analysis reveals a gap in researching the automation feasibility of end-of-life (EoL) processes compared to assembly processes. Adapting manufacturing process assessment approaches to EoL processing specifics is necessary. Focusing on component design, which directly influences EoL process complexity, can enhance EoL analyses. Assessing automation feasibility and economic viability should consider this complexity.

2 Method

In the following, the developed method for the evaluation of automation possibilities of EoL processes is introduced. The method includes the assessment of part variations, that have direct influence on the process complexity and from that derives an evaluation of process-specific automation possibilities. The costs of automated and manual process implementations are compared as well, enabling financial recommendations regarding the level of automation. The developed method answers two questions:

- To what extent is automation of the considered process possible?
- To what extent is automation more beneficial than manual disassembly?

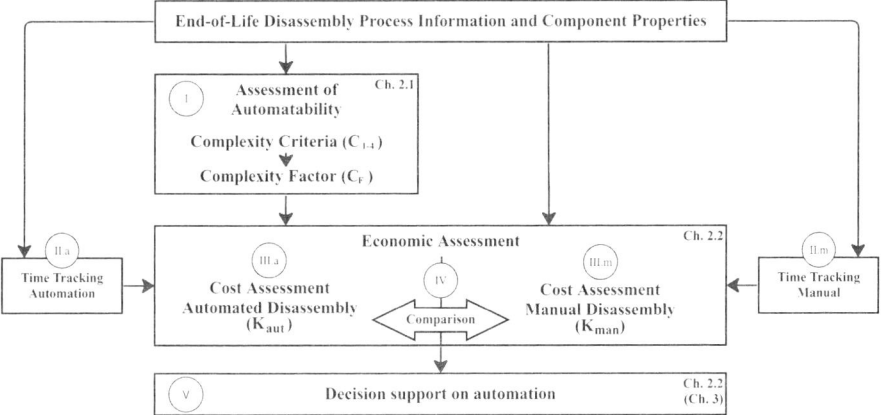

Fig. 1 Method for automation decisions in disassembly operations

Figure 1 schematically shows the logic applied to evaluate EoL processes. The method consists of 5 steps. First, the EoL disassembly process is analyzed to derive information about its complexity to evaluate the technical feasibility of automation (I). Next the time effort associated with manual (II.a) and automated (II.m) handling is evaluated that also relies on the EoL process data. The third step involves the cost assessment (III.a/III.m) of both routes that is primarily based on either duration. In the subsequent comparison (IV), the economically favorable process route is identified under the consideration of specific framework conditions. Finally, the procedure can be used as means for supporting decisions on process automation (V). This method and underlying assumptions are explained in more detail the following sections.

2.1 Assessment of Automatability of Processes

The assessment of the automatability of the considered EoL processes depends on the complexity factor, which is calculated on the basis of information on component design and disassembly processes, that are performed to remove the component.

2.1.1 End of Life Disassembly Process Information and Component-Properties

In a holistic approach the general technological and economic feasibility of an automated disassembly of vehicle parts on a level of individual processes are analyzed. Automation possibilities of processes are largely dependent on the characteristics of the targeted components [5, 13]. For a differentiated assessment of the properties of these components, various categories are considered. Based on different approaches for the evaluation of automated disassembly feasibility the proposed evaluation categories are the *variance in the characteristics of the component and its condition*, the *surfaces and their requirements* as well as *geometric and physical properties*, which include weight, size and shape as well as stiffness [4–8, 13–21]. To evaluate and quantify the processes in a comparable manner, a standardized scale based on a scoring model can be used. We propose a scale from "0"—*not feasible for automation* to "5"—*most favorable for automation*, which enables a detailed evaluation. The optimal degree of differentiation can be determined by the choice of the scale limit. For this purpose, criteria for quantification must be identified in each category or an expert assessment must be carried out. It is crucial to normalize the evaluation scale in every category to enable further calculations. The so found quantification can be utilized for further calculations of the complexity and in conclusion the economic evaluation.

2.1.2 Complexity Factors for Automation

For the assessment of the individual process automation complexity, disassembly factors are identified based on existing literature and their influence is determined. In the method proposed here, seven different evaluation criteria on process level for individual disassembly steps are considered. These criteria are used for the calculation of a complexity factor for each disassembly process.

Criterion 1 - Variance in the characteristics ($C_{1a,b}$) [4, 7, 19]: This criterion contains the *variability of the components and their positions* (C_{1a}) as well as the current *condition of the components and joining compounds* (C_{1b}). The condition is especially relevant in EoL product treatment [4]. The variances require the use of sensor technology to recognize the individual case at hand and a reacting, flexible control of the systems to adapt the processes accordingly [7]. In addition, the variety of the connection types and quantities influence the overall assessment of the automation feasibility like Sabaghi et al. indicate [19]. Within the proposed assessment method, the latter is respected indirectly. This method focusses on the assessment of individual processes. Therefore, each connection is represented separately. The respected tool changes and trajectory times add to the total sums. In this way, a high variety in this regard will be respected in the final evaluation.

To assess the variance in the characteristics and quantify it on the desired scale, a case specific system has to be installed. It can include a number of possible target item variations and/or conditions to categorize them. The assessment has to be carried out by experts.

Criterion 2 - Surface requirements (C_2) [16, 17, 20]: The surface properties of the target item have a significant influence on the automated handling. It is influencing the feasibility of different gripper technologies [16, 17, 20]. The surface requirements also influence the process. If a surface must remain in flawless condition due to optical requirements or if it is not resistant to external influences, the process must be designed according to these specific requirements and possibly the degree of automation, not only in the choice of hardware but also in the movement itself, need to be adapted.

To assess the complexity influences of the surface requirements, a normalized scale can be utilized, where the different possible surfaces are valued. This assessment is based on experts' estimation.

Criteria 3a-c – Geometric and physical properties (C_{3a-c}) [5, 6, 8, 15, 18, 21]: The geometric and physical properties of the target item are crucial for automation [5, 6, 8, 15, 18, 21], because they determine the size, layout and configuration of the disassembly system. For an adequate representation of these influences, a distinction is made between three characteristics of the target item: *Size and dimensions* (C_{3a}), *weight* (C_{3b}) as well as *stiffness* or deformability (C_{3c}).

These properties influence the choice of hardware and the handling system. It has to be capable of handling items with the corresponding characteristics. Especially the stiffness of an item strongly effects the automatability because the handling will be significantly more complex if the part changes shape, orientation or position before, during or after the handling process [18, 21]. Most handling processes are based on the principles of form or force closure, which cannot or just partly be applied to objects with a high degree of freedom [18].

A quantification and representation on a normalized scale is less complicated in the first and second category. The assessed dimensions can be described in discrete values which can be put in ratio on the aforementioned scale. The representation of the stiffness on a normalized scale remains a challenge, since it is not just depending on the material but also on the geometry of the item. If the stiffness is not quantifiable by material and geometric properties and the other geometrical and physical properties are not representable by fixed measures, an expert's estimation for the evaluation is necessary.

Criterion 4 - Accessibility (C_4) [13–15, 19]: The accessibility of a target item is relevant for the disassembly [13] but also significant for the complexity of an automation process [19]. Therefore, the accessibility is a relevant factor in the automation feasibility assessment of disassembly processes [13–15, 19]. Limited space and obstacles lead to more complex hardware and movement designs.

The expression of the accessibility can be supported by measures like a lack of the space and obstacles but has to be at least supplemented by expert estimations.

2.1.3 Complexity Factor (C_F)

The aforementioned criteria (C_{1-4}) are designed to evaluate the complexity of an automation process. An analysis of more than 150 repair guides shows that it is necessary to carry out this assessment for every single process step which is necessary to retrieve the target item. An evaluation of an item itself is not sufficient to estimate the automation effort since different joining connections and parts have to be respected in the overall retrieving process of the part.

A differentiated assessment of individual process steps also enables the result of the suggestion of a partly automated process. To create an applicable scenario, the normalized scale used in these categories is determined to be from zero to five, where a zero means a process is not automatable in this category, a one means it is automatable but very difficult and a five means the most uncomplicated automation condition. So, a higher value on this normalized scale would mean a lower complexity. To enable further calculations, the evaluated complexity of the process steps has to be determined. Therefore, the value of complexity in the individual categories for each process step can be combined to one complexity factor, which is the quotient of the summarized criteria with the number of categories (N_{Cat}). One way to make this factor more conclusive is to weight the total—similar to a weighted sum model approach like e.g. Sabaghi et al., James et al. or Tang et al. proposed [6, 7, 19]. This will be represented by an individual weighting factor (g_i) for each category. Equation 1 shows this approach to calculate the complexity factor for each process step.

$$C_F = \frac{\left(\sum_{i=1}^{N_{Cat}} c_i * g_i \right)}{N_{Cat}} \tag{1}$$

When one or multiple categories show that one characteristic is preventing any automation attempts, a zero at the specific assessment indicates that an automation is not advised. This factor can be utilized in different following steps. The first would be a general evaluation of the feasibility or possibility to realize an automation. The second utilization of this factor is an adjustment of the estimated automation costs. A more complex system means usually more necessary effort and therefor it usually means a more expensive development, realization and operation.

2.2   Cost Assessment of Automation

To assess the profitability of automation, it is essential to compare the costs of the automation project with the costs of manual dismantling. Not only the operational costs, but also the initial costs incurred by acquiring the necessary tools that enable automation need to be considered. The cost analysis therefore includes the initial or investment costs, the specific fixed costs and the variable costs. For the sake of simplified modeling, jump fixed costs are not considered separately.

In cost analysis, the utilization of approximations is paramount due to the intricacies and impracticality associated with recording exact costs comprehensively. The pragmatic use of estimates proves to be a central strategy that takes into account the great complexity of cost evaluation. Furthermore, the delimitation of a focused time frame is of central importance in the cost assessment, especially in the determination of the initial depreciation costs and the variability of the costs depending on the number of units.

2.2.1 Costs of Automated Disassembly (K_{aut})

The costs of automation are created by a combination of various elements. Equations for calculating the investment costs, the fixed costs and the variable costs are presented in the following.

Investment Costs (K_{inv}): Initially, the processes must be defined in order to be able to implement automation (step III.a). Secondly, the necessary hardware has to be derived. The choice and number of machines and systems is dependent on their flexibility, the number of considered items and their quantity. More complex and flexible processes lead to more complex hardware, programming and implementations.

Another crucial point for automation costs and the automation feasibility are the operating times [13, 15] (step II.a), which include the time for preparation, moving, disassembly and post-processing [15] or can be expressed as the Total Time for Disassembly as proposed by Gungor and Cupta [22]. These process times can be recorded in different ways, such as simulations, standardized values or measurements in real environments.

In addition to the process time, a loading or a changeover time between the vehicles (T_c) can be taken into account. On the assumption that industrial robots are predominantly utilized for automation, the number of robots required (R) can be determined using the cumulative operation times of each process (COT) required for the disassembly of a vehicle with the average operating time of a robot (F_{aot}) in the period under consideration (PUC). For this assumption, the number of planned processes must be known, which in turn can be approximated using the assumed number of vehicles (V) to be processed within the same time. The initial investment required for the robots (K_R) can then be calculated by multiplying the acquisition price (P_R).

Alongside the robots required, the end effectors (EE) have to be purchased as well. Their number can be derived from the investigated processes for automation and their acquisition price (P_{EE}) has to be considered. Also, the costs for the periphery must be included (F_{peri}). The interaction between robots and humans within a semi-automated working environment also influences the initial costs, as safety measures have to be implemented (F_{saf}). The average complexity factor will be added as a factor to consider the influence of complexity on costs. When considering a defined quantity of vehicles, a fixed time frame is to be set. To derive the investment costs, the timeframe has to be set against depreciation times for the investments. If the PUC exceeds the depreciation time, this might be disregarded. K_{inv} can then be calculated as shown in Eq. 2, assuming linear depreciation, with a number of individual processes (n).

$$K_{inv} = \frac{\left(\left(\frac{COT*V}{F_{aot}}\right)*P_R\right)*\left(1 + F_{peri} + F_{saf} + \frac{\sum_1^n C_F}{n}\right) + \left(\sum(EE * P_{EE})\right)}{depreciation\ time} * (PUC)$$

(2)

Fixed Costs (K_{fixA}): The fixed costs for an automated or semi-automated workspace differ from an entirely manual disassembly. The hardware for an automated process will cause another need for space. Also, buffer and necessary storage space can vary from case to case. Derived from the analysis of the processes this will result in a necessary area for automated processes (AR_A). The required area has to be considered with a certain square meter price (P_{area}). In addition, the technical building services may vary in volume and the energy demand is expected to be different, which has to be respected with a dedicated cost factor (F_{tbaA}). This results in different property costs (K_{propA}). Another area of fixed costs to note are security and insurance costs. The not directly disassembly ascribed labor costs (K_{laA}) in terms of over-head and maintenance staff can also have an influence on the profitability assessment. The fixed costs, calculated as shown in Eq. 3, can only be calculated after determining the number of required robots as space requirements and, subsequently, property and not directly ascribed labor costs directly depend on the number of robots deployed.

$$K_{fixA} = ((P_{area} * AR_A) + (F_{tbaA} * AR_A)) + K_{laA} \tag{3}$$

Variable Costs (K_{varA}): The variable costs for an automated or semi-automated process are proportional to the quantity of operations. Therefore, the variable costs include any costs which are directly related and scaling with each processed item. To estimate these costs, the time a specific hardware element is used for a specific process (t_{aut}) is considered. If combined with a factor of energy demand (F_{endA}) and energy prices (P_{en}) the energy costs for each process are calculated. All necessary processes have to be summed up in the next step. Also, an action rate (F_{ar}) can be applied for each hardware item which can combine an average of different costs which at least statistically increase per use like maintenance costs or operation resources. The costs of demanded energy (K_{energy}) could also be calculated by multiplying the cumulated energy demand by the energy price (P_{en}).

$$K_{varA} = \left(\sum (t_{aut} * F_{endA} * P_{en}) \right) + F_{ar} \tag{4}$$

It is crucial to multiply the variable costs with the number of processed items within the period under consideration (V) when calculating the overall costs.

$$K_{aut} = K_{inv} + (K_{fixA} * PUC) + (K_{varA} * V) \tag{5}$$

2.2.2   Costs Assessment of Manual Disassembly (K_{man})

To obtain a reference value for assessing the economic benefits of automation, the manual execution of the processes, must also be assessed (step III.m).

Personnel costs (K_{perM}): Regarding personnel costs for deployed disassembly workers (K_{perM}), each process step requires assignment to a qualification level (Q_n). This assignment must be based on company data and experts' estimations. To determine the number of required employees of each qualification level (NE_{Qn}), the processing times of manual disassembly are considered. Equally to the time tracking of automated disassembly processes, the processing times during the manual disassembly need to be determined (step II.m) as the processing times directly influence the number of disassembly workers required to execute all disassembly processes. These processing times can be recorded by simulations, standardized values or measurements. To calculate the number of necessary employees (NE_{Qn}) meeting the cumulated time requirements of each process in the respective qualification level, the total processing times of processes assigned to a qualification level are multiplied by the assumed number of vehicles (V) requiring disassembly. The resulting total processing time per qualification level is then divided by the average workhours of one disassembly worker in the PUC (F_{man}). The personnel costs (K_{perM}), shown in Eq. 6, are the product of the number of disassembly workers of each qualification level (NE_{Qn}) and the employment costs per worker of each qualification level (K_{empQn}), which include the salary and the employer's social security and general surcharge.

$$K_{perM} = \sum_{Qn=1}^{Qmax} \left(\left(\frac{\sum Processtime_{Qn} * V}{F_{man}} \right) * K_{empQn} \right) \tag{6}$$

Fixed Costs (K_{fixM}): The fixed costs for the manual disassembly (K_{fixM}) are calculated analogously to the fixed costs for the automated disassembly. They consist of the property costs (K_{propM}), not directly disassembly ascribed labor costs (K_{laM}). These costs can be calculated as the labor costs of a number of employees (NE_{NDDQn}), set as a percentage of the number of necessary employees per qualification level (NE_{Qn}). Fixed costs (K_{fixM}) are calculated based on the required assembly workers, factoring in space requirements. Property and labor costs are directly tied to the number of workplaces needed in the PUC. The costs for the necessary tools (K_{tool}) can also be regarded as an element of the fixed costs since they are not directly related to the number of items worked on, but to the number of workplaces that require the specific tools.

$$K_{fixM} = K_{propM} + K_{laM} + K_{tool} \tag{7}$$

Variable Costs (K_{varM}): In manual disassembly, personnel costs constitute a significant portion of variable costs. Similar to the automated process, variable costs for manual disassembly are directly proportional to the quantity of operations, encompassing all expenses directly linked and scaling with each processed vehicle. To estimate these costs, the time (t_{man}) a specific hardware element (i.e., tools) is in use for a specific process is combined with the energy demand (F_{endM}) and energy prices (P_{en}), whereof the energy costs for each process can be calculated. As in the automated process, an action rate (F_{arM}) can be applied for each hardware item, which can combine an average of different costs, which at least statistically increase per use, like maintenance costs or operation resources.

$$K_{varM} = \left(\sum (t_{man} * F_{endM} * P_{en}) \right) + F_{arM} \tag{8}$$

To derive the total costs for the manual disassembly (K_{man}), the variable costs need to be multiplied with the estimated or known number of processed vehicles within the period under consideration (V).

$$K_{man} = K_{perM} + (K_{fixM} * PUC) + K_{varM} * V \tag{9}$$

3 Use Case

An exemplary process to apply the method is derived from the analysis of workshop guidelines created for manual vehicle disassembly in the course of repairing or replacing a part. Although not being designed for automated disassembly, the guidelines contain process chains that are suitable to demonstrate the application of the methodology. Some assumptions have to be made for external factors and input data, that have to be provided by the company and its environment when applying the method. In this case these factors rely on estimations to compensate a lack of data and are shown in Table 1, 2 and 3.

The methodology is presented for the disassembly of a propeller shaft with the necessary process steps: *Unscrewing the nuts, removing the cover, unscrewing bolts* and finally *removing the propeller shaft*. Results of necessary calculations according to the equations presented in Sect. 2 are shown in Tables 1, and 3. Table 1 and 2 display the estimated process complexity and time recordings, whereas in Table 3 the automated and manual disassembly is compared. The threshold value (C_{FT}) indicating the limit for inadvisable automation is set at 0.4. None of the studied processes fall below this threshold, categorizing all as automatable. Additional factors and values are detailed in Table 3.

Table 1 Propeller shaft disassembly process assessment, complexity.

Propeller shaft process complexity			Categories: 7					$ØC_F = 0.45$
Weighting factor [g_j]	0.17	0.17	0.05	0.11	0.05	0.22	0.23	C_F
Unscrewing nuts	4	3	4	4	5	5	3	0.55286
Cover removal	4	3	5	2	5	3	1	0.40000
Unscrewing bolts	2	4	4	3	5	4	1	0.41571
Propeller shaft removal	2	4	2	4	5	3	2	0.41857

For the manual disassembly it is estimated that all processes listed in Table 1 and 2 can be executed by employees of qualification level one (Q_1). It is assumed that the processes take the same time as an automated treatment. Further, for administrative, planning, supervising and other overhead tasks at least one further person of qualification level 2 (Q_2) is necessary for two-thirds of the operating time.

The comparison shown in Table 3 and Fig. 2 display that automated dismantling is economically advantageous for the processes under consideration, independent of the number of vehicles. This can be explained by the high personnel costs, that primarily influences the manual disassembly route. In contrast, the variable costs appear to be significantly higher for automated disassembly.

One reason for the high influence of personnel costs is the fact that they are subject to rounding up inaccuracies as only whole integers for the number of personnel can be considered. In general, economies of scale and synergy effects can be seen, that show an increase of economic feasibility of robot utilization with the number of processes and products to be treated. In the case of a component where only some of the disassembly

Table 2 Propeller shaft disassembly process assessment, operating times and cumulated energy demand.

Operating times [h] automation	Quantity of processes	Cum. Energy demand [kWh/h]	Product (of process time * quantity)	Sum of operating time	Robot 1	Screw driver	Clip gripper	Tool changer	Time_{total} for $V = 2.500$ [Total operating time * V]
		5.480			0.685	0.220	0.065	0.400	1.712,5
Unscrewing nuts	2	0.320	0.040	0.020	0.020	0.020			Time_{total} + T_C for $V = 2.500$
Cover removal	1	0.800	0.100	0.100	0.100		0.050	0.050	2.500
Unscrewing bolts	6	3.840	0.480	0.080	0.080	0.030		0.050	
Propeller shaft removal	1	0.520	0.065	0.065	0.065		0.015	0.050	2.337,5

Table 3 Estimations (*) and calculations of automated and manual propeller shaft disassembly.

General factors		Automated disassembly				Manual disassembly				
Number of indiv. processes	4	K_{aut} = 296,714 €				K_{man} = 438,551 €				
Cum. of all processes	10	Variables	Units	Robot/general	End effectors	Variables	Unit	NE_{Q1}	NE_{Q2}	NE_{Q3} General
P_R*	65K [€]	F_{saf}*	[%]	10		Q_n total	[h]	0.685	0	0
P_{EE}*	15K [€]	R	[h]	0.640		F_{man}	[h]	1,337	1,337	1,337
P_{tool}*	750 [€]	COT	[h]	0.685		t_{man}	[h]			0.685
P_{area}*	30.00 [€/m²/mon.]	F_{aot}	[h]	2,674		Q_n OH	[h]	0.00	0.46	0.00
PUC	1 [Year(s)]	EE			3	NE_{Qn}		1.28	0.85	0.00
V*	2,500 [1/PUC]	F_{peri}*	[%]	60	60	F_{endA}*	[kWh]			1
T_C*	0.25 [h]	AR_A	[m²]	95		F_{arM}*	[€]			0.10
P_{en}*	0.30 [€/kWh]	F_{tbaA}*	[%]	100		AR_M	[m²]			71.25
Depreciation time	10 [Year(s)]	t_{aut}	[h]	s. table 2	s. table 2	F_{tbaM}*	[%]			75
Adm. Surcharge*	40 [%]	F_{endA}*	[kWh/h]	7		$F_{NENDDQn}$	[%]	10	15	10
Social security surcharge*	20 [%]	F_{ar}*	[€]	0.70	0.15	NE_{NDDQn}		0.13	0.13	0.00
Q_1 Net annual salary (NAS)*	72K [€ p.a.]	K_{laA}	[€]	192K		**Additional variables**				
Q_2 NAS*	88K [€ p.a.]	K_{varA}	[€]	7.14		Q_n OH = Sum(Q_n OH rate * quantity of process *				
Q_3 NAS*	104K [€ p.a.]	OH Q_1		0		process time)				
Workdays per year*	251 [Days]	OH Q_2		1		Q_n OH rate = Allocated process time share at Q_n				
Working hours per day*	7 [h]	OH Q_3		1		F_{tbaM} = Tech. building services costs factor, manual				
Robot operation time per day*	14 [h]	F_{ar} cum. EE	[€]		2.70	$F_{NENDDQn}$ = Percentage of NE_{Qn}, overhead task rate				
Downtime / comp. Days*	60 [Days]	F_{ar} cum. R	[€]	2.80		NE_{NDDQn} = Sum($F_{NENDDQn}$ * NE_{Qn}), = Number of				
Annual working hours	1,337 [h]	K_{energy}	[€]	1.64		not directly disassembly allocated employees				

processes are considered as automatable based on the complexity factor, both the calculation steps of the automatic and manual analysis are used in the calculation for the respective assigned processes.

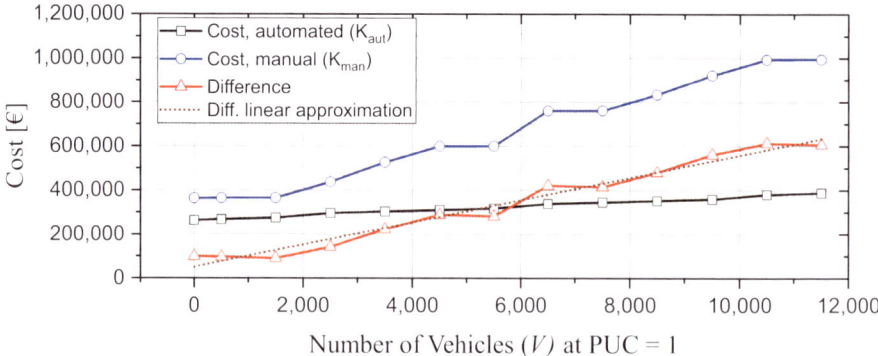

Fig. 2 Comparison of K_{aut} & K_{man} with increasing number of vehicles (V) at $PUC = 1$ year

4   Conclusion, Limitation and Outlook

The study introduces a method for evaluating automation possibilities within vehicle disassembly processes. In a first step, the complexity of a disassembly task is evaluated which demonstrates the extent to which automated disassembly is recommended or even possible. In the second step, a comparison between manual and automated process layouts is undertaken. This comparison is based on the analysis of complexity and time measures, which serve as the basis for economic considerations. In summary, quantifying the economic benefits of EoL process automation supports decision-making. Its successful application in an Excel-based tool confirms its viability for practical use in industrial settings. The case examined shows that complexity and process time determine the economic advantage of automation as a function of the number of processes. It also suggests that a synergy of different processes can reduce the costs per process.

Key areas that require further investigation are highlighted in the study. The development of normalized scales for the subjective assessment of complexity is demanded. Additionally as some studies suggest, the integration of forces to be applied for joint separation lead to a more comprehensive evaluation [6, 8, 12, 13]. This could enhance the complexity factor calculation. Challenges persist in time tracking due to uncertainties, primarily due to process planning and operational variability in the field of disassembly. Additionally, diverse factors, e.g., maintenance rates, can be introduced to enhance the precision. Other factors like the not directly disassembly related labor costs (K_{laM}), which are derived as a percentage of the disassembly ascribed labor, can enhance the precision of the analysis.

The principle of only considering full-time positions can also be disputed. In the use case, described in this publication, rounding inaccuracies might be present, that have an influence on the overall results. This influence would decrease if more processes and therefore more employees are considered. Furthermore, the process times for manual disassembly are approximated based on the times of automated disassembly, which should be replaced by recorded disassembly times in the future.

The method can be strengthened by integrating life cycle assessment (LCA) results for environmental considerations, providing a broader perspective on sustainability. Furthermore, decision-making between disassembly cell and line layouts, that are impacted by vehicle numbers and logistical considerations, could be explored for their influence on an automation scheme's economic impact. Moreover, extending this method's application beyond automotive disassembly to other fields emerges as an avenue for future research.

Acknowledgements. The researchers express their particular gratitude to Thorsten Schrader, Audi AG, for the cooperation and provision of data sets used to apply and validate the model as well as the expertise that contributed to the creation and application of the methodology presented. They also express their thanks to Florian Gorges for his support during the literature research.

References

1. Deutsches Institut für Normung e.V. (DIN), Deutsche Kommission Elektrotechnik Elektronik und Informationstechnik, (DKE), and Verein Deutscher Ingenieure e.V. (VDI), "Deutsche Normungsroadmap Circular Economy," (2023. [Online]. Available: https://www.din.de/de/forschung-und-innovation/themen/circular-economy/normungsroadmap-circular-economy
2. Temmes, J. and Ranta, J., "Utilisation of modern production automation: Some technological, economic and social impacts," *Robotics*, vol. 3, no. 1, pp. 89–94 (1987).
3. Gomes, O. and Pereira, S., "On the economic consequences of automation and robotics," *JEAS*, vol. 36, no. 2, pp. 134–153 (2020).

4. Foo, G., Kara, S., and Pagnucco, M., "Challenges of robotic disassembly in practice," *Procedia CIRP*, vol. 105, pp. 513–518 (2022).
5. Herrmann, C., Mennenga, M., Raatz, A., Schmitt, J., and Andrew, S., "Assessment of Automation Potentials for the Disassembly of Automotive Lithium Ion Battery Systems," in *Leveraging Technology for a Sustainable World*, D. Dornfeld and B. Linke, Eds., Springer, Berlin, Heidelberg (2012).
6. James, A. T., Gandhi, O. P., and Deshmukh, S. G., "Development of methodology for the disassemblability index of automobile systems using a structural approach," *Proceedings of the Institution of Mechanical Engineers, Part D: Journal of Automobile Engineering*, vol. 231, no. 4, pp. 516–535 (2017).
7. Tang, Y., Zhou, M., Zussman, E., and Caudill, R., "Disassembly modeling, planning, and application," *Journal of Manufacturing Systems*, vol. 21, no. 3, pp. 200–217 (2002).
8. Mok, H. S., Kim, H. J., and Moon, K. S., "Disassemblability of mechanical parts in automobile for recycling," *Computers & Industrial Engineering*, vol. 33, 3-4, pp. 621–624 (1997).
9. Hancock, P. A., "Automation: how much is too much?," *Ergonomics*, early access.
10. Weber, S. F., "A Modified Analytic Hierarchy Process for Automated Manufacturing Decisions," *Interfaces*, vol. 23, no. 4, pp. 75–84 (1993).
11. Mital, A., Motorwala, A., Kulkarni, M., Sinclair, M., and Siemieniuch, C., "Allocation of functions to humans and machines in a manufacturing environment: Part I—Guidelines for the practitioner," in *Allocation of functions to humans and machines in a manufacturing environment: Part I—Guidelines for the practitioner* (Elsevier Ergonomics Book Series), pp. 33–59, Elsevier, (2000).
12. Desai, A. and Mital, A., "Evaluation of disassemblability to enable design for disassembly in mass production," *International Journal of Industrial Ergonomics*, vol. 32, no. 4, pp. 265–281 (2003).
13. Kroll, E. and Hanft, T. A., "Quantitative evaluation of product disassembly for recycling," *Research in Engineering Design*, vol. 10, no. 1, pp. 1–14 (1998).
14. Go, T. F., Wahab, D. A., Rahman, M., Ramli, R., and Azhari, C. H., "Disassemblability of end-of-life vehicle: a critical review of evaluation methods," *Journal of Cleaner Production*, vol. 19, no. 13, pp. 1536–1546 (2011).
15. Yi, H.-C., Park, Y.-C., and Lee, K.-S., "A study on the method of disassembly time evaluation of a product using work factor method," in *SMC'03 Conference Proceedings. 2003 IEEE International Conference on Systems, Man and Cybernetics. Conference Theme—System Security and Assurance (Cat. No.03CH37483)*, Washington, DC, USA, pp. 1753–1759 (2003).
16. Seliger, G., Szimmat, F., Niemeier, J., and Stephan, J., "Automated Handling of Non-Rigid Parts," *CIRP Annals*, vol. 52, no. 1, pp. 21–24 (2003).
17. Seliger, G., Gutsche, C., and Hsieh, L.-H., "Process Planning and Robotic Assembly System Design for Technical Textile Fabrics," *CIRP Annals*, vol. 41, no. 1, pp. 33–36 (1992).
18. Sanchez, J., Corrales, J.-A., Bouzgarrou, B.-C., and Mezouar, Y., "Robotic manipulation and sensing of deformable objects in domestic and industrial applications: a survey," *The International Journal of Robotics Research*, vol. 37, no. 7, pp. 688–716 (2018).
19. Sabaghi, M., Mascle, C., and Baptiste, P., "Evaluation of products at design phase for an efficient disassembly at end-of-life," *Journal of Cleaner Production*, vol. 116, pp. 177–186 (2016).
20. Poss, C., Irrenhauser, T., Prueglmeier, M., Goehring, D., Zoghlami, F., and Salehi, V., "Perceptionbased Intelligent Materialhandling in Industrial Logistics Environments," in *Proceedings of the 2019 11th International Conference on Computer and Automation Engineering*, Perth WN Australia, pp. 146–151 (2019).

21. Nguyen, H. G., Kuhn, M., and Franke, J., "Manufacturing automation for automotive wiring harnesses," *Procedia CIRP*, vol. 97, pp. 379–384 (2021).
22. Gungor, A. and Gupta, S. M., "An evaluation methodology for disassembly processes," *Computers & Industrial Engineering*, vol. 33, 1–2, pp. 329–332 (1997).

Fabrication of Media-Tight Connections Between Metal and Plastics Using HyJOIN® Technology

Annett Klotzbach[✉], Mosarouf Hossain, Dieter Luong, and Philipp Götze

KIST + ESCHERICH GmbH, Lockwitzgrund 100, 01257 Dresden, Germany
a.klotzbach@kist-escherich.com
http://www.kist-escherich.com

Abstract. Thermal direct joining with hyJOIN® enables quick and reliable connections between metal and plastics without any additional material like screws, rivets or adhesives. Due to its reversibility, it facilitates the implementation of new repair and recycling strategies while at the same time minimizing the consumption of resources.

During joining the parts are pressed together. At the same time, electromagnetically induced eddy currents heat the metal. The plastics softens in the contact area and binds to the metal surface during cooling. As plastics often adhere poorly to metal, the surface is structured or coated during a pre-process.

Within the paper the hyJOIN® process is applied to generate media-tight connections between metal and plastics. To analyze the connection performance a burst pressure test specimen, made of an aluminum plate and a glass fiber reinforced PA6 plastic container, was developed. During the feasibility study the influence of interface design and joining process parameter to tightness and burst pressure level was evaluated. The main results will be shown in this paper. Finally, the transfer of the process knowledge to new, compact and easy to recycle hybrid part designs will be discussed. hyJOIN® systems thus offers the possibility for innovative and smart production.

Keywords: Thermal joining · Hybrid part · Metal plastic connections · Lightweight design

1 Introduction

Modern lightweight construction no longer relies solely on the use of lightweight and functionally integrated fiber composite structures. The targeted selection of the right material in the right place is becoming increasingly important. As a result, the demands on new joining technologies are growing to produce complex hybrid assemblies from different material classes. While productive welding processes are used for connections of the same material type, multi-material connections between metal and plastics are often screwed or glued.

K. Dröder and T. Vietor (Eds.): CD 2024, Proceedings, pp. 32–41, 2025.
https://doi.org/10.1007/978-3-658-45889-8_3

For some years now, there has been an alternative to traditional adhesive bonding of metal to plastics: thermal direct joining or press joining. In this process, the metal partner is heated in the joining area. Subsequently or simultaneously, the thermoplastic part is pressed together with the metal. The plastics reaches the glass transition temperature and melts at the contact area. After cooling the plastic solidifies and adheres onto the metal surface. To achieve a strong and media-tight connection, it is recommended to roughen or coat the metal surface in the joining area. In this way, a microscopic or macroscopic form and force fit can be achieved. The scientific and technical principles have been and are being developed in various research projects [1–6].

In addition, the first industrial applications and suitable system technology are already available on the market, e.g. KIST + ESCHERICH with hyJOIN® [7].

In general, the hyJOIN® process involves an ablative pre-treatment of the metal. The assembly is pressed together. During pressing the metal is heated locally by inductive eddy currents so that the plastic melts and can anchor itself to the metal surface (Fig. 1).

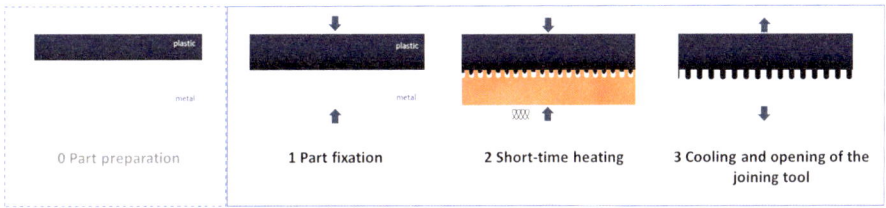

Fig. 1. Schematic diagram of the process steps for metal plastic connections made by hyJOIN®.

Although the process is being used successfully in initial series applications, there is still a need for intensive research to develop a comprehensive understanding of the process. One of the reasons for this is that the technology can be used for an extremely wide range of applications. A major area of application is the implementation of compact thermal management solutions such as actively cooled housings for energy storage, sensor technology or electronics. In this area of application, the thermal direct joining of a metallic heat sink with a plastic injection molded part offers advantages: screws and additional seals can be removed. It is also often possible to reduce the width of the joining flange and therefore make the design more compact. At the end of the life cycle, the connection can then simply be thermally separated.

A generic hybrid cooler was developed to meet customer requirements for know-how protection. This publication therefore presents initial research results on the fabrication of media-tight connections using hyJOIN®.

2   Materials and Methods

2.1   Burst Pressure Container

To evaluate the media tightness of metal plastic connections, a test specimen design was developed and fabricated as a first step (Fig. 2). The test specimen consists out of a flat metal plate with sufficient rigidity. In this case, a material thickness of 3–4 mm was used for the aluminum alloys (wrought alloy or die-cast aluminum). An injection molding tool was created for the plastic cover. Different melting rib geometries were realized by exchanging individual segments. Three different geometries were used in the comparative study (Fig. 3). The test specimen also has a connection for the test media (water or pressurized air). [8].

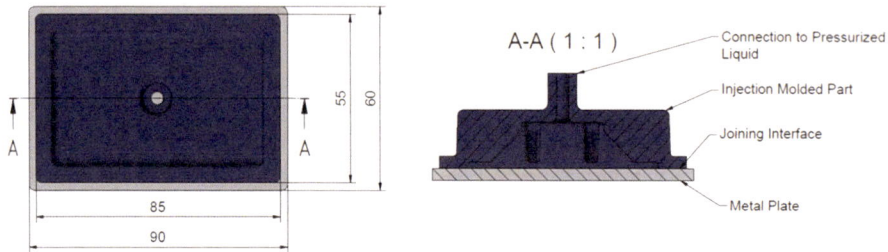

Fig. 2. Burst pressure test specimen.

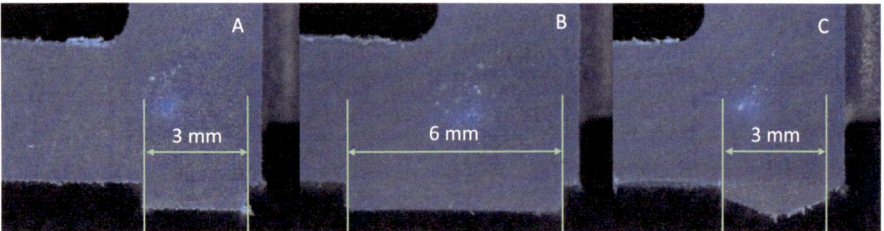

Fig. 3. Melting rib geometries (series A, B, C).

2.2 Materials and Part Preparation

The following materials were used for the tests:

Metal	AlMg3	90 mm x 60 mm x 4 mm with pretreatment
Plastics	PA6-GF30	85 mm x 55 mm; melting rib geometry: 3 mm, 6 mm, 3 mm tip

To achieve good joint strength, the metallic surface must be pre-treated before joining. Various ablative or coating processes can be used [9]. In this case, the pre-treatment was carried out on both sides using a chemical etching process [10]. In this (electro-) chemical process, individual alloying elements are leached out of the area close to the surface, resulting in micro-roughness and micro-anchoring on the surface. This 3D nanoscale-sculptured surface ensures that the melted plastic can adhere well to the aluminum.

The plastic cover was not pretreated, precleaned, dried or conditioned before joining.

2.3 Test Procedures

To compare the joint properties achieved, the tests listed in Table 1 were carried out on all test specimens. In the bubble leak test, the test specimen is inflated with air and held under water. The air bubbles rising to the water surface within one minute are then counted (Fig. 4). In the pressure loss test, the container is pressurized with compressed air and the air supply is closed after a rest period. A pressure gauge is used to determine the pressure drop after a predefined time (Fig. 5).

Finally, the samples are subjected to burst tests, where the samples are destroyed (Fig. 6). The container is filled with water. A manually operated piston pump is used to increase the water pressure until the container fails. The maximum pressure achieved is documented. [8].

Table 1. Overview of test procedures.

Test routine	Test specs	Evaluation
Bubble leak test	Part under water, air 2 bar, 60 s	Number of air bubbles
Pressure drop testing	Air 5 bar, 10 min	Pressure difference
Burst test with piston pump	Water, time till bust: approx. 5 s	Value [bar]

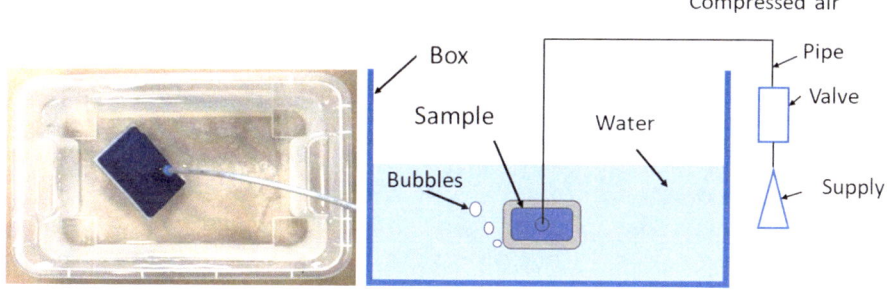

Fig. 4. Air bubble leak test. [8]

3 Experimental Setups

3.1 Joining Tool and Machine

Two different joining tools were designed and manufactured for the fabrication of the test specimens. Basically, such a tool consists of an upper part and a lower part. The two parts fix the joining partners to each other and enable the two parts to be pressed together. To heat the metal in the joining area, a geometry-adapted inductor is integrated either in the upper movable pressure die (Fig. 7 left) or in the lower tool (Fig. 7 right). In both cases, inductive heating takes place through the flange of the plastic part. This concept will also enable the plastic part to be attached to larger and more complex metal housings in the future.

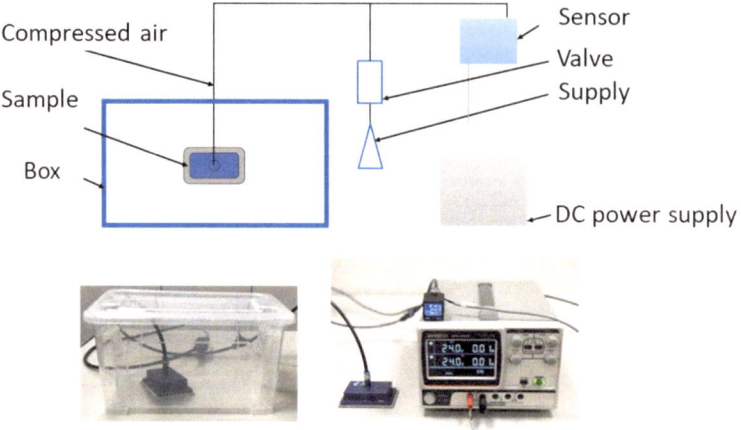

Fig. 5. Pressure drop test. [8]

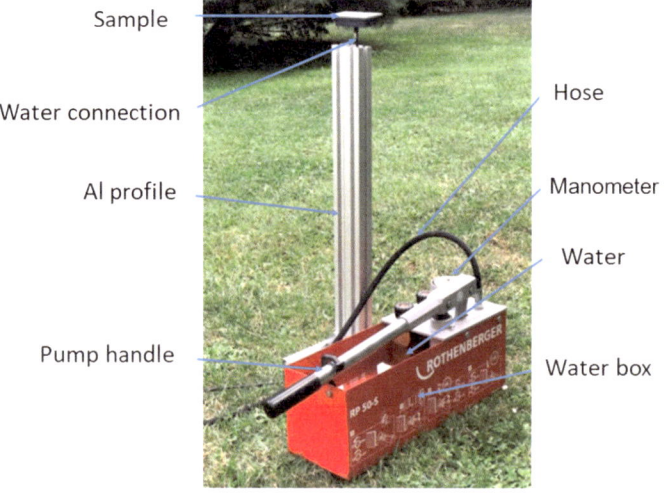

Fig. 6. Burst pressure test. [8]

In both experimental setups, a servo drive enables force and path control during the joining process. In this way, the joining force as well as the displacement distance[1] can be controlled and monitored during joining. An induction generator with a maximum output of 30 kW at medium frequency was used for the inductive heating of the metal plate.

---

[1] The displacement distance is the value by which the assembly is pressed together during heating. Depending on the component tolerances, a typical displacement distance of 0.2 mm to 0.5 mm is typical for assemblies with a flat joining surface. Observing the displacement distance is essential for the dimensional accuracy of the joined assembly.

The tests carried out in the publication were performed exclusively on the laboratory joining station (Fig. 7 right), as this setup allowed simplified thermal field monitoring.

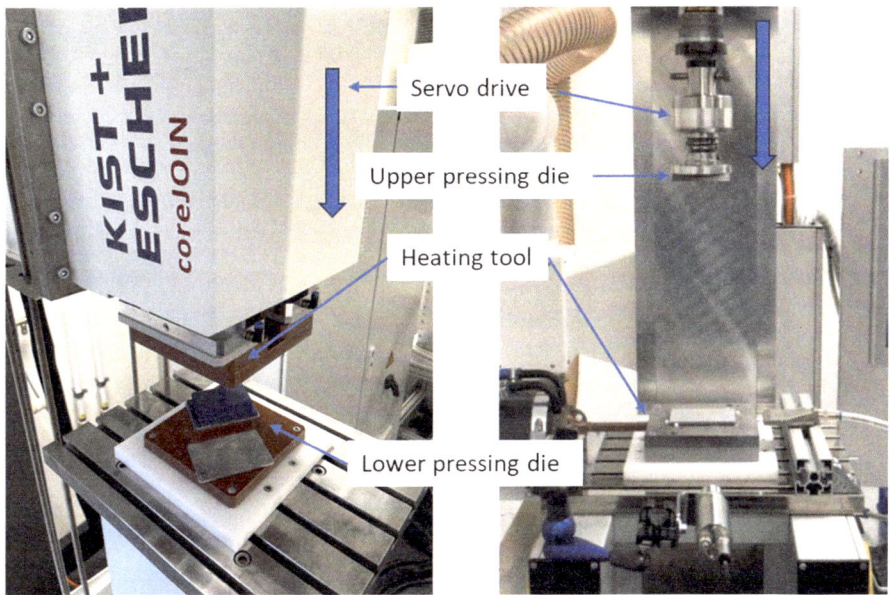

Fig. 7. Experimental joining setup: coreJOIN® modular press (left); Lab joining station (right).

3.2 Joining Matrix

The results of the thermal direct joining process are influenced by a variety of parameters. An important influencing factor is the plastic material, its composition and the dimensional tolerances. The following parameters can be used as setting parameters for the joining process:

– Joining force [kN]: Force with which both parts are pressed together during heating
– Joining path or displacement distance [mm]: Change in path due to the melting of the joining rib
– Heating power [%]: Generator power in percent
– Heating time [s]: Heating time of the part

In preliminary tests, the basic parameters were selected so that the melting ribs could bond to the metal over the entire circumference. Due to the distortion of the injection molded parts, 0.4 mm was defined as the minimum required displacement distance. At the same time, care was taken visually to ensure that the melt did not escape to the outside. Based on these framework conditions, the heating time was varied for the fixed parameters: joining force (3 kN, 6 kN), heating power (50%, 100%) and rib geometry. To validate the results, 5 test specimens were joined with the identical parameters.

4   Results and Discussion

As part of the comparative tests, 60 test specimens were produced and tested as described in Sect. 2.3. It was found that only one specimen from test series A4 (3 mm melting rib, 50% power, 6 kN joining force) exhibited a slight leak. Table 2 shows the parameters used in the individual test series as well as the test results.

Table 2. Overview of parameter setup for different trials and test results.

Serie	Rib geometry	Induction power [%]	Joining force[kN]	Heating time[s]	Bubble leak test	Pressure drop test
A1	3 mm	100	3	5,8	no	100%
A2	3 mm	100	6	5,8	no	100%
A3	3 mm	50	3	13	no	100%
A4	3 mm	50	6	12,4	*30 bubbles	*32%
B1	6 mm	100	3	6,3	no	100%
B2	6 mm	100	6	5,6	no	100%
B3	6 mm	50	3	13,3	no	100%
B4	6 mm	50	6	12,7	no	100%
C1	3 mm tip	100	3	6	no	100%
C2	3 mm tip	100	6	5,7	no	100%
C3	3 mm tip	50	3	12,2	no	100%
C4	3 mm tip	50	6	12,2	no	100%

*one of five test specimens

As described in 2.3, all test specimens were initially tested for leaks. 59 test specimens successfully passed the bubble leak test (no air bubble) and the pressure drop test (no pressure loss = 100%). Only one test specimen was leaky (Table 2). After the leak test, a destructive burst pressure test was carried out.

Figure 8 shows the bursting pressure strength determined for the individual test series. Test series B (melting rib 6 mm) were generally able to achieve average burst pressures between 20 and 27 bar. It was observed that the test series with a joining force of 3 kN showed the highest burst pressures. A direct proportionality between the melting rib width and the maximum bursting pressure could not be proven. Comparable failure thresholds were determined for test series A (3 mm) and C (3 mm tip).

In general, it can be stated that tight and stable connections can be produced both at high joining forces (6 kN) and moderate joining forces (3 kN). A higher joining force tends to result in a slightly lower necessary heating energy. The test evaluation showed that it is possible to create highly resilient joints even with short joining times of less than 10 s. This was done with 100% generator output. By using field concentrators, the coupling of the eddy currents into the aluminum could be optimized so that the melting temperature of the plastic could be reached within a few seconds.

Since a displacement distance of 0.4 mm was aimed for in each of the tests, the different melting rib geometries result in significantly different melting volumes. For this reason, the heating time had to be determined experimentally for each parameter set individually. Since, in addition to the displacement distance, the recognizable melt exit was also visually evaluated and considered when determining the heating time, there is a degree of uncertainty which must be considered in a detailed error analysis. To evaluate the joining efficiency, the heating energy applied was therefore set in relation to the volume to be melted. The 6 mm rib geometry shows the lowest energy input per volume. This is due to the large contact surface to the metal. The heat generated can be transferred quickly and evenly into the plastics. In comparison, test series C (melting rib 3 mm tip) show a significantly higher energy input per melting volume. The risk of overheating and thus damage to the plastics is increased.

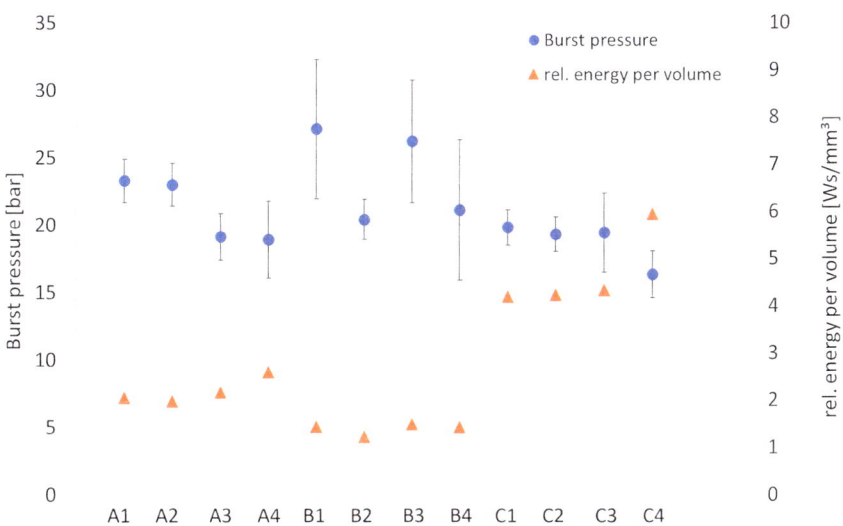

Fig. 8. Results of burst pressure and energy input per volume of different melting rib geometries

After bursting of the samples, the fracture surfaces were analyzed at a light microscope. The results showed a predominantly adhesive failure in the joint (Fig. 9 right). Some samples also showed localized fracture in the plastics. Figure 9 left shows a typical cross-section through the joint. A continuous bond without air inclusions or gaps is recognizable. Excess melt is pushed laterally out of the joint.

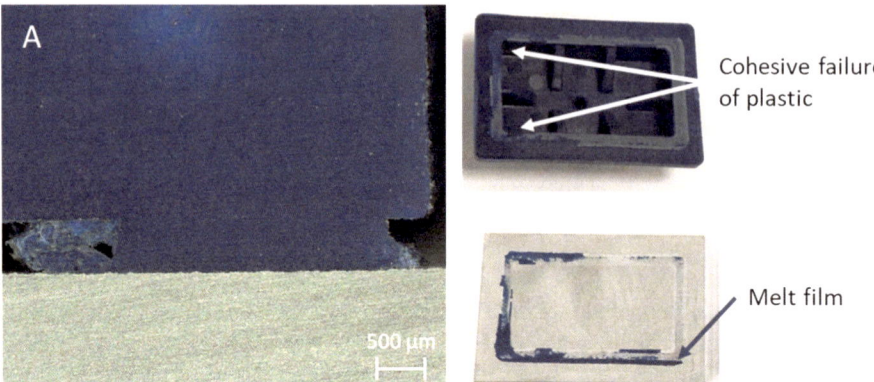

Fig. 9. Crosscut of hybrid connection with 3 mm melting rib (left); Breaking surface of hybrid connection with 6 mm melting rib (right).

5 Conclusions

The publication shows the results of basic tests on the thermal direct joining of aluminum with glass fiber reinforced PA6 using hyJOIN® technology. A burst pressure test specimen was developed with which airtight and watertight joints can be produced. Depending on the melting rib geometry and the joining parameters, failure pressures between 15 and 32 bar could be achieved. Almost all the joined and destructively tested hybrid specimens were completely airtight up to 5 bar. It has been shown that tight joints can be produced with melting rib widths of just 3 mm. The successfully tested parameter settings show that there is a large process window in which media-tight connections can be created.

The test specimen geometry makes it possible to carry out reproducible comparative tests with other material combinations or surface treatments. As the process is reversible, the container can also be used to develop repair strategies. The assembly is heated in the joining area with the help of the joining tool. As soon as the plastic begins to melt, the plastic part is removed by applying gentle pressure. Any plastic residue remaining in the structure is then melted again when a new plastic part is attached and does not impair the joint performance, or only to a small extent.

In summary, the work proves that metal plastic joints produced with hyJOIN® are media tight. The process can therefore replace the classic screwing of assemblies with an additional seal. Through the substitution of additional materials and process reversibility, the technology enables material resources to be saved and alternative repair and recycling concepts to be implemented in the future.

Acknowledgements. Part of this research was funded by the "Federal ministry of education and research" within the collaborative project: JOASIS: Development of a 30% lighter seat frame for electric vehicles using HPCi® technology (FKZ: 01DR21023B).

References

1. https://www.Photonikforschung.de/projekte/lasertechnik/projekt/laserleichter.html, last accessed 2024/01/04.
2. https://www.werkstoffplattform-hymat.de/Group/TailoredJoints/Pages, last accessed 2024/01/04.
3. https://flexhyjoin.jimdo.com/description/. last accessed 2024/01/21.
4. https://www.iws.fraunhofer.de/de/technologiefelder/trennen-und-fuegen/kleben/thedi.html. last accessed 2024/01/21.
5. Spancken, D., Van der Straeten, K., Beck, J., Stötzner, N.: Laser Structuring of Metal Surfaces for Hybrid Joints. Lightweight Design worldwide (11), 16–21, https://doi.org/10.1007/s41 777-018-0028-6 (2018).
6. Dröder, Klaus (Hrsg.): Prozesstechnologie zur Herstellung von FVK-Metall-Hybriden. Results of BMBF-collaborative ProVorPlus. Springer Nature 2020. -ISBN: 978–3–662–60679–7
7. Thermal Direct Joining Allows Plastic and Metal to Be Fused together. Plastics Insights. https://en.kunststoffe.de/a/specialistarticle/thermal-direct-joining-allows-plastic-an-5164351, last accessed 2024/01/24.
8. Hossain, M.: Untersuchungen zum thermischen Direktfügen von Hybridbauteilen mittels induktiver Erwärmung. Diploma Thesis. Hochschule für Technik und Wirtschaft Dresden. Fakultät Maschinenbau. Studiengang Produktionstechnik (2023).
9. Vogt, S., Götze, P., Klotzbach, A., Möller, M., Neugebauer, Ch., Decker, D., Weiler, M., Gabdyl, J.: Experimental investigations of laser structuring for metal surfaces in metal-plastic joints. Procedia CIRP 111 697-700 (2022).
10. Baytekin-Gerngross, M.: Nanoscale sculpturing of metals and its applications. Christian-Albrechts- Christian-Albrechts-University Kiel, Faculty of Engineering Functional Nanomaterials, Dissertation (2018).

Automated Production of Mycelium-Based Composite Products Using the Fibre Injection Moulding Process

Simon Mangold$^{(\boxtimes)}$ (iD), Malte Mehner, Moritz Ströhle, Sebastian Kist, Florian Kößler, and Jürgen Fleischer

Wbk Institute of Production Science, Karlsruhe Institute of Technology, Kaiserstraße 12, 76131 Karlsruhe, Germany

`{malte.mehner,florian.koessler,Juergen.fleischer}@kit.edu`
`https://www.wbk.kit.edu`

Abstract. The production of plastics has increased exponentially in the past two decades, with packaging making up a large portion of usage. Environmental concerns have surged due to the disposal of plastic waste, with only a small percentage being recycled while the rest is either burned or ending up in landfills. As a result, there is a growing need for alternatives, including biodegradable materials like mycelium, the structural part of fungi. While mycelium holds promise for technical applications, the manufacturing process is mainly manual, which hampers efficiency and scaling to large scale series applications. Therefore, automating the manufacturing process of mycelium-based products has the potential to improve precision, efficiency, and cost-effectiveness of the manufacturing. One potential method of automation is the Fiber Injection Molding (FIM) process. The characteristics of this process is the injection of a mixture of structural fibers and a thermoplastic binder into a mold by means of a large volume airflow. This study aims to investigate the applicability of the FIM process for manufacturing natural products bound by mycelium and the required adaptions to an existing plant, using two different types of fungal spores, namely Trametes Versicolor and Pleurotus Ostreatus. The spores were grown on fibers of hemp, straw, and wood and form-filling was conducted via FIM. The results of the study include a summary of the challenges faced when using FIM for manufacturing mycelium-based products, optimized process parameters and concepts for adapted machinery equipment. The study found that FIM is a suitable method for producing mycelium-based products, and the optimal process parameters varied depending on the type of fungal spores and fiber used. However, the study also identifies some challenges, such as the transportation of materials in the large airflow. In conclusion, the FIM process can be used to manufacture mycelium-based products effectively.

Keywords: Fibre Injection Moulding · Mycelium Composite · Biologicalisation

K. Dröder and T. Vietor (Eds.): CD 2024, Proceedings, pp. 42–52, 2025.
https://doi.org/10.1007/978-3-658-45889-8_4

1   Motivation

The issue of plastic waste has gained considerable attention in recent years due to its negative impact on the environment. Consumer goods are among the main sources of plastic waste. As a result, the packaging industry has been seeking alternatives to traditional plastic packaging. One of the promising solutions is the usage of bio-based compounds as custom-molded elements [1–4]. The mechanical integrity of these composites can be achieved by mycelium, the vegetative lower part of fungi. It grows due to its symbiotic relationship with the materials that feed it, forming entangled networks of branching fiber [5–7]. These properties qualify the material for applications where waste materials could be used as primary material, such as packaging. The materials have the advantage of after usage being able to rot without causing any negative impact on the environment. Mycelium compounds have additionally been adopted in a broad variety of applications other than packaging due to their benefits in sustainability. However, the manufacturing process is primarily manual in today's state of the art, which hinders large-scale applications [8, 9]. To address this issue, a study is conducted to examine the applicability of the so-called Fiber Injection Molding (FIM) process for the automated manufacturing of mycelium compounds. FIM is a technology from the field of textile fiber processing for direct processing of short to medium long fibers (length of 80 to 100 mm). Fehler: Verweis nicht gefunden (a) contains a schematic visualization of the included process steps. At first, the fiber material is prepared, as a mixture of structural and binder fibers. It is then transported via an airflow and blown into an air-permeable mold. The binding is activated by inducing heat and a pressing in form, resulting in a thermal consolidation. This generates a contour-accurate 3D preform that is manageable and available for further use in for example the resin transfer molding process chain. The main advantages of FIM include the almost waste-free component production, the possibility of producing different wall thicknesses and the wide range of materials that can be used. The method has only a few process steps and offers many degrees of freedom with regard to the use of material and the part geometry. Depending on the planned application and the required number of components, sliding table systems (for small series and pilot plant operation) and rotary systems (for series applications) can be used.

The approach followed within this contribution, to use FIM for mycelium compound manufacturing relies on maintaining the mold filling and as many of the other process steps as possible alike. The mycelium is intended to replace the thermoplastic binder.

2   State of the Art

The contribution of Elsacker et al. [10] gives a comprehensive summary of the manufacturing pipeline for mycelium-based products. The authors condense it as a seven-step manufacturing process.

(1) The mycelium is initially grown on agar plates, in grain-substrate, in a liquid nutrient solution, or in the pre-grown homogenised substrate;
(2) The substrate is autoclaved or pasteurised to eliminate any type of already present microorganisms on the substrate and thereby preventing contamination during the growth and incubation process;

(3) A specific amount of the mycelium tissue is added to the substrate. If the substrate was not humidified before autoclaving (step 2) an amount of sterile water is added. To improve growth, a sterile solution of nutrients can also be added;

(4) The inoculated substrate is hand-packed in a sterilized mould which has the desired shape. The mould is sealed with a filtered air-permeable cover to maintain a micro-climate;

(5) The mycelium grows through the substrate in a controlled environment. The material can be grown in two phases, first in a mould to bind the fibres, and secondly outside the mould to solidify the outer skin of the material during a period;

(6) The grown material is heat-treated at a specific temperature for several hours to end the growth process and dehydrate the material;

(7) A coating or post-processing can be applied to the material to improve its properties.

Fabrication method of mycelium-based composites per Elsacker et al [10].

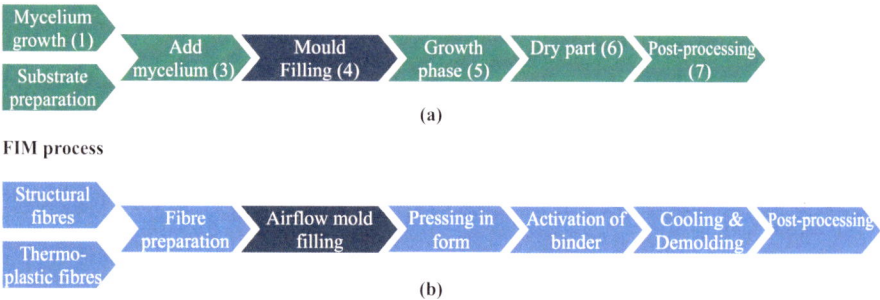

Fig. 1. Comparative visualisation of the regarded process chains

Figure 1 shows both manufacturing process chains in comparison. The similarity of the approaches is thereby underlined. Consequently, the emerging research question can be formulated to: What changes to the plant technology and what process variables are necessary to enable FIM plants for the manufacturing of mycelium-based products?

Jiang et al. and Houette et al. [11, 12] present similar fabrication steps to retrieve qualitative acceptable mycelium compound parts. The adaptions to the plant technology primarily depend on the manufactured product, since it determines the type of substrate and fungal genus. In order to develop suitable components and technologies for the adaption of the plant technology, current industrial applications of mycelium compounds are analyzed along with the respective manufacturing technology. Nowadays, multiple industrial operators on the field of mycelium composite products have established. Applications of mycelium-based products include the construction sector with insulating panels, acoustic tiles and architectural elements. One of the companies operating in that field is MOGU S.r.l. from Italy. The design sector includes applications of leather-like material combinations or other textiles used in fashion. For packaging materials, the major companies are Ecovative Design LLC, Antalis Verpackungen GmbH and Krown-Design BV. Typical packaging products today include structural packaging elements and form packaging elements.

With currently available technology, a large part of the manufacturing process consists of manual labor. This includes above all the material preparation, mold filling and the in-house handling of molds. The selection of substrate and fungal spores mainly are subject to intellectual property restrictions and therefore material combinations are not known to the public. The current applications in industrial sector underline the selection of mycelium-based products for packaging applications as a suitable use-case. For the used material mixes, one can assume the manufacturability of wood chips and hemp fibers. Besides the characterization of material properties, academia has also addressed questions in the domain of production science. Appels et al. have conducted a study, in which various fabrication variables were analyzed in relation to the resulting product's properties [13]. The study produced nine material samples by growing *Trametes Multicolor* on sawdust and straw, with or without heat pressing, and *Pleurotus Ostreatus* on cotton with heat pressing, cold pressing, and no pressing, as well as on straw with heat pressing, cold pressing, and no pressing. The findings indicate that different mushroom strains have varying effects on the colonization level and thickness of the air-exposed mycelium, which influences the stiffness and water resistance of the resulting materials. Heat pressing increases the stiffness, strength, and homogeneity of the materials, causing them to transition from a "foam-like appearance" to a "cork- or wood-like" appearance. Additionally, the materials' visual appearance significantly changes as a result of these various parameters.

Furthermore, researching possible material combinations has perceived attention. Materials under research include kenaf, hemp, corn stover, Biotex Jute, Biotex Flax, [14] and hemp hurd, wood chips, hemp mat, hemp fibers, non-woven mats [15].

A first approach in examination of the application of FIM for the manufacturing of Mycelium compounds has been presented by Mangold et al. [16]. The study finds that FIM is conceptually applicable without giving further information on necessary adoptions on the manufacturing plant or process parameters. The process consists of some preprocessing of the fibers before injection. Firstly, the fibers have to be inoculated with fungi spores and the mold has to be sterilized. In the FIM process these fibers get automated transported into the mold via airflow. Later on, the mold is closed and fibers get pressed into it. After a growth time the mycelium composite gets heated up to stop mycelium growth and activate the natural binder. At last postprocessing like pressing to achieve a certain density is possible.

Summarizing the state of the art, one can state that currently available manufacturing methods lack of automation which enable for large scale application of the products. At the same time, FIM appears as promising candidate for the automated manufacturing of mycelium compounds.

3 Material Assessment

To qualify FIM plants for the automated production of mycelium compounds, some requirements have to be met by the material systems and the environment. A list of requirements on the material system as derived from a use-case analysis and an extraction of it is presented in Table 1.

For an economically feasible application, a fast growth rate in the mold (Step (5) in Fig. 1) is required. This aspect reduces the number of mandatory molds for the manufacturing in operative environments. In order to identify promising fungal spores that meet these requirements, a literature review was conducted. The review revealed *Trametes Versicolor* and *Pleurotus Ostreatus* as promising candidates for usage in automated manufacturing processes. In addition to identifying suitable fungal spores, the review also analyzed different potential substrates for the fungal spores. Hemp fibers and straw fibers were identified as qualifying substrates for the processing by FIM. Several material probes were prepared inside the laboratory scale FIM plant with the derived material combinations. Ambient humidity and temperature during the growth phase was approximately 50% and 21 °C. Probes were grown for 7 days before drying in the above described ambient.

Table 1 Extract of the requirements as from analysis.

Requirement	Minimum	Optimum	Maximum
Growth time in mold	3 d	5 d	7 d
Fiber Lengths	5 mm	50 mm	100 mm
Transportable by airflow (Pneumatic conveying)		Yes	
Heat resistance	60 °C	80 °C	100 °C
Good separability from mold after growth		Yes	

To further qualify the materials, specimens were manufactured and tested for their ultimate tensile strength according to ISO 527-1 and ISO 527-2. The prepared geometries were according to DIN EN ISO 20753, with slight adoptions due to shrinkage in the drying phase. Tests were conducted on a ZwickRoell test machine with a controlled strain rate via an extensiometer in a closed loop (type A1).

The results of the ultimate strength tests show a brittle material behavior of the probes. Figure 2 shows two selected stress-strain curves of the different material mixes, bound by mycelium of *Trametes Versicolor*.

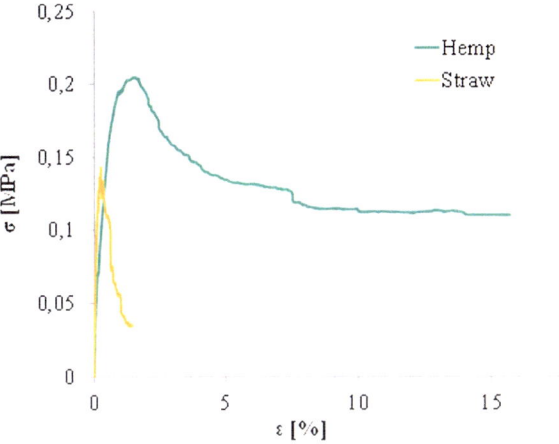

Fig. 2. Stress-strain-curves

The ultimate tensile strengths of four successfully measured probes were averaged and calculated to 0.148 MPa for straw fibers and 0.278 for hemp fibers. Thus, the material characteristics are comparable to those of silicone rubbers, polyurethane foam materials or expanded polystyrene – the chosen benchmark. For determining the manufacturing properties of the material by FIM process, a reference geometry was selected. It is depicted in. Fig. 6. (a). The selected geometry is a form, originally made from Polystyrene material used to upholster corners of an easily damageable good. It was modelled in CAD environment and molds for the FIM plant were designed and manufactured.

4   Preparation of Fim Plant

For the study, a lab-scale FIM plant designed for middle-scale series production is used, as depicted in Fig. 3. The plant was developed and tailored to the examinations in laboratory environment, as described by Moll [17]. In said prior work, especially the mold filling with conventional fibers and binder materials was examined. Therefore, an image processing system was included in the plant. In order to use the new fungi-based material system, several adaptations are necessary, which are generalizable to other similar FIM plants. The key adaptation required is the material preparation unit, consisting of the cutting unit and the fiber supply. The material preparation unit is an essential component and the preparation plays a critical role in ensuring the proper material delivery to the mold. The fiber supply unit of the material preparation unit is responsible for loading the pre-grown substrate material into the FIM plant. It is designed to provide a consistent and steady flow of material, which is crucial for ensuring that the FIM parts have uniform properties and dimensions.

To use straw and hemp materials, larger cutting forces than processing carbon fibers are required. Additionally, when processing fungi material systems, the handled materials are soaked and thus partly adhesive. This makes existing material preparation modules unsuitable. To overcome these challenges, a concept of a newly cutting and portioning unit is introduced in Fig. 4.

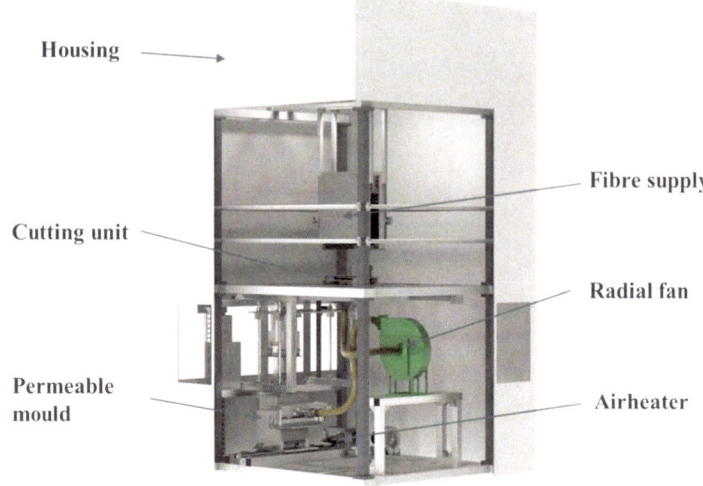

Fig. 3. Rendering of the utilized and adopted FIM plant [8]

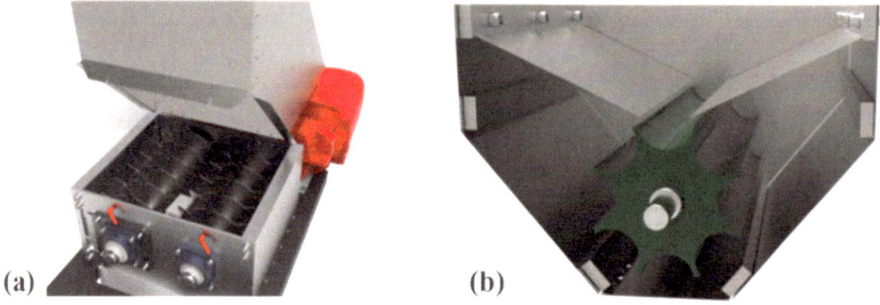

Fig. 4. (a) Cutting blades of the material preparation unit; (b) portioning unit.

Mounted to the feeding unit, the preparation unit employs rotary blades to cut the material into smaller portions. This cutting process is crucial, as it allows the material to be more effectively mixed and fed into the mold. The motor driving the rotary blades is electrically actuated, ensuring smooth and precise cutting operations. Once the material has been cut, it falls into the portioning unit. This unit is responsible for feeding a precisely measured amount of substrate into the radial fan of the FIM plant. To accomplish this, a rotary portioning element is employed. This element ensures that a consistent amount of material is fed into the radial fan, which facilitates the injection of the material into the mold. Overall, the fiber preparation unit enables the process controller to set specific values for the process control.

Secondly, the air-permeable mold used in the FIM process requires adaption according to the dimensions of the processed fibers. The inhomogeneous lengths of the biological fibers pose special challenges to the system. In order to find suitable dimensions, the molds were tested in practical experiments. Perforated plates with round holes in staggered rows with a hole diameter of 3 mm proved to be suitable for the material combinations used. It stopped the fibers in the mold but allowed for a consistent airflow. The form utilized in this study comprises a series of five individual molds arranged in a linear configuration. It is specifically designed as a two-part structure to facilitate the demolding, allowing also for an easy removal of the form from the system to accommodate the insertion of new, and empty molds. Mold filling takes place from the top, where the material is introduced into the mold. For material growth under pressure, each mold is equipped with springs that enable it to be firmly secured and braced. This ensures that the material within the mold is subject to the necessary pressure for successful growth (Fig. 5).

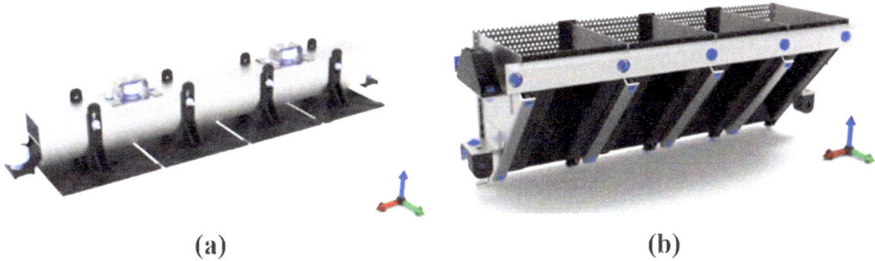

(a)   (b)

Fig. 5. Separable form consisting of upper part (a) and lower, air-permeable part (b)

Thirdly, process parameters, such as set target points for the press and the set airflow require updates. The best mold filling was achieved at a rotational speed of the radial fan of 2500 rpm. Too low airflow speeds result in fiber batches clinging to plant parts before the mold and too high speeds lead to a rather inhomogeneous distribution of fibers. Manufactured parts were dried in an external oven as the needed temperatures are more efficiently achieved there. Having these adoptions of the plant at hand, the manufacturability of the reference geometry was determined by means of practical experiments.

5   Results

For the practical test runs on the adopted plant, the qualified material systems are used. Figure 6 shows a selection of the manufactured parts, which have non-grown mycelium areas. During manufacturing of these parts, some challenges occurred, which were unforeseen, yet not specific to the FIM plant and thus require further examinations in future applications.

To transport the material into the mold by airflow, the material passes through the radial fan. If the set speed is too large, material is pushed to the housing of the fan and sticks due to its adhesive properties. Future applications need to consider that behavior when working with fans primarily intended to transport dry materials. Furthermore, blowing the material in the air-permeable form reduces the humidity in the material.

Thus, chances hold that the subsequent growth stage of the material may be disturbed since the fungi requires a certain amount of humidity for growing. One solution could be in the integration of moisture sensors into the mold. In the presented study, no sensor was used and just by experimental data, the process was tested. The variety in moisture is also considered to be the reason for the lack of mycelium growth in some areas of the part (Fig. 6).

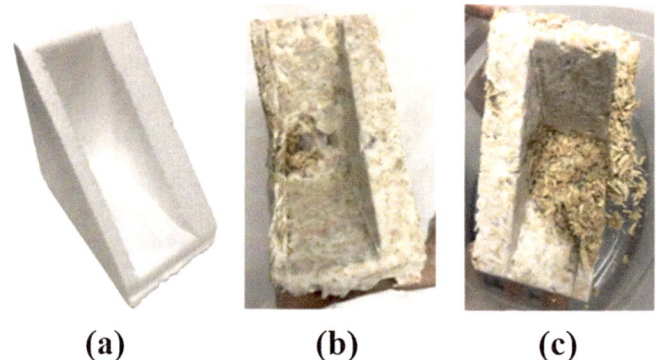

(a)   **(b)**   **(c)**

Fig. 6. Reference part (a), samples of the manufactured mycelium-composite parts with almost fully-grown mycelium (b) and partly not grown mycelium (c)

Another challenge to the FIM process is the actual growth time of mycelium itself. It would be beneficial if pre-grown mycelium parts could be blown into a mold. Also, new geometries require a new mold and plant parameter sets have to be adjusted again.

On the contrary, the cleanliness of the plant exhibited a surprisingly favorable outcome. Employing an alcohol-based surface cleaning method on the critical points of material interaction, the prepared probes demonstrated robust growth phases with minimal impurities attributable to competing fungi within the probes. An air filter was used in addition to surface cleaning. A strict hygiene is crucial to prevent growth of mold reed. Another point to mention is the increased homogeneity of the distribution of fibers in the mold comparing it with a manual filling process. The automated natural fiber cutting to a certain length helped to build this homogeneity, thus helping to reduce the rather high differences of mechanical properties comparing it to expanded polystyrene.

6   Conclusions

The conducted study reveals detailed insights to a promising process for the automated manufacturing of mycelium compounds. The possible advantages of the application of FIM technology also provoke several restrictions and disadvantages which require further examinations and improvements before industrial applications. Among others, the growth of the mycelium compound within the form presents an economic challenge since it requires numerous expensive forms. Additionally, the form bound process has disadvantages regarding adaptability to new product geometries for it is form bound.

Therefore, using FIM for mycelium compound manufacturing raises concerns regarding its economic feasibility, which necessitates a comprehensive analysis once the process parameters and conditions are better understood. As part of the study, mold filling was carried out to explore the viability of the approach. However, the selected reference part posed significant challenges because of its geometry and successful mycelium growth was not consistently achieved in the end. Therefore, further investigation and analysis of a wider range of material systems and process parameters are necessary to overcome these limitations and explore more feasible alternatives.

On the other side an automation offered advantages in better process hygiene and a higher reproducibility of the form filling process and fiber homogenization. In the future further research will be done on optimizing the inhomogeneous moisture distributed over the mold. In the end it can be said, that FIM has its drawbacks but offers some promising advantages over a manual production of mycelium-composite parts.

Acknowledgements. The authors would like to thank the Academy for Responsible Research, Teaching, and Innovation (ARRTI) of Karlsruhe Institute of Technology for supporting the research.

References

1. Haneef, M., Ceseracciu, L., Canale, C., Bayer, I.S. *et al.*: Advanced Materials from Fungal Mycelium: Fabrication and Tuning of Physical Properties. Sci Rep *7*, p. 41292 (2017).
2. Bonfante, P., Genre, A.: Mechanisms underlying beneficial plant-fungus interactions in mycorrhizal symbiosis. Nat Commun *1*, p. 48 (2010).
3. Shakir, M. A., Azahari, B., Yusup, Y., Yhaya, M. F., *et al.*: Preperation and Characterization of Mycelium as a Bio-Matrix in Fabrication of Bio-Composite. Journal of Advanced Research in Fluid Mechanics and Thermal Sciences *65*, p. 253 (2020)
4. Yang, L., Park, D., Qin, Z.: Material Function of Mycelium-Based Bio-Composite: A Review. Frontiers in Materials *8* (2021)
5. Alemu, D., Tafesse, M., Mondal, A. K.: Mycelium-Based Composite: The Future Sustainable Biomaterial. International Journal of Biomaterials *2022* (2022)
6. Bruscato, C., Malvessi, E., Brandalise, R. N., Camassola, M.: High performance of macro-fungi in the production of mycelium-based biofoams using sawdust – Sustainable technology for waste reduction. Journal of Cleaner Production *234*, p. 225 (2019).
7. Aiduang, W., Jatuwong, K., Jinanukul, P., Suwannarach, N. *et al*: Sustainable Innovation: Fabrication and Characterization of Mycelium-Based Green Composites for Modern Interior Materials Using Agro-Industrial Wastes and Different Species of Fungi. Polymers *16*, p. 550 (2024).
8. Javidan, A., Le Ferrand, H., Hebel, D., Saeidi, N.: Application of Mycelium-Bound Composite Materials in Construction Industry: A Short Review
9. Shanmugam, V., Mensah, R., Försth, M., Sas, G. *et al*: Circular economy in biocomposite development: State-of-the-art, challenges and emerging trends. Composites Part C: Open Access 5 (2021)
10. Elsacker, E., Vandelook, S., van Wylick, A., Ruytinx, J. *et al.*: A comprehensive framework for the production of mycelium-based lignocellulosic composites. Sci Total Environ *725* (2020).
11. Jiang, L., Walczyk, D., McIntyre, G., Bucinell, R. *et al.*: Bioresin infused then cured mycelium-based sandwich-structure biocomposites: Resin transfer molding (RTM) process, flexural properties, and simulation. Journal of Cleaner Production *207*, p. 123 (2018).

12. Houette, T., Maurer, C., Niewiarowski, R., Gruber, P.: Growth and Mechanical Characterization of Mycelium-Based Composites towards Future Bioremediation and Food Production in the Material Manufacturing Cycle. Biomimetics 7, p.103 (2022).
13. Appels, F.V., Camere, S., Montalti, M., Karana, E. *et al.*: Fabrication factors influencing mechanical, moisture- and water-related properties of mycelium-based composites *161*, p. 64 (2018).
14. Jiang, L., Walczyk, D., McIntyre, G., Bucinell, R. *et al.*: Manufacturing of biocomposite sandwich structures using mycelium-bound cores and preforms *28*, p. 50 (2017).
15. Lelivelt, R., Lindner, G., Teuffel, P., Lamers, H., 2015. The production process and compressive strength of Mycelium-based materials, p. 1 (2015).
16. Mangold, S., Kist, S., Friedmann, M., Fleischer, J.: Faserblasverfahren für biologische Bauteile *164*, p. 67 (2022).
17. Moll, P.: *Ressourceneffiziente Herstellung von Langfaser-Preforms im Faserblasverfahren*, Karlsruhe (2021).

Cross-Process X-ray Inspection Strategy
in Battery Cell Assembly

Steffen Masuch[1,2][✉], Sophie Gräfnitz[2], and Klaus Dröder[1]

[1] Institute of Machine Tools and Production Technology, Technische Universität Braunschweig,
Langer Kamp 19B, 38106 Braunschweig, Germany
{s.masuch,k.droeder}@tu-braunschweig.de
[2] PowerCo SE, Industriestraße Nord, 38239 Salzgitter, Germany
https://www.tu-braunschweig.de/iwf, https://www.powerco.de

Abstract. The production of lithium-ion battery cells (LIBs) for electric vehicles requires a considerable amount of energy and raw materials. Due to the increasing added value along the production chain, it is essential to recognise deviations in the intermediate products as early as possible using 100% inline measurement processes. For this reason, LIBs are inspected during cell assembly using plain radiography to qualitatively check their internal geometry. For safety-critical features, such as the anode-cathode-overhang (AC-overhang) in the composite, measuring methods are required that are not fulfilled by plain radiography. The use of computer tomography (CT) offers a solution for this demand. However, to realise a 100% CT inspection, the scan time of conventional systems and the number of applications along the cell assembly must be reduced.

This research presents an approach to overcome this limitation in the form of a cross-process X-ray inspection strategy based on a technical and economic analysis of the cell assembly. To reduce scan time and the number of X-ray applications, multiple internal features are measured in a single CT scan prior to electrolyte filling. An initial feasibility assessment of this inspection position is being investigated using a state-of-the-art metal-jet tube and photon counting detector. This provides an inspection time of 1 s, enabling detailed inspection to be combined in a time-efficient process. However, moving the CT inspection downstream would contradict the key objective of early defect detection. To overcome these challenges, an upstream vision system was developed. This vision system detects outliers immediately after the stacking process by detecting the position of the external feature separator in the electrode separator composite (ESC) without X-rays and at a much lower cost than CT. This approach aligns with the primary objective of early defect detection in the production chain through the cross-process inspection strategy.

Keywords: Inspection strategy · Battery production · X-ray inspection · Vision system · Fast CT

© The Author(s) 2025
K. Dröder and T. Vietor (Eds.): CD 2024, Proceedings, pp. 53–66, 2025.
https://doi.org/10.1007/978-3-658-45889-8_5

1 Introduction

The production of LIBs for electric vehicles is very resource intensive. Early defect detection is therefore crucial to ensure resource efficiency. In addition to ecological and economic aspects, the properties of the intermediate product must be quantified during production to guarantee function and safety. Zero-defect production is the goal, to be achieved through the extensive use of 100% inspection processes. With the current testing strategy in LIB production, cost-intensive test equipment is used to check the quality of the product and to continuously optimise production processes. Not all desired requirements can be fulfilled in mass production, as economic aspects are weighted more heavily. This applies to the use of X-ray technology in high-throughput cell assembly. In this production stage individual mechanical parts are assembled to form a component, whereby the inspection of the internal and external geometry has a central role. Inline-capable, measuring non-destructive visual inspection can be used at various inspection points to check the external geometry. In contrast, the inspection of internal geometric features requires the use of X-rays. Compared to visual inspection technology, the line integration of X-ray technology is more complex due to the radiation protection and the acquisition costs are many times higher.

Another important factor for an inline X-ray application is the data acquisition time of the method. Plain radiography is a method for the assessment of the quality of a region of interest of the test object with a single image in less than half a second. However, the precision and accuracy of the process is in many cases insufficient due to the laws of central beam projection [1]. This is why the use of CT is mandatory for measurement tasks. In CT, several projections are combined to form a 3D reconstruction. Depending on the number of projections required, CT scans can take a considerable amount of time. For this reason, until recently they have only been used for sample testing. Manufacturers are promising dramatic reductions in the time required for a CT scan with new innovations, particularly the use of liquid metal-jet tubes [2]. As these systems have only been in use for a short time, the feasibility of an inline application for LIB hast to be evaluated.

The inspection of features with X-rays can be carried out at various inspection points in the production chain due to the principle of operation. A feature can either be inspected in the intermediate product or in the finished LIB at the end of the process chain. The selection of the position along the production chain of X-ray inspections is a complex challenge due to the interdependence of economic and technical factors. In this context, the challenges for the implementation of CT in high-throughput cell assembly are presented in this research using a manufacturing scenario. A new cross-process X-ray inspection strategy is then derived based on technical and economic factors. In this strategy, two approaches are pursued to reduce the measurement effort of CT. The first strategy is based on the inspection of critical inspection features using a fast CT scan at a later stage in production. The second is to add a vision system without X-ray after the stacking process to detect outliers in the cell assembly as early as possible.

2 X-ray Inspection Strategy in Cell Assembly

This chapter evaluates the inspection options in cell assembly, focusing on X-ray technology. For the economic analysis of LIB production, the data from the baseline manufacturing plant of [3] is used with the example of Battery 1 from the BatPaC model. With a capacity of 6 GWh, the production is classified as medium-sized, which has the advantage of providing a reference point for smaller and larger factories. The NMC622-G battery has 67 Ah and the total cost can be estimated at 113 $/kWh. In the scenario, 22 million prismatic cells are produced per year on 300 production days with 3 shifts of 8 h each. In cell assembly, the systems are designed with an overproduction of 37%. If 8 production lines are assumed, a cycle time of 6 s/cell must be realised. This process chain is analysed in more detail below in order to derive an appropriate X-ray inspection strategy.

2.1 Technical Analysis of Test Equipment and Value Stream Analysis of Process Chain

Figure 1 shows the production chain of the baseline manufacturing plant from [3] with the individual steps of cell assembly and possible inspection points with plain radiography. In addition, a qualitative Sankey diagram illustrates the material flow of the good and bad parts that arise in the production steps and are identified in the inspection processes. With plain radiography, there is a risk of good parts being classified as bad (false negatives). There is also a risk of bad parts being classified as good (false positive) [1]. In addition, Fig. 1B allocates the value stream along the process chain by specifying the material and production costs. The manufacturing costs of the product consist of material and production costs. For the final LIB 20% of the manufacturing costs are related to production costs and 80% to material costs. In this analysis, the overhead costs were allocated to the individual process costs [3]. Additionally, the factors general sales, administration, research and development, profit and warranty were not considered. The breakdown of production costs for each process step is in accordance with the cost overview in [4].

The output of electrode production is individual sheets of anode and cathodes consisting of a substrate coated on both sides with active material. Due to the complex manufacturing process and the use of cost-intensive raw materials in electrode production, 61.6% of the total manufacturing costs are already achieved in electrode production [3]. Further analysis of the value and cost increase in cell assembly assumes good material (true positives) is transferred from electrode and separator production to cell assembly. Cell assembly comprises the individual steps of stacking, contacting, housing and filling. All processes are carried out in a dry room atmosphere. Dry room operation is one of the major cost drivers in cell assembly, so the usable area of systems has a particular impact on production costs. In the first production step of cell assembly, anodes, cathodes and the separator are combined to form an ESC. The separator, which is ceramic-coated on both sides with a central polyethylene layer, is within this analysis considered as a purchased material and accounts for 15.8% of the total material costs of the finished LIB [3]. In the ESC, the separators typically exhibit a circumferential overhang to the anodes (SA-overhang) and the anodes in turn overhang the cathodes circumferentially

(AC-overhang). The SA-overhang is implemented to prevent electrical contact between the anode by the cathode. The AC-overhang is relevant to ensure full coverage of the cathode. The SA-overhang is implemented to prevent an electrical contact between the anode by the cathode to prevent lithium plating. The position of the separators and electrodes in the process can currently be checked with a visual inspection during the stacking process [5]. However, simultaneous X-ray detection of the separators and electrodes in the ESC after the stacking process can only be achieved with long CT scan times due to the very different X-ray attenuation of the respective materials. For this reason, there is still no inline test for the inspection of the SA-overhang. Furthermore, the only way to measure the AC-overhang in all four corners of the ESC is to perform a CT. According to recent publications, a conventional CT takes several minutes [6]. For this reason, a vision system during stacking combined with an additional plain radiography is currently the only option in high-throughput production for quality control.

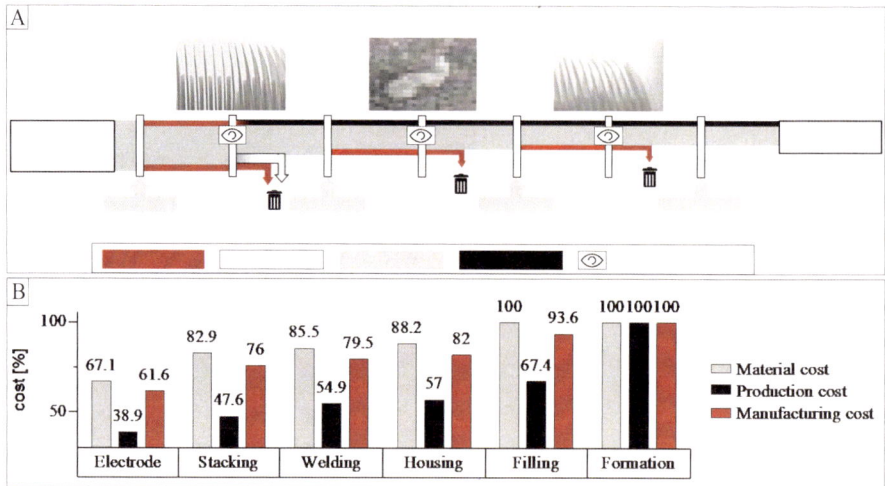

Fig. 1. **A** – Cell assembly process chain of the baseline scenario with possible options for plain radiography (RG) to test internal features of a LIB; **B** – Break-down of production, material, and manufacturing costs by individual production steps (based on [3, 4]).

In the following ultrasonic welding process, the tabs are connected to the arrester. The tabs are relatively inexpensive nickel-coated copper sheets. The requirements for the inside of the weld are the absence of pores to achieve a homogeneous current flow and a mechanically stable connection [7]. At this point, 100% defect inspection is possible using plain radiography. Plain radiography is considered highly reliable (all true positive or negative) due to its performance in defect detection, as seen in analogous tests for weld pore inspection in semiconductor manufacturing [8]. However, to measure the size and position of the pores in 3D, it is again necessary to perform a CT scan. Compared to the inspection of the AC-overhang, the requirements of the inspection without CT are fulfilled by plain radiography.

During housing, the ESC is inserted into a can. Due to the narrow tolerances between the ESC and the housing, there is an increased risk of damaging the ESC in the corner area when inserting it. It is essential to maintain a defect-free housing to prevent damage to the ESC and consequently the migration of particles and short circuits [9]. Due to the required defect detection in the corner area, tests using plain radiography fulfil the requirements. However, as with the measurement of pores in the weld seam, CT can reveal more information about the damage and the origin of defects. CT at this point would allow further analyses to be carried out over a large data set.

The final process step in the cell assembly stage is electrolyte filling, in which liquid lithium hexafluorophosphate salt, dissolved in a carbonate-based solvent system, is filled in under vacuum. At 11.8% of the manufacturing costs, the electrolyte is a comparatively expensive component of the LIB [4]. No other mechanical parts are added, and the LIB is finally sealed. Therefore, no new internal features are added at this stage.

In the next production stage, formation and ageing, the LIB is being electrochemically activated and the self-discharge rate is determined, followed by the end of line tests. In proportion, the increase in total value between cell assembly and delivery is 7.4% due to high manufacturing costs resulting from long storage times and expensive equipment [3, 4].

2.2   Cross-Process X-ray Inspection Strategy

Based on the technical and economic analysis carried out, a proposal for the use of 100% measuring X-ray inspection in high-throughput cell assembly is given below. Taking into account the factors influencing the choice of inspection technology, the inspection position and the harmonisation of the requirements of individual inspections along the entire cell assembly, the concept developed is referred to as a cross-process inspection strategy. This is illustrated below and visualised in Fig. 2.

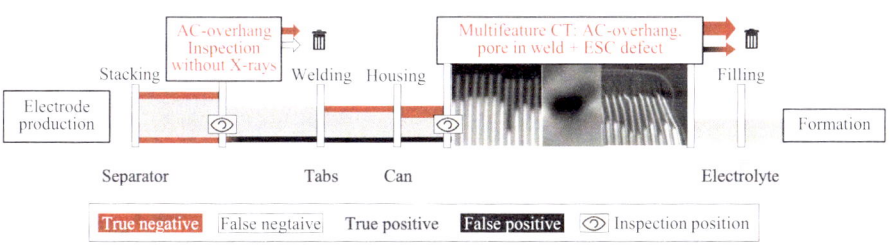

Fig. 2. Cross-process X-ray inspection strategy for implementing a multifeature CT in cell assembly.

The technical challenges of inspecting internal features were discussed. After three process steps, X-ray technology can be applied. Plain radiography can be applied after three process steps to generate individual X-ray images to visualise defects. CT scans with longer inspection times are required for measurement tasks such as determining the position of electrodes. One approach to directly reduce the measurement time is to inspect several features in one CT and accelerate the acquisition of each projection. This inspection is called multifeature CT and has the advantage of minimising the risk of false test decisions. (false positive / negative). The idea of multifeature CT is to inspect the AC-overhang, weld defects and the position of the ESC after housing. A CT in the cell assembly should not be carried out at the end of the process chain, as the value stream analysis shows a high level of total value added during the electrolyte filling and conditioning process. It is recommended to inspect after the housing, as the increase in total value due to welding and housing is only about 6% (see Fig. 1B). This test position is also advantageous from a technical point of view, as the probability of a downstream deviation in the characteristics is estimated to be low. Despite the mentioned benefits, the concept of zero-defect production is at odds with the idea of multifeature CT at a later stage in the production chain, since rejects should be identified as early as possible. However, later inspection is possible if a robust production chain has been demonstrated or other upstream quality assurance methods can be used to achieve an early identification of good parts. For this reason, an alternative control should also be implemented after the stacking process, which identifies bad parts and must be developed in coordination with the multifeature CT. For this reason, the following chapter describes the development of a low-cost alternative inspection method without X-rays after the stacking process.

3   Ac-Overhang Inspection Without X-rays after Stacking

This section analyses whether a deviating position of the electrodes of an ESC can be identified without X-rays. To develop a new inspection process, the causes of a deviating electrode or separator position in the manufacturing process are first analysed. A new test criterion is defined based on the root cause analysis. This allows indirect control of the electrode position via the separator position. A vision system for recording the test criterion is then set up and tested.

3.1   Root Cause Analysis of Stacking and Definition of a New Test Criterion for AC-Overhang

In this example, the single sheet stacking process in the Research & Development Line of PowerCo SE is analysed. Typical position deviations of the sheets (anode A, cathode K, separator S) are shown in Fig. 3A and B. In the process, the electrodes and the separator are placed on the stacking table as individual sheets, each with a vacuum gripper. The stacking table must move to four positions to place the different sheets. The "single position error" can occur if the sheets are placed in such a way just the top sheet moves. This deposit error can occur primarily due to incorrect gripper handling. In the process itself, a vision system can be used above the stacking table after stacking to identify this typical stacking error [10]. Once a sheet has been deposited, the stack is fixed vertically by a clamping system. The position of sheets already placed in the stack may be affected during the ESC build up due to friction and transverse forces applied by the gripper, the clamping system or by the acceleration of the stack. Typical error patterns are referred to as "diagonal force errors" and can be categorised as "single step" or "multi step". This error is not detected by the vision system above the stacking table, so a further check must be carried out at to identify this failure.

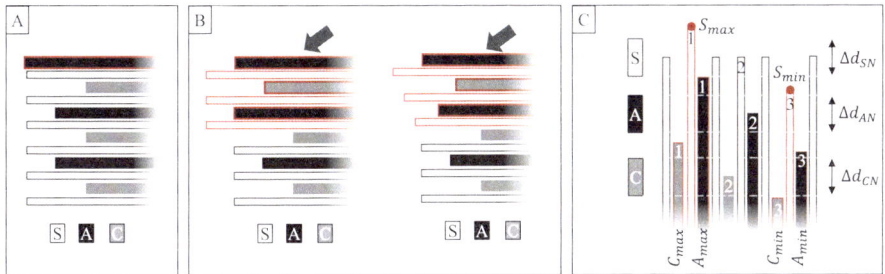

Fig. 3. **A** – Single position error; **B** – diagonal force error (left type "single step"; right "multi step"); **C** – Test criteria separator position and definition of group 1, 2, 3 of separator, anode and cathode.

For the "diagonal force error", it can be assumed that the oblique force applied to the ESC causes the adjacent sheets to move. In this scenario, or in the presence of other disturbances causing systematic displacement of adjacent sheets, the electrode position could be inferred from the position of the separators in one spatial direction. A new definition of the test criterion is shown in Fig. 3C. If there is a correlation between the separator, anode and cathode position, an outlier could be identified by defining a maximum permissible distance between separator in one spatial direction: $\Delta d_{SN} < \Delta d_S$ and $\Delta d_S = S_{max} - S_{min}$. For this outlier, it would apply that there is also a position deviation of the anodes $\Delta d_{AN} < \Delta d_A$ where $\Delta d_A = A_{max} - A_{min}$ and / or $\Delta d_{KN} < \Delta d_C$ where $\Delta d_C = C_{max} - C_{min}$ due to the large deviation of the separator position determined.

3.2 Vision System for Detection Separator Position

The separator position is an external feature which could be inspected from the lateral side using a vision system. The first step in the development of a vision system is the definition of the requirements of the object to be inspected. This data is then used to make a selection and alignment of the system components. The test object is an ESC with a total of 31 anodes (graphite), 30 cathodes (NMC622) and 61 separators. SA- and AC-overhang are 1.5 mm. The theoretical stack thickness is 11.5 mm. The sheet corners must be placed in a two-dimensional tolerance zone $\Delta d_{SN} = \Delta d_{AN} = \Delta d_{CN}$ with a width of 800 μm in one spatial direction. The smallest feature is the thickness of the separator, which is 15 μm, so according to the Shannon theorem, a digital resolution of at least 7.5 μm must be achieved [11].

The setup illustrated in Fig. 4A is developed based on these requirements. The resulting lateral image of the ESC is shown in Fig. 4B. A telecentric lens and a 10-megapixel area sensor with a pixel pitch of 3.45 μm are used for the inspection task. With the selected aperture (f-number = 10), an optical resolution of 100 lp/mm is achieved. With the magnification of the lens of 0.55 and the pixel pitch, this results in an effective resolution of 6.27 μm/pixel. The required contrast between the separator tip and the background is achieved by positioning the collimated blue illumination at an angle of 45° to the optical axis and an exposure time of 10 ms. The separators stand out as thin white lines against the black background. Manipulation of the shape of the separator can result from contact between the anode and cathode edges. In addition, an electrostatic charge can lead to attraction between individual separator sheets. As a result of these influences, a bend to the right can be seen in all the separator sheets. It would be possible to correct the separator position by spline interpolation followed by bend correction with the length of the spline. Due to the qualitatively small differences in the magnitude of the bend, the position of the separator tip is used as the spatial position without further corrections (see Fig. 4C).

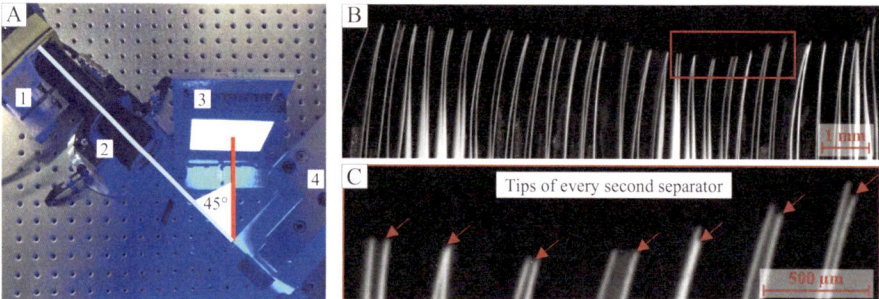

Fig. 4. **A** – Setup of vision system optical axis (white) and central beam of illumination (red), 1: camera, 2: telecentric lens, 3: collimated illuminator, 4: ESC; **B** – Resulting image of separator with high contrast of separator tips; **C** – Tips of every second separator are marked in the red section of image B.

3.3 Investigation of the Correlation Between Separator, Anode and Cathode Position

There must be a strong linear correlation in one direction between the position of the separators and the position of the electrodes in the bad ESC to use the vision system for outlier detection. For the correlation analysis, bad parts (true negative) are removed from production after the single sheet stacking process described above. No good parts are analysed because it is assumed that the position of the sheets correlate when they are correctly stacked, as the relative position differences of all sheets are small.

The bad parts are identified by CT of the corners of the ESC using the conventional system (see chapter 4, resolution 15 μm/voxel). Using this CT system, 29 ESC corners were identified which had a position distance of the long edge of the anode with $0.8\text{mm} \geq \Delta d_A \leq 3\text{mm}$ and cathodes with $0.8\text{mm} \geq \Delta d_C \leq 3\text{mm}$. . The tip position of these corners in the direction of the long edge is then recorded with the developed vision system. Finally, the tips of the separators and electrodes are labelled using the ImageJ program (version 1.52a). The pairings are numbered as shown in Fig. 3C. A labelled example images are shown in Fig. 5A. Based on the resolution of the CT and vision system, the actual position of all sheets is calculated before comparison. The main result across all pairs of anodes, cathodes and separators is shown in the diagram Fig. 5B and C. The distribution of the Pearson correlation coefficient (PCC) r across all pairs in the corner images is centred around 0.95 for the majority of the data. There are only a few outliers, which means that this experiment demonstrated a strong correlation between the positions of adjacent separators, anodes and cathodes in one spatial direction for the stacking process investigated.

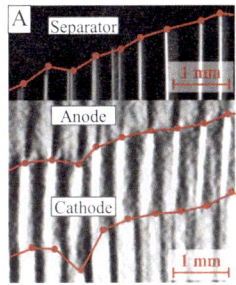 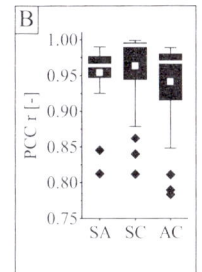 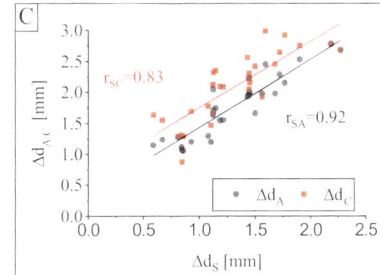

Fig. 5. **A** - Example section of labelled image data set from the vision system (top) and the CT slice (bottom); **B** - The diagram shows the PCC for the pairs of separator-anode (SA), separator-cathode (SC) and anode-cathode (AC). Over all 29 corner images the data shows a strong correlation. **C** - In the diagram, the maximum distance between the separators is plotted against the maximum distance between the anodes and cathodes of all 29 corners. The diagram visualises the linear correlation between the distances Δd_S, Δd_A and Δd_C. Compared to the difference between the anode and the separator, the difference between the cathodes is greater.

For the bad parts the criterion $\Delta d_{SN} = 0.5mm < \Delta d_S$ can serve as an indicator for an outlier, so that for anode $\Delta d_{AN} = 0.8mm < \Delta d_A$ and cathode $\Delta d_{CN} = 0.8mm < \Delta d_C$ also applies in the ESC. This is illustrated by the proof of a correlation between Δd_S, Δd_A und Δd_C which is shown in Fig. 5C. In the long term, the system must be tested further to evaluate how high the proportion of incorrect test decisions is (false positive, false negative).

4   Fast Multifeature Computed Tomography

The requirement for the realisation of a fast CT before electrolyte filling is the recording of the fastest possible projections during the rotation of the object in the X-ray beam. This requires a high tube power with a small focal spot in order to enable short exposure times for the detector [12]. As manufacturers are continuously working on this challenge, great technical progress has been made in recent years with the development of liquid metal-jet tubes and photon-counting detectors. In contrast to conventional tubes, which have a solid metal anode with air or water cooling, liquid metal-jet tubes enable considerably higher tube outputs. The increased and stable performance with a small focal spot can be achieved by the greatly improved heat dissipation through a liquid metal target made of gallium or tungsten [2]. In conventional systems, established flat-panel detectors (FPD) are also used, which convert the X-rays into an electrical signal in two steps with a scintillator layer and a photodiode. In contrast, photon-counting detectors convert X-rays in a single step, often resulting in improved image quality with higher resolution and contrast sensitivity [13]. In the following, a fast, multifeature CT with such a metal-jet system is tested before electrolyte filling. The first part of this chapter describes the data acquisition procedure. Subsequently, CT reconstructions are presented and compared to measure several features.

4.1 Methodology

Two systems are available for perform a CT scan. The technical data and the parameters for setting the CTs are shown in Table 1. Details on the components can be found in the manufacturers' data sheets. As defined in the introduction, system 1 is referred to as a conventional system, as a closed microfocus tube (max. Power 39 W) and a FPD are installed. System 2 is defined by the use of a liquid metal beam tube (max. Power 1000 W) and a photon counting detector. The first step is to set the placement of the object and the required magnification. The test object is an ESC housed in pouch foil without electrolyte, which is shown in Fig. 6. The ESC has the same characteristics as described in Chapter 3. The results of the pouch cell are assumed to be applicable to a prismatic cell due to the relatively low X-ray absorption of the aluminium housing material compared to the ESC material. The inspection area is selected for the analysis as this is where the features to be inspected are located. An object is chosen which has all production defects to evaluate the visualisation capability in the different CT scans. For all CT scans, the object is positioned identically in the X-ray beam with the central beam between the anodes and cathodes in the corner area and the electrode edges mirror-symmetrical to the axis of rotation.

As mentioned in Sect. 2.1, the required CT scan times of less than 6 s cannot be achieved with the conventional system. For this reason, the scan with the conventional system should be of high quality and not as fast as possible. The reconstruction data is used as a comparison data set. The combination of exposure time, acceleration voltage and current is set to maximise the grey scale spectrum. A focal spot of 16.5 µm at 130 kV and 0.1 mA is used for acqusition. A stepwise exposure is performed with 720 projections and an exposure time of 200 ms. The waiting time between projections is 1 s to reduce the influence of vibrations during image acquisition. The time required to acquire the projections is 144 s. The total CT scan time, including waiting time and excluding reconstruction, is 864 s.

With the metal-jet system, two CT scans are performed to acquire data as quickly as possible. To evaluate the dependence of the quality of the CT reconstruction on the acquisition speed, two fast CT scans are performed with a total scan time of 10 s (scan 2) and 1 s (scan 3). The reconstruction time is not included in this study. In both scans, 1000 projections are acquired at 160 kV, a focal spot size of 30 µm and 4.4 mA. Scan 3 uses the detector's minimum exposure time of 1 ms. The reconstructions of all the scans are carried out with a filtered back projection and have a resolution of 20 µm/voxel. The reconstructions are aligned equally, and three slices are compared for all defect cases, which can also be seen in Fig. 6.

4.2 Results and Discussion

The evaluation of fast CT with the metal-jet system is based on a qualitative comparison of the visibility of the features and various comparative measurements with conventional

Table 1 Overview of system and CT parameters.

Parameter	Conventional System	Metal-jet System	
Tube	Hamamatsu L9181–06	Excillum MetalJet E1 + Dectris Eiger 4M	
Detector	Rad-icon 3030		
Scan	1	2	3
Goal	High quality CT	Fast CT	Fast CT
Scan time	864 s	10 s	1 s
Power	13 W	700 W	700 W
Spot size	16.5 μm	30 μm	30 μm
Voxel size	20 μm	20 μm	20 μm
Exposure time	200 ms	10 ms	1 ms
Projections	720	1000	1000

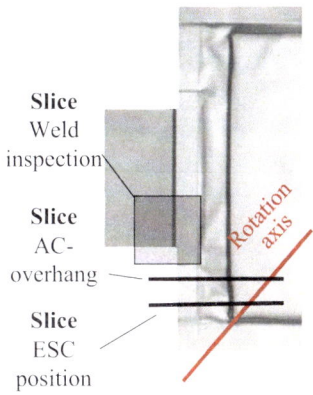

Fig. 6. Image of the test object and inspection position of slices

CT. Figure 7 shows the different raw slices of scans 1, 2 and 3. The AC-overhang (Fig. 7A), the weld seam quality (Fig. 7B) and the position of the ESC in the pouch foil (Fig. 7C) be evaluated in the slices. Basically, there is a lower signal-to-noise ratio in the fast CT scan. However, the features can still be easily identified in all scans in the manual evaluation. This is particularly evident in the qualitative analysis of the images. The position of the electrode, the weld pore, the cut in the arrester tab and the bent anode tips in the edge area can be identified in each dataset.

When the AC-overhang in the images in Fig. 7A is evaluated quantitatively using ImageJ software, it becomes clear that there are slight deviations in the position and characteristics of the features between the slices. The manual comparison measurements with three repetitions exhibit an increased difference of the mean value between the scans of the two systems. The main influencing factors are assumed to be the deviations in the beam hardening artifacts in the anode tip area in the slices of the conventional system and a deviation in slice position due to a different alignment of the reconstructions. In contrast, the measurement deviations between scans 2 and 3 are low. In addition, the standard deviation of the manual repeat measurements is small for all measurements. Furthermore, the setting of a static threshold filter in Fig. 7C shows that an automated evaluation of all slices could be possible, since path tracking for the anodes is possible.

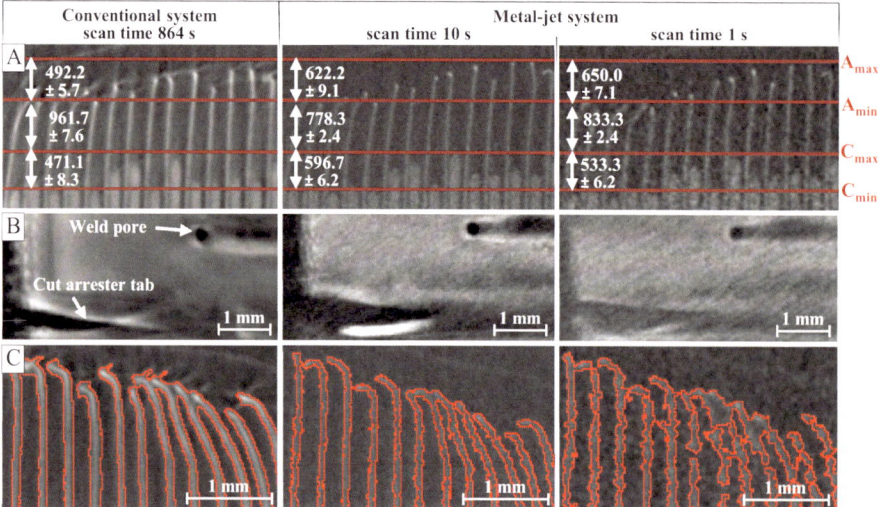

Fig. 7. A – Comparison of AC-overhang measurement; **B** – Slice of weld defects (cut in arrester tab and weld pore); **C** – Visualisation of anode tracing for detect bended tips

The comparison has shown that with a metal-jet system, the acquisition time of the projections for a multifeature CT can be reduced by using a liquid metal-jet tube and a photon-counting detector. Even in the shortest scan, which takes only 1 s, all inspection features can be identified. Further tests must investigate whether the required precision and accuracy can be achieved with this fast CT. The initial results show that this system can be used in the future to implement a 100% multi-feature CT inspection with the proposed cross-process inspection strategy in LIB mass production.

5   Conlusion and Outlook

In this research, a technical and economic analysis was carried out to derive a cross-process inspection strategy. The strategy replaces the multiple use of plain radiography with one application of 100% CT scan of internal features of a LIB in cell assembly. The prioritised inspection position before electrolyte filling is identified as advantageous, as all relevant inner features can already be inspected here and the economic loss between the stacking process and electrolyte filling due to the processing of bad parts is low at 6% of the manufacturing costs. Despite the advantages of using multifeature CT, two challenges for the feasibility of this concept need to be analysed. Firstly, the cross-process strategy must ensure a low proportion of defective parts that are processed after the stacking phase. Therefore, at this stage, a cost-effective additional inspection without X-ray systems must be developed to identify outliers. These requirements are met by the vision system investigated in Chapter 3. As an alternative inspection technique, it was found that the position of the separators can be used to infer misaligned electrode positions due to the high correlation between adjacent sheets in an ESC. Secondly, conventional CTs are unable to achieve the inspection times required for 100% inspection in high throughput cell assembly. The investigation of a fast CT in Chapter 4 provides an initial benchmark. Using a liquid metal-jet tube and a photon counting detector, a fast CT scan was performed to measure AC-overhang, detect weld defects and damages of the ESC.

Further investigations must initially focus on the testing and optimisation of the vision system and the fast CT. When used in mass production, it is very important for both systems that the image data can be analysed automatically. Above all, when implementing fast CT, not only the acquisition times of the projections must be considered, but also the downstream reconstruction times up to the final inspection decision. Due to the amount of data generated, task-specific and time-optimised algorithms must be developed. In the future, the intelligent networking of the individual tests will lead to further innovations in a cross-process inspection strategy in LIB production. The developed cross-process inspection strategy can provide a crucial basis for implementing further approaches to increase efficiency in the use of cost-intensive measuring equipment.

References

1. S. Masuch, P. Gümbel, N. Kaden, and K. Dröder, "Applications and Development of X-ray Inspection Techniques in Battery Cell Production," *Processes*, vol. 11, no. 1, p. 10, 2023.
2. A. Adibhatla, P. Takman, E. Espes, and U. Lundstrom, "X-ray sources for high throughput and extreme resolutions using liquid MetalJet technology and Tungsten targets (Conference Presentation)," in *Advances in Laboratory-based X-Ray Sources, Optics, and Applications VII*, 2019.
3. P. A. Nelson, S. Ahmed, K. G. Gallagher, and D. W. Dees, "Modeling the Performance and Cost of Lithium-Ion Batteries for Electric-Drive Vehicles", *Third Edition*, 2019.
4. Y. Liu, R. Zhang, J. Wang, and Y. Wang, "Current and future lithium-ion battery manufacturing," *iScience*, vol. 24, no. 4, p. 102332, 2021.
5. R. Leithoff, A. Fröhlich, and K. Dröder, "Investigation of the Influence of Deposition Accuracy of Electrodes on the Electrochemical Properties of Lithium-Ion Batteries," *Energy Technol.*, vol. 8, no. 2, p. 1900129, 2019.
6. Niedermeier, J., Kopp, A., Schmidt, J., Schmidt, P., Bernthaler, T., & Schneider, G., Ed., "Metrologische Computertomografie zur seriennahen Anwendung an großformatigen Batteriezellen zur Qualitäts- und Funktionsbewertung," *DGZfP Jahrestagung*, 2018.
7. E. Rohkohl, M. Kraken, M. Schönemann, A. Breuer, and C. Herrmann, "How to characterize a NDT method for weld inspection in battery cell manufacturing using deep learning," *Int J Adv Manuf Technol*, vol. 119, 7-8, pp. 4829–4843, 2022.
8. T. Tyystjärvi, I. Virkkunen, P. Fridolf, A. Rosell, and Z. Barsoum, "Automated defect detection in digital radiography of aerospace welds using deep learning," *Weld World*, vol. 66, no. 4, pp. 643–671, 2022.

9. G. Qian et al., "The role of structural defects in commercial lithium-ion batteries," *Cell Reports Physical Science*, vol. 2, no. 9, p. 100554, 2021.
10. S. Michaelis, J. Schütrumpf, J. Zienow, and A. Persichetti, "Battery Production Equipment 2030: Roadmap," 2023.
11. F. L. Luo, H. Ye, and M. H. Rashid, *Digital power electronics and applications*. London: Elsevier Academic, 2005.
12. E. A. Zwanenburg, M. A. Williams, and J. M. Warnett, "Review of high-speed imaging with lab-based x-ray computed tomography," *Meas. Sci. Technol.*, vol. 33, no. 1, p. 12003, 2022.
13. T. Flohr, M. Petersilka, A. Henning, S. Ulzheimer, J. Ferda, and B. Schmidt, "Photon-counting CT review," *Phys Med.*, 79:126-136, 2020.

Process Chain for Functionally Integrated Structures Based on Continuous Fibre Reinforced Thermoplastic Sheets

Vicky Reichel[✉], Werner Berlin, and Klaus Dröder

Institute of Machine Tools and Production Technology, Technische Universität Braunschweig,
Langer Kamp 19B, 38106 Braunschweig, Germany
{v.reichel,w.berlin,k.droeder}@tu-braunschweig.de
https://www.tu-braunschweig.de/iwf/fup

Abstract. Thermoforming enables the efficient processing of continuous fibre reinforced thermoplastic (CFRT) such as organo sheets into geometrically complex and structurally advantageous structures; however, resulting components are often hybrized through additional processes for an increased rigidity. For this, injection moulding is a widespread option, as it allows for the precise integration of plastic ribs into the thermoformed products, leading to improved structural integrity and mechanical stiffness [1, 2]. As thermoforming and injection moulding offer synergistic advantages, especially based on many related process steps, the processes can be successfully integrated [3]. When further manufacturing steps such as cutting are combined within this integrated process, the previously required handling steps are avoided entirely and the overall process cycle time can be reduced [4].

Within the present research, the integration of multiple process steps into an integrative process chain is examined through a prototypical experimental setup. The investigated one-shot injection moulding mould (IMM) combines the thermoforming and trimming of the final contour as well as the injection moulding step within an index plate mould. This enables comparably short process cycles while maintaining part quality, such as the defined position of CFRT in edge contours. The process chain is validated and resulting demonstrator components are analyzed regarding the adhesion between CFRT and applied thermoplastics by applying tensile loads. Even in comparison to conventionally established processes, the investigated integrated process chain enables feasible results. Within the examined process parameters, the temperatures within the different system components significantly influence the overall part quality. Overall, the concept for an integrated process chain shows promise for further investigations, as it offers significant gains in overall process efficiencies through reduced handling steps and potential time savings.

Keywords: One-shot process · CFRT sheet-based structures · Injection moulding · Thermoforming

© The Author(s) 2025
K. Dröder and T. Vietor (Eds.): CD 2024, Proceedings, pp. 67–80, 2025.
https://doi.org/10.1007/978-3-658-45889-8_6

1 Introduction

Fibre reinforced plastics offer a high lightweight potential while increasing stiffness and strength characteristic compared to regular plastic components. Additionally, when using continuous fibre reinforced thermoplastics (CFRT) an optimized use of the load path is possible while a repeated formability above the glass transition temperature of the matrix is possible. As a starting material with a flat, sheet-like geometry, this semi-finished product can be formed into complex geometric shapes and thus increase rigidity and strength with fibre reinforcement, especially in flat component areas. In comparison to unfilled plastics, the wall thickness and material usage can be reduced. Another key advantage is the possibility of hybridizing the structure using other polymers through additionally processes, injection moulding is a wide-spread option [3, 5, 6]. In this case, a material bond can be created with a material of the same type, either through process-integrated heating of the CFRT by means of a melt or by applying a temperature in advance, for example from radiation fields [7, 8]. Common functionalizations are back-injected ribs for stiffening surfaces or the injection moulding of connection points. There are various suitable areas of application, such as lightweight automotive parts, large industrial machine parts, thin electronic housings for notebooks or smartphones but also shoe soles and bicycle components in sport articles are manufactured with these processes [4, 9].

However, in addition to the mentioned advantages, there are some challenges in terms of processability. A typical processing method for CFRT components is thermoforming. For this, the flat semi-finished product is heated to melting temperature and thus to a flexible state. The subsequent shaping of the semi-finished product is achieved with a high applied pressure and in a forming tool with subsequent cooling processes. The polymer is conventionally applied in a separate step, in a separate injection mould. The temperature control for both the shaping process of the CFRT and the plastic application are both selected in accordance with the melting point, as they are decisive for the resulting mechanical properties as well as the overall energy-efficiency. Due to the usually sequenced process steps, several heating and cooling cycles are conventionally required and the overall process requires a number of handling steps.

The aim of the paper is to present a mould for injection moulding and its process chain for the manufacturing of CFRT-plastic components combining the three process steps of thermoforming, trimming and plastic application. An overview of the state of the art of process chains are given and advantages and challenges are discussed. The mould and its functions is shown in detail. The investigation of the production and testing of specimen based on different CFRT materials is presented. Following on this a discussion of the results and the reference to the process concludes the paper.

2 Manufacturing Technology of Functionalized CFRT

The production of components from functionalized CFRT is subject to an interplay of trimming, forming and gating operations. To combine the three process steps in one mould is the aim of the project. Common process chains from different manufacturers are conceptually visualized in **Fig. 1**.

State of the art - sequential processes cycle

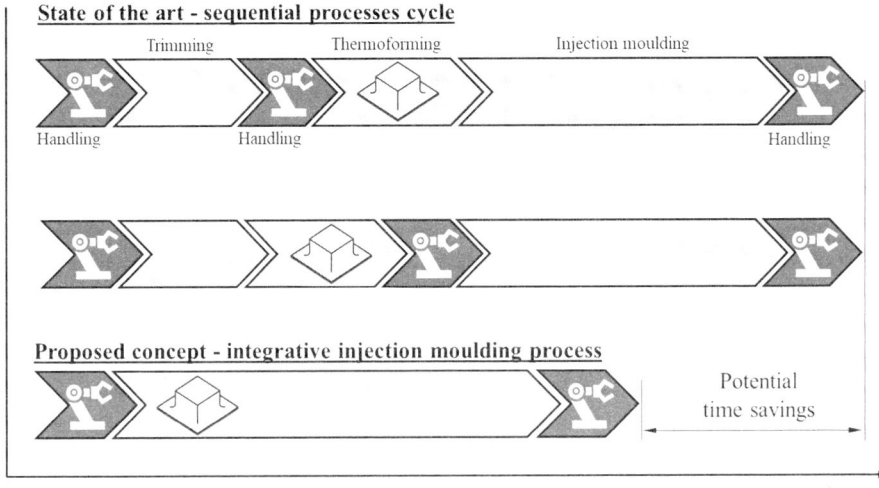

Fig. 1. Schematic process chains for CFRT sheet-based parts

In accordance with process guidelines as established by Krauss Maffei [10, 11] and Engel [12, 13], the CFRT semi-finished product is first brought into a near net shape form [2, 3, 9, 14, 15]. The blank is then transferred for example to an IR field located above the injection moulding mould (IMM) and heated. As soon as the target temperature is reached on the CFRT, it is placed in the IMM, formed and injected. The advantage of this process lies in the ability to trim complex geometries using a laser. Hardly any other trimming process can generate such precise, tool-free and dust-free cuts. The resulting advantage is the high material utilization of the expensive CFRT semi-finished products. The integration of thermoforming into the injection moulding process has the energetic advantage of process heat being used to reach the target temperature window required for the bond of CFRT and molten thermoplastic matrix [2, 15, 16].

A major disadvantage of decoupled trimming is the necessary accuracy of the positioning of the blank in the IMM. This leads to indentations at the edges of the component, which must be overmoulded to ensure a clean edge finish. The position of the blank in the edge can also not be defined, which can be disadvantageous for applications designed for high loads.

According to Dieffenbacher [17, 18], another option combines the trimming and thermoforming step. This is particularly possible with tapes, which are deposited, cut and consolidated in the target contour. A disadvantage of this process chain is the additional handling step required for moving the semi-finished part to another tool for functionalization with plastic in injection moulding.

In general, sequential processes have the advantage of decoupling and therefore greater flexibility to react to complications in production than integrative processes. This is a disadvantage when it comes to short process times in mass production. The integrated process chain suggested within the present publication addresses a solution for this conflict of goals by developing a mould and its process to combine the single steps.

3 Process Definition and Mould Design

In order to achieve the combination of the discussed process steps, a concept for a single mould enabling all manufacturing steps has to be designed. Since the plastic application can only take place in an injection mould, the trimming and thermoforming of the CFRT are integrated into it. **Fig. 2** shows a schematic sequence of the different process steps with their relative durations.

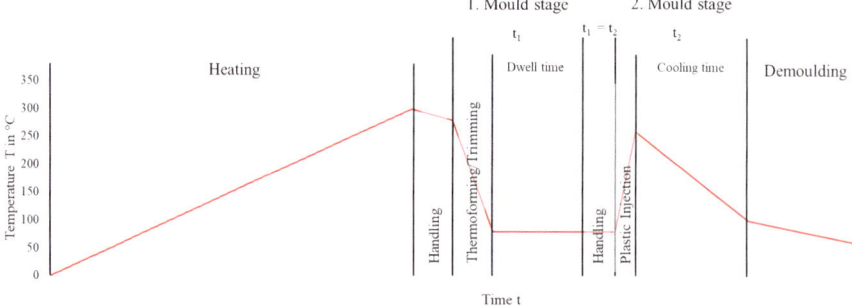

Fig. 2. Qualitative of progression of process temperature in the process chain for CFRT material based on Polyamid 6 (PA6)

Heating the CRFT is the longest step of this process. Efficient heating techniques such as infrared heating minimize this time compared to circulating air heating by oven. This is followed by the handling step, in which the heated and therefore limp textile is transferred to the opened mould. With an efficient handling system, only a small amount of heat is lost via room air convection. With the closing of the mould and the resulting full-surface contact between the blank and the mould surface for the thermoforming process, the temperature drops drastically to the temperature of the tool. This is followed by a dwell time in the first mould stage due to the time required for the plastic application in the second mould stage. The temperature remains constant until the start of the injection moulding step. Due to the plastic melt, which is heated above the melting temperature for processing, the formed and trimmed CFRT also experiences an increase in temperature, which is partially reduced again during the cooling time. During demoulding, when the part is exposed to ambient air, the temperature is reduced further and the component is finished.

Within this concept, the temperature control is a key challenge. Features such as the forming quality of the CFRT sheet, the bond between the CFRT and the plastic component and the desired time savings compared to other processes are heavily dependent on the temperature control within the mould. A high temperature in the melt is required for thermoforming the CFRT (T_O), while the contact surface in the mould has significantly lower temperatures for injection moulding (T_w). Depending on the material, this can be a delta of 170 °C for PA6 (T_O: 260 °C and T_m: 80 °C) or a delta of 120 °C for PP (T_O: 180 °C and T_m: 60 °C). Due to these high temperature differences, the thermoforming of the CFRT cannot be fully completed, which means that a higher input temperature of the CFRT must be selected. This is associated with increased time and energy requirements as well as an increased risk of degradation effects in the polymer matrix [1]. A further conflict of objectives in the second mould stage is the temperature during injection on the CFRT, which is decisive for the bond [1, 19]. Higher temperatures of the inserted CFRT are beneficial for the resulting overall stability; however, analogous conflicts regarding heating and cooling times arise, necessitating further investigations and evaluations regarding an optimal trade-off between a high efficiency and an appropriate resulting component quality.

Description of mould and functions

The aim was to reproduce the process sequence described in **Fig. 2** in an injection mould using index plate technology. This technology allows the transfer of the formed and cut CFRT through a rotation from the upper to the lower mould half, moving the ejector side to the second process step of the plastic application (See **Fig. 3**). Therefore, in the ejector side the mould form is identical for the upper and lower half of the mould where thermoforming and trimming as a first step and injection moulding is realized as a second process step.

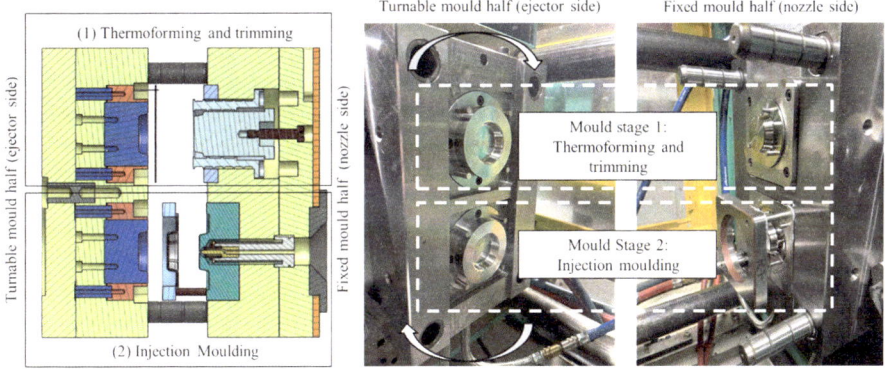

Fig. 3. Schematic design of IMM and its functions (l.) and manufactured mould (r.)

The advantage of this concept is, on the one hand, the possible decoupling of the process steps to increase quality and reduce process susceptibility. In addition, a finished part can be ejected with each opening and closing movement. The first part therefore has a full cycle time, and the output time is reduced by half with each additional part, thus maximizing the efficiency of the process.

The mould specifications cover a range of crucial functions, including the provision of a holding fixture designed for the heated CFRT. Additionally, the mould facilitates the precise thermoforming of the CFRT, employing a dedicated heater for an effective implementation of the forming punch. Following the thermoforming process, the mould incorporates a shear cut mechanism, ensuring the desired shape and dimensions. In order to facilitate the next process steps, the entire mould half is rotated into the stage of plastic application.

The schematic for the thermoforming and trimming unit is shown in **Fig. 4**. In the nozzle side (1), a spring-loaded mould half (2) is visible, which is pressed in with the closing movement of the mould and thus releases the cutting sleeve (3) after the shaping process. By moving the cutting punch (10) inside the cutting sleeve (3) the shear cut of the CFRT part is enabled after the forming mould (8) is closed completely and thermoforming is finished.

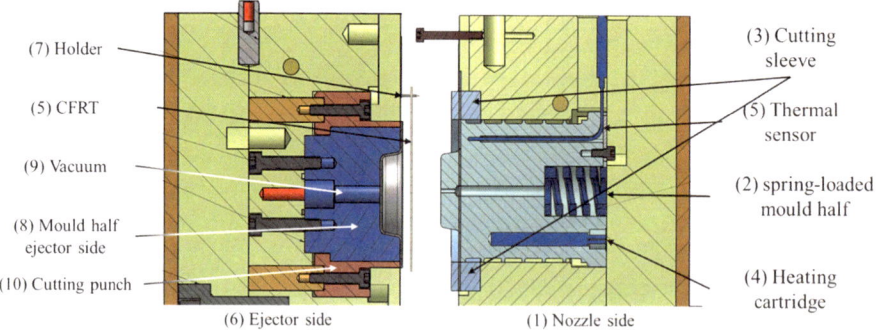

Fig. 4. Schematic mould illustration of mould stage 1: thermoforming and trimming

Shear cutting is an established process with high economic efficiency for the large-scale production process [20]. It was selected as a suitable trimming process, as it can be synergetically integrated into an injection mould, utilizing the closing movement for applying the shearing forces. The entire cutting edge of the specimen is produced in one stroke. Furthermore, the upper part of the right mould half is heated by heating cartridges (4) with 2 x 315 W in order to adjust the temperature difference between the mould and the preheated CFRT sheet more precisely than with a water-based mould temperature control. The CFRT (5) is held in position on the ejector side (6) by a holder (7). The aim after thermoforming is for the insert to remain within the mould half of the ejector side (6) for the purpose of transfer it to the lower mould half for the following injection moulding step. To ensure the remaining of the CFRT in the ejector side (8), a vacuum channel (9) is integrated. The upper and lower parts of the ejector side (6) are identical.

The structure of the second functional stage is shown in **Fig. 5**. The ejector side (6) is manufactured in the same way as the first functional stage. The nozzle side (1) contains the hot runner (11) and the injection moulding cavity (12). After the insert has been rotated into the lower half of the mould, the mould closes and the plastic is applied. The design of the cavity creates an undercut through which the finished component (15) is removed from the stripper ring (14) when the mould opens. This replaces the usual ejector pins on the ejector side (6).

The specimen consists of a circular CFRT sheet insert, which is injected on one side (see **Fig. 6**) at the shear-cut edge, an overmoulding is realized through the cavity geometry. By selecting a circular shape, the influence of the fibre angle to the cutting plane on possible fraying can be determined [21]. Possible fibre protrusions can thus be

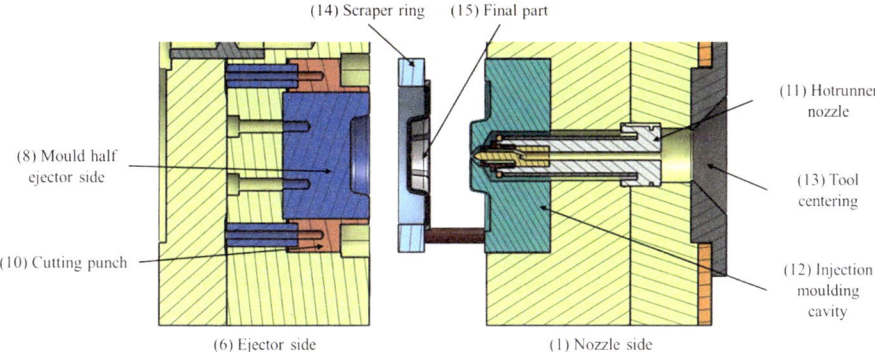

Fig. 5. Schematic mould illustration of mould stage 2: injection moulding

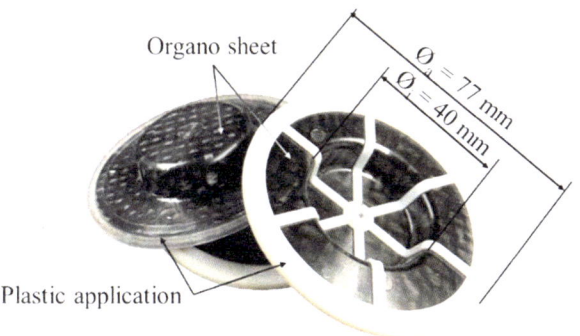

Fig. 6. Specimen geometry, size and characteristics

enclosed by the melt during injection and lead to an improved bond between the melt and the CFRT insert. The outer circle diameter of the functional demonstrator is Ø = 77 mm including a 6 mm overmoulding on the sides. The CFRT insert has a diameter of Ø = 71 mm, so that 3 mm of the edge is overmoulded all the way around. The six flow channels are arranged symmetrically at each point and have a rectangular cross-section of 2 mm^2.

4 Processing

4.1 Results on Shear Cutting of CFRT

Three different CFRT materials are investigated as shown in **Table 1**. All CFRT are based on textile fabrics in 0°/90° fibre orientation. The weight per unit area and amount of fibre are different for glass fibre (GF) based and flax fibre (FF) based material.

Overall, the specimens at the cut edge show that the hypothesis of an influence of the fibre angle cannot be confirmed. No difference can be determined at high and low fibre angles. It was found that an influence of the tilting of the cutting tools due to gravity is visible in the cut edge. The contours in the upper area of the specimen mostly result

Table 1. List of investigated CFRT materials

	Material A	Material B	Material C
Manufacturer designation	Tepex® dynalite 104-RG600(x)/47%	Tepex® dynalite 102-RG600(x)/47%	Tepex® dynalite 831-F150(x)/41%
Declaration in testing	PP	PA6	PLA
Plastic component	Polypropylene	Polyamid 6	Polylactic acid
Fibre material	Glass fibre	Glass fibre	Flax fibre
Manufacturing temperature	165 °C	220 °C	150 °C

in a frayed edge, while the lower area can be cut cleanly. Due to the tilting, the cutting clearance cannot be kept constant around the whole cutting edge when the cutting tool plunges into the opposite mould half.

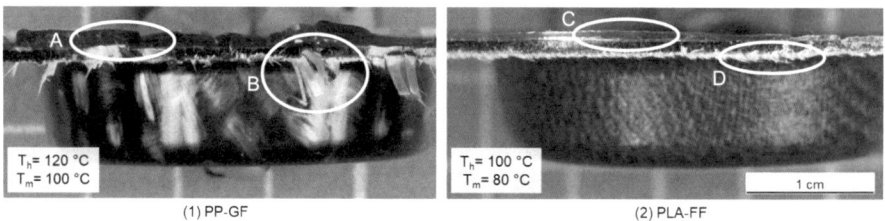

(1) PP-GF   (2) PLA-FF

Fig. 7. Result: side view of cutting-edge characteristic for different materials

The cutting clearance was chosen to be 0.05 mm due to the high demands on leak tightness in the second tool stage. Despite this small cutting clearance, the fibres are able to fill the smallest gaps can therefore not be cut but only torn. As a result, fringes become visible at the edge of the cut.

In **Fig. 7** a cutting result for PP-GF (1) and PLA-FF (2) is shown. The figure shows both fibre fringes (B, D) and matrix leakage (A, C) in the cutting cavity. This occurs as a result of high temperatures of the heating cartridge (T_h) and mould (T_m), which reduce the temperature difference between CFRT and mould and thus prolong the molten state. For the PP-GF material, higher fibre fringes and matrix leakage can be determined. The reason for this is assumed to be the better cuttability of flax fibres as well as the lower matrix content in the CFRT semi-finished product. In addition, the temperature window for processing the PLA-FF is comparably small, so that a degradation of the material can be observed rather than a reduction in the temperature difference between the tool and the material.

For PA6-GF material, this material degradation has not extensively developed due to its higher temperature stability. Resulting from a high temperature difference between material and cutting tool, the material cools down within a comparably short time and therefore fibre fringes and matrix leakage is prevented. As shown in **Fig. 8** fibre fringes cannot be entirely overmoulded by the injected melt. The presence of fibres lead as well to higher accessibility for the injected plastic melt in the sealing area where the resulting gap is filled.

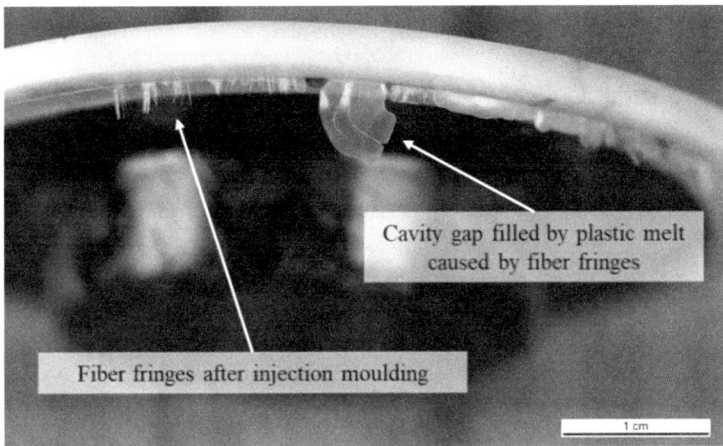

Fig. 8. Influence of fibre fringes in injection moulding step

Nevertheless, there were specimen of all materials with a low amount of fibre fringes, which could be used in the following process step of injection moulding. Even after approximately 3000 strokes, no wear on the cutting-edge tool affecting the process could be detected. By changing the fibre material an adaption on the wear resistance might become necessary.

4.2 Results on Testing the Pull-Off Force

To evaluate the functionality of the mould in the process sequence, test specimens were produced and tested for the pull-off force of the components. The test setup is shown for an exemplary test in **Fig. 9**. The test specimen is clamped at the top and bottom and subjected to tensile force. The clamping jaws allow the injected plastic to be loaded in shear to the base body of the CFRT so that the pull-off force can be determined. It is to be expected, that increased temperatures in the injection moulding process results in improved bonding quality and thus increased the resulting pull-off force.

The test plan resulting from this expectation (see **Table 2**) varies the temperatures for the individual materials. The melting temperatures differ depending on the matrix material. The results can therefore be used to design the temperature control of the overall process, in which the temperatures of the heating cartridges, mould and plastic melt are decisive.

There were ten specimen per variation manufactured and at least five were tested valid. Due to the selected testing procedure, in some specimen the CFRT sheet material was tested rather than the pull-off force (see **Fig. 10**). For the further evaluation, merely

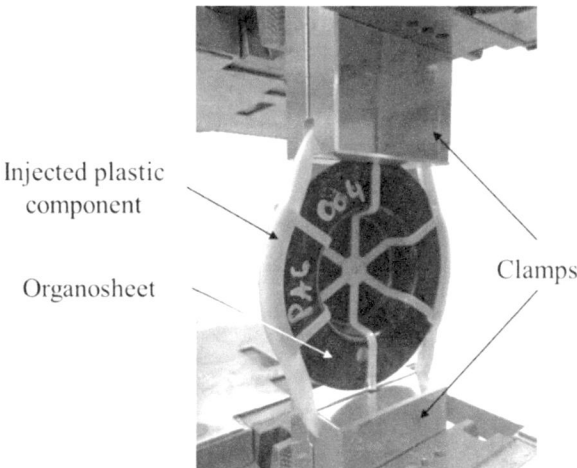

Fig. 9. Test setup for the pull-off test in terms of determining the pull-off force

Table 2. Test plan

| | Test plan 1 – Reference specimen | | | Test plan 2 | |
	PLA/Flax fibre	PP/Glas fibre	PA6/Glas fibre	PP/Glas fibre	PA6/Glas fibre
Temperature of heating cartridge T_h in °C	80	100	130	100	130
Mould temperature T_m in °C	60	60	80	100	120
Temperature of plastic melt T_p in °C	180	190	260	190	260

test results resulting in a detachment of thermoplastic and CFRT sheet were designated as "valid" and evaluated further.

The pull-off forces determined can be seen in **Fig. 11**. The highest pull-off force was determined for PP at a mould temperature T_m of 100 °C. An increase in the mould temperature T_m is accompanied by an increase in the pull-off force, an even stronger analogue influence is visible for PA6. An increase in T_m leads to improved bonding between the base material and the injection-moulded compound (see **Fig. 11** r.). For PP the pull-off force increased by + 34% and for PA6 by + 57% while the mould temperature difference to the reference value was increased by $\Delta = 40$ °C.

Fig. 10. Test results for valid testing (l.) and invalid testing (r.)

The forces determined are higher for PLA and PP samples compared to the mechanically superior PA6. This is partly due to the temperature control and the better sample quality of low-melting polymers due to the lower temperature difference between melt temperature T_p and mould temperature T_m. The temperature input in the bonding surface takes place almost exclusively via the temperature of the melt.

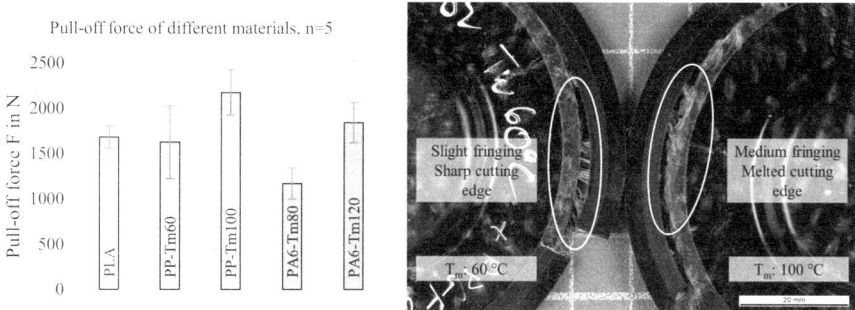

Fig. 11. Results for pull-off force (l.) and examples for specimen with different mould temperatures T_m (r.)

By melting the surface of the CFRT sheet insert, welding can be achieved for bonding. Despite the visible standard deviation, the data basis is sufficient to show a significant influence of the mould temperature T_m.

5 Conclusions

With the process presented within the presented research, the possibility of manufacturing small, mass-produced components with an integrated process chain is successfully demonstrated. Through an integration of trimming and thermoforming within the first mould stage distortion and undefined alignments of the CFRT part can be avoided. As the formed part is left within the mould half without demoulding and rotated into the position required for injection moulding, intermittent handling steps are avoided entirely. As the following injection moulding as second mould stage is parallelized and uses the same tool stroke as the thermoforming and trimming, a scalable high-throughput integrated process chain was achieved. The overall process was validated with a demonstrator geometry and a combination of various conventionally used material combinations. Similar to most established processes for injection moulding CFRT parts, the pull-off of thermoplastic and CFRT part is mostly dependent on suitably high temperatures during moulding. As a result, the achieved forces during pull-off force test are comparably low. This could be compensated through additional heating strategies in line with other approaches for injection moulding from the state of the art in future experiments. However, possible individual process optimizations for the sub-process of injection moulding does not diminish the overall effectivity of the presented process chain. With the validated integrated processes, the basis for a functionally integrated, scalable and highly efficient thermoforming, trimming and injection moulding is established. Especially the reduction of required handling tasks as well as heating/cooling cycling times offers promising future approaches for a more efficient and productive process chain for functionalized CFRT components.

Acknowledgements. This study was carried out within the project HyStamp - "Functional integration in the mould through the process combination of casting and injection moulding in the production of thermoplastic hybrid components" (KK5052902HD1, 2021–2023) which was funded by the Central Innovation Programme for small and medium-sized enterprises (SMEs) as a programme of the German Federal Ministry for Economic Affairs and Climate Action. We kindly thank our project partner **febana - Feinmechanische Bauelemente GmbH, Sömmerda** for the innovative ideas, support and cooperation.

References

1. B.-A. Behrens et al., "Numerical Modelling of Bond Strength in Overmoulded Thermoplastic Composites," J. Compos. Sci., vol. 5, no. 7, p. 164, 2021, https://doi.org/10.3390/jcs5070164.
2. A. Hürkamp, T. Ossowski, and K. Dröder, "In-mould assembly of functionally integrated structures: A surrogate model for fast quality assessment," CIRP Annals, vol. 71, no. 1, pp. 37–40, 2022, https://doi.org/10.1016/j.cirp.2022.04.009.
3. M. Schuck, "Kombination von Materialleichtbau mit konstruktivem Leichtbau," Lightweight Design, vol. 5, no. 2, pp. 54–59, 2012, https://doi.org/10.1365/s35725-012-0090-7.
4. Lanxess, Tepex - Fields of application, https://lanxess-performance-materials.com/en/Products-and-Solutions/Tepex/Fields-of-Application, last accessed 2024/01/19
5. M. Bouwman, Donderwinkel T., and S. Wijskamp, "Overmoulding–an integrated design approach for dimensional accuracy and strength of structural parts. ITHEC Proceedings," in ITHEC 2016: 3rd International Conference and Exhibition on Thermoplastic Composites: 11–12 October 2016, Congress Center Bremen, Germany: conference proceedings, 2016.
6. M. A. Valverde, R. Kupfer, T. Wollmann, L. F. Kawashita, M. Gude, and S. R. Hallett, "Influence of component design on features and properties in thermoplastic overmoulded composites," Composites Part A: Applied Science and Manufacturing, vol. 132, p. 105823, 2020, https://doi.org/10.1016/j.compositesa.2020.105823.

7. J. P. Beuscher, R. Schnurr, A. Müller, M. Kühn, and K. Dröder, "Introduction of an in-mould infrared heating device for processing thermoplastic fibre-reinforced preforms and manufacturing hybrid components," 2017, https://doi.org/10.24355/DBBS.084-201906241 541-0.

8. C. Bruns, J.-C. Tielking, H. Kuolt, and A. Raatz, "Modelling and Evaluating the Heat Transfer of Molten Thermoplastic Fabrics in Automated Handling Processes," Procedia CIRP, vol. 76, pp. 79–84, 2018, https://doi.org/10.1016/j.procir.2018.01.011.

9. P. Egger, "Spritzgießtechnik steigert Effizienz in Faserverbundfertigung," Lightweight Design, vol. 7, no. 1, pp. 58–62, 2014.

10. KraussMaffei, FiberForm-Technologie, https://www.kraussmaffei.com/de/unsere-verfahren/fiberform-technologie, last accessed 2024/01/19

11. Andreas Burkert, Krauss-Maffei bringt das FiberForm-Verfahren auf Trab. [Online]. Available: https://www.springerprofessional.de/mechatronik/umformen/kraussmaffei-bringt-das-fiberform-verfahren-auf-trab/15784462, last accessed 2024/01/15

12. Engel Austria, Organomelt, https://www.engelglobal.com/de/de/unternehmen/media-center/news-presse/integrierter-thermoplastprozess-fuer-niedrigste-stueckkosten-im-composite-leichtbau, last accessed 2024/01/15

13. Engel Austria GmbH, Organomelt Verfahren erreicht neue Dimension. [Online]. Available: https://plasticker.de/Kunststoff_News_35275_Special_k19_Engel_Organomelt_Verfahren_erreicht_neue_Dimension?special=k19, last accessed 2024/01/15

14. C. Brecher, M. Emonts, A. Kermer-Meyer, H. Janssen, and D. Werner, "Herstellung von belastungsoptimierten thermoplastischen Faserverbundbauteilen," in Leichtbau-Technologien im Automobilbau, W. Siebenpfeiffer, Ed., Wiesbaden: Springer Fachmedien Wiesbaden, 2014, pp. 70–75.

15. A. Hürkamp, S. Gellrich, A. Dér, C. Herrmann, K. Dröder, and S. Thiede, "Machine learning and simulation-based surrogate modeling for improved process chain operation," Int J Adv Manuf Technol, vol. 117, 7-8, pp. 2297–2307, 2021, https://doi.org/10.1007/s00170-021-070 84-5.

16. R. Giusti and G. Lucchetta, "Modeling the Adhesion Bonding Strength in Injection Over-molding of Polypropylene Parts," Polymers, vol. 12, no. 9, 2020, https://doi.org/10.3390/polym12092063.

17. Dieffenbacher, FibreForge, https://dieffenbacher.com/de/forming/produkte/tapeverarbeitung/fiberforge, last accessed 2024/01/15

18. Dieffenbacher, Erste DIEFFENBACHER Fiberforge für Südkorea, https://dieffenbacher.com/de/unternehmen/news/detail/erste-dieffenbacher-fiberforge-fuer-suedkorea, last accessed 2024/01/15

19. K. Dröder, A. Hürkamp, T. Ossowski, B.-A. Behrens, H. Wester, and R. Lorenz, Integrierte Prozesssimulation von Thermoformen und Spritzguss. Hannover: Europäische Forschungsgesellschaft für Blechverarbeitung e.V. (EFB), 2022.

20. W. Volk and J. Stahl, "Shear Cutting," in CIRP encyclopedia of production engineering, S. Chatti and T. Tolio, Eds., Berlin, Heidelberg: Springer Berlin Heidelberg, 2019, pp. 1–9.

21. V. Zal, H. M. Naeini, A. R. Bahramian, and B. Abbaszadeh, "Experimental evaluation of blanking and piercing of PVC based composite and hybrid laminates," Adv. Manuf., vol. 4, no. 3, pp. 248–256, 2016, https://doi.org/10.1007/s40436-016-0147-4.

Adaptive Unscrewing Automation for Electric Motors

Patrick Alexander Schmidt[✉], Florian Schmutzler, David Löffler, Christian Fritsch, and Uwe Frieß

Fraunhofer Institute for Machine Tools and Forming Technology IWU, Reichenhainer Straße 88, 09126 Chemnitz, Germany

{patrick.alexander.schmidt,florian.schmutzler,david.loeffler, christian.fritsch,uwe.friess}@iwu.fraunhofer.de

Abstract. This paper presents the development of solutions for the autonomous disassembly of electric motors, a vital component in the realm of e-mobility. With the escalating growth in electric vehicle (EV) production, exemplified by the registration of 10.8 million EVs worldwide in 2022 and the substantial sales of specific electric motor models, the importance of efficient recycling and disassembly processes has become more pronounced. Recognizing this, our project "Zirkel" aims to address the challenges of circular economy and resource efficiency through innovative disassembly techniques.

Our methodology involved comprehensive pre-disassembly workshops to familiarize ourselves with the electric motor as a component, understanding its structure, and identifying potential challenges in the disassembly process. The workshops provided crucial insights into the specificities of electric motors used in popular EV models like the Volkswagen ID.3 and ID.4, paving the way for the development of an autonomous disassembly system.

The technical challenges in developing this system were manifold, tackled through a multi-staged approach that integrates both sensorial and mechanical processes. This approach allowed us to efficiently navigate the complexities associated with the disassembly of electric motors, ensuring a balance between precision and practicality.

In the demonstrator section of the paper, we elucidate the step-by-step process of overcoming these technical hurdles, detailing the development stages and the innovative solutions we employed. This includes the use of advanced automated techniques, adaptive designs, and automation of previously manual processes.

Keywords: Autonomous disassembly · Screw detection · Design for recycling · Circular economy · Electric motors

1 Introduction

In the current era of rapid technological advancement, it is increasingly important to embrace a circular economy. This approach focuses on optimizing the utilization of products and materials, minimizing waste and decreasing the demand for raw resources.

K. Dröder and T. Vietor (Eds.): CD 2024, Proceedings, pp. 81–93, 2025.
https://doi.org/10.1007/978-3-658-45889-8_7

The automotive industry exemplifies significant challenges, particularly in relation to vehicles and their components often becoming prematurely obsolete before being used in a second or third life cycle. This results in discarded vehicles being shipped overseas as used cars before ultimately ending up in scrapyards, depleting resources like rare earth metals, copper, and high-quality alloys. The global scale of this issue is underscored by having exported 23 million used vehicles worldwide in the period between 2015 and 2020. Approximately 66% of these vehicles were traded to developing and transition countries. The European Union (EU) emerges as the main exporter with around 11.5 million units being traded of which 42% were exported outside the EU [1].

Today's global landscape is characterized by increasing competitive pressures in various industries, including resource scarcity, rising raw material costs, waste disposal challenges, pollution concerns, carbon emissions, energy scarcity, and unpredictable cost increases. In response to these challenges, we introduce "Zirkel": a project addressing the core principles of the circular economy, resource efficiency of e-mobility components. Zirkel is an innovative initiative featuring a multifaceted strategy that includes advanced automated disassembly techniques for electric motors, adaptive design and automation of previously manual disassembly processes [2].

In this paper, the necessary steps towards an autonomous disassembly of electric motors are discussed. The electric motor is a pivotal component, reflecting the expanding significance of electric vehicles (EVs). In 2022, a notable 10.8 million EVs were registered worldwide, almost doubling from 6.7 million in 2021. [3] Significantly, the electric motor model we analyzed—a permanent magnet synchronous motor in hairpin design from Volkswagen—was sold 253,000 times in 2022 and incorporated into models such as the ID.3 and ID.4. [4].

This paper is structured to initially provide a comprehensive overview of the current state of dismantling technologies on electric motors, followed by a detailed explanation of our methodology. Subsequently, we delve into the development of our demonstrator. The process of developing this autonomous disassembly system was intricate and multi-staged, necessitating the overcoming of several technical challenges regarding sensor technology and the mechanical side. These challenges and their resolutions are explained in the demonstrator section of this paper.

2   State of the art

Reducing the criticality and increasing the sustainability of rare earth materials are challenges that are also being tackled by politics, for example in the United Nations Sustainable Development Goals. [5] On the part of the European Union, the Green Deal [6] the Circular Economy Act and the Battery Ordinance are guidelines that oblige OEMs, e.g. of batteries and drives, to create take-back systems, ensure reuse or dispose of their products in an environmentally friendly manner. The EU BattVO also introduces a digital battery passport, labeling requirements and documentation of the CO_2 footprint.

In order to counter the disadvantage for the German industry on the global raw materials market, a closed recycling loop, especially for rare earth materials, must be established along the value chain of EVs. These metals are playing an increasingly important role in society in view of the growing importance of renewable energies and e-mobility. The reuse and recycling of traction motors, along with traction batteries, has one of the greatest leverage effects for this project. When electric motors are scraped, the rotor and stator are separated. The focus when recycling electric motors is on the copper of the stator. The rotors are classified as mixed scrap as there is a lack of recycling on an industrial scale. Both components are currently shredded separately. The shredded materials are then sorted through various filter systems (e.g. magnetic separators). The economically strategic rare earth magnets (e.g. neodymium (Nd) or samarium (Sm)) contained therein are not extracted and end up in steel recycling. [3, 4].

To this day, no large-scale industrial magnet recycling system exists. Only a few processes are shown to be scalable but are not jet scaled up. There are also research projects for recycling rare earth materials from electric motors, which are running worldwide. [7, 8].

The realization of a material cycle usually fails due to economic and technical obstacles. Be it the number of parties involved, a large heterogeneity in terms of specifications or the OEM's reluctance to provide information. As there is still no standardization for Nd magnets in electric motors and the variety of alloys used is large, industrial recycling is currently uneconomical. Additionally, the disassembly process of e-motors is mostly performed manually. However, as the growing waste streams from old electrical appliances and wind turbines are expected to be the main source of rare earth materials such as Nd over the next two decades, a functioning circular economy should be the goal. [3, 6, 7].

Investigations into the dismantling of electric motors have been or are currently underway in the following projects [9]:

- MORE, 2011–2014, recycling of electric motors [10]
- DeMoBat, 2019–2023, industrial disassembly of battery modules and electric motors [11]
- ZirkulEA, 2022–2025, circuit capability of the electric drivetrain [12]
- AgiProbot, 2019–2024, agile production system using mobile learning robots [13]
- DemoSens, 2020–2023, digitization of automated disassembly [14]

In the MORE project, the reuse of the magnets from the rotor was tested, as they are not damaged during operation. In a specially developed process, the magnets were removed non-destructively from the laminated core, demagnetized and freed from adhesive residues. The recyclates obtained in this way can generally be reused after renewed magnetization. The magnets were also produced from recycled Nd powder, but these have a 3% lower remanence. [3].

As part of the Fraunhofer Society, the Fraunhofer IFAM has developed a process for recovering recyclates from magnetic materials and magnetic material mixtures. The recycling process is based on a combination of pyrometallurgical and wet-chemical process steps. The Fraunhofer IWKS is also researching the establishment of a circular economy for functional components for e-mobility at its Center for Dismantling and Recycling for Electromobility (ZDR-EMIL). [4, 9].

Detachable and debondable connections, such as screw connections, are an important means of designing products in a circular way. However, there are many challenges that need to be overcome when automating disassembly:

Automated screwhead detection can be challenging due to the small size in comparison to mere object identification and potential deformation of screwheads, requiring increased accuracy. [15] To meet this challenge, a systematic pre-processing method is used that links edges in the images using a sampling rate adjuster. In addition, the application of a Gaussian blur filter improves performance by removing noise, and the Canny edge detection algorithm acts as an edge isolator, removing everything but edges in the image. Various detection methods are used for screws on flat surfaces. The Hough Line Transformation identifies the screw thread as two parallel lines but does not provide information on the tip direction. It has a low error rate on isolated screws but a higher one when there are distracting features in the background. On the other hand, boundary-following algorithms offer precise orientation but face challenges in close-proximity scenarios. To detect overlapping screws, a template-based technique can be used, although it may have a longer processing time. [16].

To classify different types of screws, a two-step approach is employed. It uses detectors such as R-CNN or YOLOv3, which balances speed and accuracy. First, the screw head is localized, and then it is classified into discrete classes. Efficiency is improved through network design optimizations. Comparing YOLOv5 models, it was found that a larger classifier maintains comparable accuracy but takes twice as long to evaluate the image. To address false alarms, an additional 'None' class has been introduced. The second step involves robot motion to configure a tool for disassembly. A tool recommendation system, which combines the screw image extraction mechanism and an EfficientNetV2-based image classifier, ensures adaptability to real-time robotic disassembly with fewer parameters. [15, 17].

3   Methodology

The process of autonomous disassembly can be fundamentally divided into overarching planning tasks (IT level) and their implementation by the mechatronic system (OT or field level). The process is described below.

Ideally, at the beginning of the disassembly, the planning is carried out as a form of "cognition" (Fig. 1, step 1). Semantic information is queried from a database or cloud environment to retrieve the information model of disassembling a specific component. This model contains all the necessary information for the disassembly process. Based on this, the state of the component to be dismantled is initially captured at the disassembly station (Fig. 1, step 1). The real state, both mechanical (actual geometry) and electrical, as well as the position and condition of the connecting elements are recorded. These data form the populated information model representing the entire actual state.

In the analysis of Volkswagen's electric motors, a comprehensive disassembly guide was developed using CAD data from the rear axle machine. The guide effectively segmented the disassembly process into 14 manual steps, providing detailed information on components, joining connections, and specific procedures. Alternative sequences were considered during development.

During practical disassembly workshops, the overall condition of the motor (Fig. 1, step 2.4), mechanical components (Fig. 1, step 2.1), and connectors were assessed (Fig. 1, step 2.3). Photos were taken to document signs of aging, wear, and variations in connection technology methods.

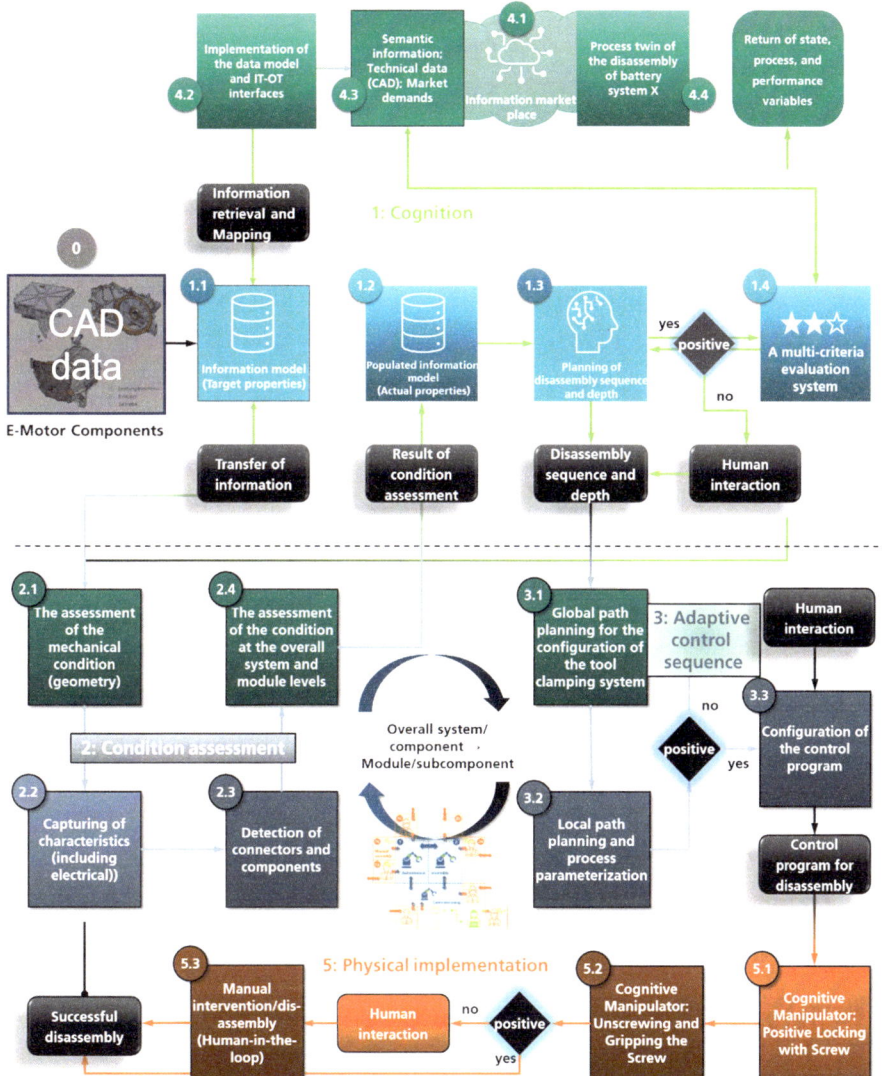

Fig. 1. Generalized, information flow-based process sequence for disassembly: five thematic core topics

Differences between CAD data and the actual motor were noted, and release torques of screw connections were measured for designing the automated unscrewing process. Observations during manual disassembly provided insights for a more disassembly-friendly design, including tool changes, screw accessibility, and gripping points for loosening components. These observations helped adjust the disassembly sequences and improve the overall process. Overall, the disassembly guide and practical assessments contributed to a better understanding of the electric motor's condition and informed the development of more efficient disassembly procedures.

The practical implementation of the disassembly process and its documentation serves as the basis for populating a database for automated cognition. However, the cognition itself (Fig. 1, step 1) and its sub-steps, which are relevant for recycling companies, cannot be fully determined at present. This is due to the lack of available data for economically viable component selection, as well as the absence of a (multicriteria) evaluation system for decision-making and the preferred dismantling route, including associated dismantling steps. The exploration of the dismantling route was conducted manually in the described dismantling workshops. A specific dismantling process was defined, and an unscrewing process was chosen. The decision to recover the permanent magnets, for example, was based on a quantity-based analysis of raw material prices for the overall assembly group. The cognitive process (Fig. 1, step 1) was thus analogized, and its fundamental characteristics were developed.

The project scope does not allow the development of a fully automated disassembly of the complete motor, but rather only that of one disassembly process. The basis for selecting the process was the compilation of materials used in the motor and their value. Here, alongside the copper of the stator's hairpins, the permanent magnets in the rotor are particularly noteworthy. To access these magnets in the rotor design provided by Volkswagen, four long-slot screws need to be loosened. Subsequently, the balance discs and the laminated cores with the embedded magnets can be pressed off the rotor shaft. In manual disassembly, the magnets themselves were driven out of the laminated cores with a hammer and chisel, requiring minimal effort.

The automation of the selected unscrewing process should be able to react adaptively to a wide range of variants and interference parameters. The control sequence (Fig. 1, step 3) is based on detailed planning, involving the control of various disassembly variants through generalized operations and automated hardware and software reactions. The VDI 1990 standard provides guidance in this regard. Adaptivity encompasses parameters such as infeed depths, disassembly paths, forces, moments, and times. Zirkel addresses the adaptive challenge by treating each rotor as unknown and adjusting path planning, screw finding, and screwdriving in real-time. The disassembly process begins with sensor-based coarse positioning of the end effector, using camera depth data. A sensor is then used to capture the precise position of the connection element, enabling fine positioning. After successful positioning (Fig. 1, step 3.3), the control program is configured with individual PLC sub-cycles. The physical unscrewing process itself involves establishing form closure (Fig. 1, step 5.1) and subsequently unscrewing and gripping the disassembled screw (Fig. 1, step 5.2). If unsuccessful, manual intervention is required.

Upon completion of the process, data including time and torque from each step is stored in a database or cloud environment, creating a digital twin of the disassembly process (Fig. 1, step 4.4).

4   Demonstrator

The necessity and methodology of disassembly for permanent magnet synchronous motors in the Zirkel project have been rationalized into an automation concept. As already mentioned, a manageable approach, considering financial and time commitments, has been chosen. Multiple processes with different technical requirements are necessary to disassemble an electric motor. To handle the complexity, the motor is manually disassembled until the rotor is exposed, as the highest value is in the permanent magnets. It took 14 disassembly steps, as per DIN 8580, to expose the rotor and test the automatic unscrewing procedure for shaft bolts. The preceding 14 disassembly steps involved levering off covers, removing retaining rings, and pulling out shafts. Currently, there is no accessible digital infrastructure to retrieve component properties and preferred dismantling routes. Insights from manual dismantling workshops serve as the data basis. This means that there is no "information marketplace" (Fig. 1, step 4.1), but it needs to be created and filled in the long term with data from manual disassembly workshops. In the future, the "cognition" (Fig. 1, step 1) can be automated without manual preliminary considerations.

The overall target is to automate the disassembly of the four long screws M6 x 125 mm (distinction from the state of the art, see e. g. "DeMoBat") that secure the rotor stack and thereby ensure the cohesion between the laminated stacks with embedded permanent magnets. As a distinctive feature, the entire process is designed to operate in an adaptive manner, which will be further elaborated. Three core challenges were identified concerning potential challenges during the planning phase:

1) **detection challenges:** Changes to the part(s) and joints due to wear, contamination, and material aging effects are expected. Practically, the screw head and its screw head profile must be reliably recognized so that both the verification and the position determination can take place.
2) **high loosening torques**: Due to load-induced material distortion, contamination, material aging/compaction of gap fillers/adhesives it is not guaranteed that former assembly parameters can be adopted for reference torque-values for disassembly.
3) **tool accessibility**: It cannot be assumed that all parts and joining connections are accessible without elaborate component handling. Narrow spaces and a wide variety of standard parts continue to complicate tool accessibility.

These challenges are mainly addressed by three coping strategies:

1) **detection challenges:** A two-stage detection process is being implemented to improve process reliability. The first stage involves a coarse-focused camera process which is followed by a fine-detection process for cross-verification. This enables the disassembly process to adapt to different screw arrangement patterns and varying numbers of screws, accommodating deviations in part and process parameters. Currently, tests are being conducted for screws with TX30 screw head drives, and the test series will be expanded to include screws from TX25 to TX40.
2) **high loosening torques**: Using a three-axis system (GANTRY) limits the automation system regarding its degrees of freedom but simultaneously it provides more stiffness by limiting torsional flexibility. The measured breakaway torque of 50 Nm can be handled effectively.

3) **tool accessibility**: The market and state-of-the-art research revealed that existing solutions for screwing technology are, on one hand, heavily focused on assembly and, on the other hand, come with disproportionately large installation space requirements in comparison to the prevalent application. The screw connections on the rotor can all be reached unidirectionally with bits that have a minimum length of 50 mm. There are no interfering contours that would hinder the accessibility.

The considerations above led to a partly realized automated unscrewing process-flow that can be broken down into four crucial steps. Figure 2 depicts the process flow within the disassembly cell. The sensor-verified grip of a screw represents the final step in the disassembly chain and simultaneously the first step in the automation loop. The coarse alignment of the tool, step 2a, is only required once for each part, as the rotor is not moved within the disassembly cell. Afterward, the steps 2b-4 can be iterated in a loop until all four screws are unscrewed.

In the following, reference is made at appropriate points to the general methodology presented in Chap. 3. As the unscrewing of the four long shaft screws is a specific use case that has neither the necessary database (material or value stream-related data twins) nor the required interconnectivity (digital mapping of the dismantling route including interface management to cloud management levels), it is not possible to address all points in a meaningful way.

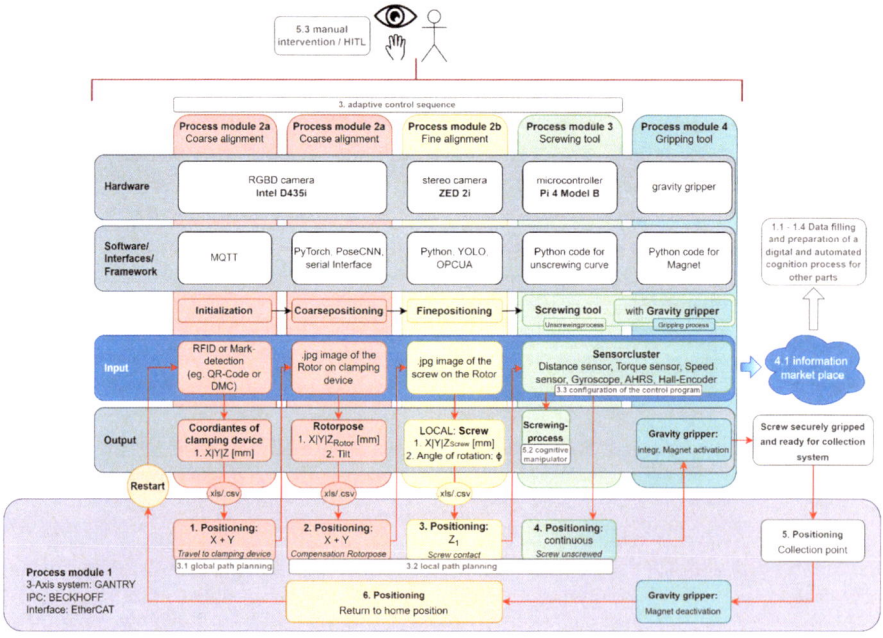

Fig. 2. Graphical overview of the specific process flow within the disassembly cell

The disassembly process starts with the detection of the motor's cartesian position in the working space of the demonstrator. (Fig. 1, step 2.1) This is accomplished by scanning a QR-code that is attached to the clamping device of the rotor. The primary objective of this initial step is to ensure convenience, as there is no longer a need for a costly setup: Approximately aligning the rotor's screw joints intended for disassembly towards the tool, and by that guaranteeing accessibility, is now sufficient and ensures adaptive positioning of the tool towards alternating screw joint positions [18, 19]. (Fig. 1, step 3.2).

To determine the global position of the rotor, cost-effective RGBD cameras are utilized, offering advantages over expensive laser scanners. RGBD cameras are well-suited for indoor use with controlled lighting conditions, making them practical for various applications such as indoor robotics, gaming, 3D modeling, and human-robot interaction [15, 16]. For coarse detection, a deep learning model called PoseCNN Pytorch is trained to handle occlusion robustly. The following steps were undertaken to train the model:

1) Dataset Creation: Synthetic datasets are generated using CAD models of the objects of interest. These datasets include 3D models with known poses.
2) Rendering: The CAD models are rendered from different viewpoints and under various lighting conditions to simulate realistic scenarios. This synthetic dataset, consisting of rendered images and corresponding ground truth poses, is used to train the CNN model.
3) Real Data Validation: After training on synthetic data, the model is validated using real-world data to ensure its generalizability to unseen, non-synthetic scenarios (as seen in the GIF below).
4) Pose Estimation on Real Images: Once trained, the CNN model can estimate the 6D poses of objects in real-world images, even though it was primarily trained on synthetic data.

The major challenges encountered during the model training can be summarized as follows:

1) Computational Intensity: Advanced pose estimation methods, particularly those based on deep learning, can be computationally intensive, resulting in lengthy error-fixing processes.
2) Symmetry and Repetitive Patterns: Objects with symmetrical or repetitive patterns can pose challenges, as the system may struggle to differentiate between different poses. The ambiguity in symmetry makes it difficult to determine the correct orientation uniquely.
3) Noisy Sensor Data: Noisy sensor data, such as inaccurate depth measurements or sensor noise, can introduce errors in pose estimation.

In the next step, after the endeffector with the screwing tool was moved roughly near the screws, the head of the screws are more precisely recognized by using the 3D camera ZED 2i locally attached at the screwdriver end effector. (Fig. 1, step 2.3) A fine positioning process is mandatory for aligning the center of the driver bit with the center of the screw. This precautionary process prevents collisions in the third step: the screwing tool proceeds to advance the screw until tactile contact is being recognized by a cluster of auxiliary sensors. As a deep-learning model is in use, the same procedure for training and implementing the model as it has been done for the coarse part detection is necessary. The neural network-based object detection model YOLOv8 was chosen due to its speed and accuracy. The only difference lies in higher demands regarding resolution. The disadvantage of this approach of being resource intensive was solved by choosing this camera as it can take photos individually with the left and right lens, hence giving data with two different angles in the same image. Moreover, the images were captured by changing the resolution (2k, 1080p, 720p, VGA) and with different brightness and contrast values. Overall there are 1503 images on which the model will be trained initially. Furthermore, to increase the accuracy of the trained model data, augmentation techniques will be used.

Once the coordinates for the fine-positioning process have been transferred, the two-staged unfastening process begins. First, the fastening torque of the screw must be overcome, that may has increased due to aging, pollution or material stress leading to material distortion. After recognizing the torque breakdown, the unscrewing parameters are adjusted in terms of productivity what especially results in high rotational speeds of the tool. (Fig. 1: step 3 "adaptive control sequence") For the screwing tool itself, a commercially available impact wrench solution is chosen which offers the desired breakaway and tightening torque for the process. The necessary intelligence is achieved by adding sensors to the tool. The inclusion of sensors like an encoder, gyroscope, inertia measurement unit, depth-, and proximity sensor(s) and a temperature sensor improve the tool by adding an adaptive control. The processing of sensor data and derivation of process-adapted actuator commands are handled by a tool-integrated microcontroller. Again, all relevant process data will be gathered and processed by the IPC. The unscrewing process itself marks the physical realization of the disassembly. (Fig. 1, step 5).

During the process of unscrewing, the screw loses its mechanical guidance as it exits the thread, potentially leading to unstable and challenging-to-control motion behavior. The implementation of an adaptive gripper, capable of clamping various screwheads, while also providing support to the threaded rod during unscrewing with mechanical guiding surfaces represents the final component of the automation process [20]. After examining gripping tools on the market, it became evident, that an integrated gripper design within a screwing tool is not available. Yet, simultaneous engineering is hardly recommended for designing an efficient and space-saving solution. At this stage, three concepts have been generated. Recognizing the effectiveness of an electromagnetic function for this task, a concept was developed. The upcoming task involves implementing the electromagnetic gripper, simulating its functionality, and subsequently conducting tests.

Figure 3 contains labeled images of the current status of the disassembly demonstrator. (Fig. 1: step 5 "physical implementation").

5   Conclusion

Through Zirkel's innovations, we aim to significantly improve resource use and strengthen supply chains, fostering a sustainable and resilient production sector. Additionally, our ambition is to develop industrial-grade automation solutions for unscrewing processes, as they currently have a niche existence compared to the vast selection of options available for screwing solutions. This research presents a pioneering method for

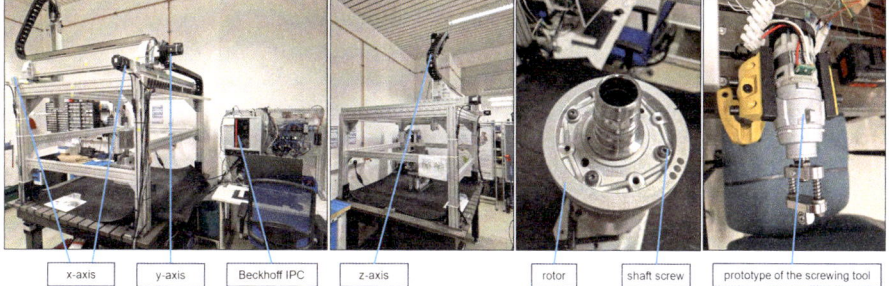

| x-axis | y-axis | Beckhoff IPC | z-axis | | rotor | shaft screw | prototype of the screwing tool |

Fig. 3. Visualization of the disassembly demonstrator, from left to right: depiction of the disassembly cell, close-up of the rotor, prototype of the un-screwing tool

the autonomous disassembly of electric motors, an essential component in the realm of e-mobility. With the increasing sales of electric vehicles, efficient recycling and disassembly processes have gained paramount importance. Our approach, which synergizes sensor-driven feedback with mechanical precision, effectively tackles the complexities of electric motor disassembly. This not only proves the technical feasibility of our method but also enhances the economic and environmental aspects of e-mobility recycling. Moreover, our work offers substantial implications for future electric motor designs. By applying 'Design for Recycling' principles, our research advocates for products that consider end-of-life recycling at the design stage, promoting more sustainable and efficient recycling processes. This approach is critical in aligning with global efforts towards environmental sustainability and resource efficiency in the emerging e-mobility sector. As the automotive industry rapidly evolves, our research contributes to the broader discussion of sustainable practices. It underscores the importance of integrating circular economy principles into vehicle design and manufacturing, addressing the technical, economic, and environmental dimensions of electric motor recycling. Our study thus serves as a catalyst for more sustainable practices in e-mobility and sets a precedent for future research and policymaking in this vital area.

Acknowledgement. The results presented in this paper originated from a current BMBF-funded project "Zirkel" (funding code 02J21E041).

References

1. United Nations, Ed., "Used Vehicles and the Environment: A Global Overview of Used Light Duty Vehicles: Flow, Scale and Regulation," environment programm, 2021.
2. E. Schneider, "Germany's Industrial strategy 2030, EU competition policy and the Crisis of New Constitutionalism. (Geo-)political economy of a contested paradigm shift," *New Political Economy*, vol. 28, no. 2, pp. 241–258, 2023, https://doi.org/10.1080/13563467.2022.2091535.
3. Statista Research Department, *Anzahl der Neuzulassungen von Elektroautos weltweit von 2012 bis 2022*. [Online]. Available: https://de.statista.com/statistik/daten/studie/406683/umf rage/anzahl-der-verkaeufe-von-elektroautos-weltweit-prognose/

4. Juan Felipe Munoz, *Motor1 Numbers: Die weltweit meistverkauften Elektroautos 2022: Tesla dominiert aktuell das Geschehen, doch China sollte man nicht unterschätzen ...* [Online]. Available: https://de.motor1.com/news/666693/elektroautos-verkaufszahlen-weltweit-2022/

5. United Nations, Ed., "Transforming our world: the 2030 Agenda for Sustainable Delevopment," Department of Economic and Social Affairs, Sep. 2015.

6. European Commission, Ed., "The European Green Deal," Brüssel, Dec. 2019. Accessed: Jan. 17 2024. [Online]. Available: https://eur-lex.europa.eu/legal-content/EN/TXT/?uri=COM%3A2019%3A640%3AFIN

7. U. Blast, R. Blank, T. Elwert, F. Finsterwalder, G. Hörnig, T. Klier, S. Langkau, F. Marscheider-Weidemann, J.-O. Müller, Ch. Thüringen, F. Treffer, T. Walter, "Recycling von Komponenten und strategischen Metallen aus elektrischen Antrieben: MORE (Motor Recycling)," Bundesministerium für Bildung und Forschung, Aug. 2014.

8. M. Schönfeldt *et al.*, "Magnetic and structural properties of multiple recycled and sustainable sintered Nd-Fe-B magnets," *Journal of Alloys and Compounds*, vol. 939, p. 168709, 2023, https://doi.org/10.1016/j.jallcom.2023.168709.

9. F. Hansen, D. Hochstädt, A. Sauter, and J. H. Seelig, "Automatisierte Demontage elektrischer Antriebsaggregate: Herausforderungen und Lösungsansätze," *Chemie Ingenieur Technik*, vol. 95, no. 12, pp. 1978–1987, 2023, https://doi.org/10.1002/cite.202300085.

10. U. Bast *et al.*, *Recycling von Komponenten und strategischen Metallen aus elektrischen Fahrantrieben: Kennwort: MORE (Motor Recycling).* [Online]. Available: https://publica-rest.fraunhofer.de/server/api/core/bitstreams/a89b5bc4-5b49-422b-8c08-a6f8e05fd71c/content

11. A. A. Assadi, *Industrielle Demontage von Batteriemodulen und E-Motoren DeMo-Bat.* [Online]. Available: https://www.ipa.fraunhofer.de/de/referenzprojekte/DeMoBat.html (accessed: Mar. 7 2024).

12. wbk Institut für Produktionstechnik, *Projektbeschreibung ZirkulEA: Kreislauffähigkeit des Elektro-Antriebsstrangs durch intelligente Demontage und Nachverfolgung* (accessed: Mar. 7 2024).

13. wbk Institut für Produktionstechnik, *Projektbeschreibung AgiProbot: Agiles Produktionssystem mittels mobiler, lernender Roboter mit Multisensorik bei ungewissen Produktspezifikationen.* [Online]. Available: https://www.wbk.kit.edu/wbkintern/Forschung/Projekte/AgiProbot/ (accessed: Mar. 7 2024).

14. RWTH Aachen University, *DemoSens: Verbundprojekt erforscht Demontage und Recycling von E-Mobil-Akkus.* [Online]. Available: https://www.pem.rwth-aachen.de/cms/pem/forschung/projekte/~pkazx/demosens/ (accessed: Mar. 7 2024).

15. X. Zhang, K. Eltouny, X. Liang, and S. Behdad, "Automatic Screw Detection and Tool Recommendation System for Robotic Disassembly," *Journal of Manufacturing Science and Engineering*, vol. 145, no. 3, 2023, https://doi.org/10.1115/1.4056074.

16. E. Ransed, "A heuristic approach to screw pose detection," vol. 2022.

17. S. Mangold, C. Steiner, M. Friedmann, and J. Fleischer, "Vision-Based Screw Head Detection for Automated Disassembly for Remanufacturing," *Procedia CIRP*, vol. 105, pp. 1–6, 2022, https://doi.org/10.1016/j.procir.2022.02.001.

18. N. Gard, A. Hilsmann, and P. Eisert, "Combining Local and Global Pose Estimation for Precise Tracking of Similar Objects," in *Proceedings of the 17th International Joint Conference on Computer Vision, Imaging and Computer Graphics Theory and Applications*, Online Streaming, --- Select a Country ---, 2022, pp. 745–756.

19. W. Chen, X. Jia, H. J. Chang, J. Duan, and A. Leonardis, "G2L-Net: Global to Local Network for Real-Time 6D Pose Estimation With Embedding Vector Features," in *2020 IEEE/CVF Conference on Computer Vision and Pattern Recognition (CVPR)*, Seattle, WA, USA, 2020, pp. 4232–4241.

20. Z. Samadikhoshkho, K. Zareinia, and F. Janabi-Sharifi, "A Brief Review on Robotic Grippers Classifications," in *2019 IEEE Canadian Conference of Electrical and Computer Engineering (CCECE)*, Edmonton, AB, Canada, 2019, pp. 1–4.

Design and Simulation

Enhancing Automotive Production Through Integrated Data Analysis and Simulation

Enno Bublitz[1,2(✉)] and Kai Warsönke[1,3]

[1] Volkswagen Aktiengesellschaft, Berliner Ring 2, 38440 Wolfsburg, Deutschland
{enno.paul.bublitz,kai.warsoenke}@volkswagen.de
[2] Institute of Machine Tools and Production Technology, Technische Universität Braunschweig,
Langer Kamp 19B, 38106 Braunschweig, Deutschland
[3] Institute for Engineering Design, Technische Universität Braunschweig, Hermann-Blenk-Str.
42, 38108 Braunschweig, Deutschland

Abstract. In the context of automotive production, the variability in component tolerances poses a significant challenge, often leading to production downtimes. Deviations in dimensional accuracy and geometry can disrupt automated assembly processes, emphasizing the critical link between input quality and assembly outcomes. As automation gains prominence, robust process monitoring becomes imperative for quality assurance and maintenance cycle optimization. However, a prevalent issue persists data fragmentation and underutilization in production. Our approach is centred on leveraging existing production data to optimize automotive production and streamline final assembly processes. We investigate the interplay of upstream process influences on downstream assembly processes within large-scale production environments. Real-world manufacturing data is analysed and validated through finite element analysis (FEA) simulations coupled with Monte Carlo simulations. Focusing on one specific automotive assembly process—the automated assembly of cockpits—we examine the influence of various body concepts on assembly automation compatibility. Data analysis from two assembly lines identifies body concepts conducive to efficient assembly, subsequently validated using our innovative coupled simulation method. This holistic approach identifies production-suitable product concepts and integrates our novel simulation method into the design of future products, bridging the gap between production data fragmentation and optimized assembly in the automotive industry. By connecting real manufacturing processes with product concept suitability, we enhance production efficiency, advance data mining and simulation methods, and improve overall production quality.

Keywords: Automotive production · Assembly · Data analytics · Fem-simulation · Industry 4.0

1 Introduction

The automotive industry's process chain encompasses a wide spectrum, starting from crafting individual parts in the press shop, assembling these into units, culminating in the formation of a vehicle body through intricate joining and welding processes in the

K. Dröder and T. Vietor (Eds.): CD 2024, Proceedings, pp. 97–109, 2025.
https://doi.org/10.1007/978-3-658-45889-8_8

body shop. Following this, the vehicle undergoes coloration and anti-corrosion treatment in the paint shop before final assembly, which involves accommodating detailed customer specifications encompassing add-on and integrated components. The demand for process-reliable, production-oriented, design-oriented and function-oriented car body construction requires the targeted definition of individual part and assembly tolerances, whereby the assemblies created from the individual parts are subject to a much larger tolerance range [1–3].

In the modern design of vehicle structures, tolerances are therefore determined on the basis of individual part analyses and empirical values. However, as automotive production is a complex system, the subsequent manufacturing processes are mutually dependent. Deviations in dimensional accuracy or geometry can lead to disruptions in downstream assembly processes. Parts cannot be assembled, for example, due to distortions or incorrectly positioned connecting parts. This phenomenon is even more apparent in automated assembly processes that rely on compliance with specified tolerances, as they cannot react as flexibly as human workers to unforeseen changes [4]. The significance of this problem is further intensified by the increasing level of automation, particularly in final assembly. New developments in computer technology, robotics, gripping and handling technology are steadily increasing this degree of automation [4, 5]. Accordingly, the assembly quality depends largely on the quality and dimensional accuracy of the components provided. It is possible to regulate interdependent processes through targeted interventions at the point of origin and thus react to errors [6]. Such and other quality control loops enable potential cost savings of 10%–20% [7]. Assembly is always the reservoir for all errors in upstream processes, meaning that this is where the effects are most frequently felt [8]. With the increasing automation of manual final assembly processes, continuous process monitoring is being introduced. The aim of monitoring the key process parameters is to ensure the manufactured product quality and to optimize maintenance cycles. An evaluation of this process data and higher-level production data enables the increasing complexity of production to be controlled. This is a possible application of data mining [9]. By evaluating the data, weak points can be identified and rectified through targeted maintenance or control measures. However, data mining methods have only rarely been used in production [9, 10].

The greatest potential lies in avoiding additional costs for reworking in production, avoiding unplanned machine downtimes and increasing production output [11]. The process data generated in automated assembly processes, which is usually used to control the process workflow, remains predominantly at the place where it is generated and is not merged with other data and information [12]. However, there is great potential here, for example to quantitatively map the effects of different component tolerances on subsequent assembly processes and, based on this, to carry out product optimizations or learn for future product generations. The aim of this paper is therefore to develop an approach that leverages this potential.

2   Approach

A new simulation method that addresses common assembly process disturbances is presented in this section. These disruptions often arise from a failure to fully consider the complex interactions within production processes. Even traditional methods that incorporate advanced control variables and feedback mechanisms often fall short, especially when evaluating the robustness of the body structure in production environments. The analysis reveals a significant issue in the way production data is utilized in the design of new vehicles, resulting in inefficiencies and errors during assembly processes [13].

The prevailing method of widening the tolerance zone with increasing assembly complexity is often insufficient to compensate for the manufacturing influences between interacting components. Additionally, the approach of simulating complex body structures as rigid bodies is inadequate if manufacturing impacts on dimensional accuracy are considered.

On the contrary, the varying component behaviour in relation to individual components can compensate for deficits in dimensional accuracy through optimized handling, clamping and spotwelding. To address these challenges, our approach utilizes a comprehensive stochastic simulation that combines the Monte Carlo method with the linear finite element method. This method combines both structural topology and material deformations to analyse error propagation across various components. It is particularly useful for assembling thin sheet metal parts that are prone to deformation under external forces and exhibit varying stiffness when combined. The Monte Carlo algorithm varies influencing parameters according to a symmetrical gaussian distribution. It also conducts a sensitivity analysis of the main contributors to identify relevant functional and connection areas.

We avoid traditional simulations that focus on isolated components or limited sections of the assembly process. Our methodology provides a comprehensive view of structural and tolerance analyses by examining the mechanical effects of clamping and welding and considering the integration of sheet metal parts within a complex car body structure. It improves theoretical modelling and provides a practical tool for designing and assembling vehicle bodies. This holistic approach enhances simulation accuracy, leading to better-designed vehicles and more efficient production lines [14].

3   Validation

This section describes the procedure for validating the idea of the coupled finite element analysis simulation. To ensure the plausibility of the method a specific structural component is selected that leads to assembly problems in two different vehicle concepts in production, whereby other external factors are limited as far as possible. In large-scale production with modular construction kits, there are variants of substructures that were developed at different times and consist of different evolutionary stages. This enables a direct comparison of product concepts that fulfil the same function for the customer. To select such a comparison case, the entire production and fault data of an assembly line was evaluated over a period of 3 years and the automatic failure messages and manual error comments that occurred were classified using an natural language processing

(NLP) algorithm. The causes of faults were then assigned to the individual grouped production faults and the associated production processes. This enables production errors to be identified that are attributable to specific upstream processes. The following table shows the TOP 5 errors that can be traced back to upstream errors or exceeded tolerances in the body shop by linking errors in assembly. In total, over 1.4 million data records of system faults in the assembly line were evaluated and classified. In addition, the data first had to be prepared for evaluation by structuring the data, eliminating implausibilities and removing outliers (Table 1).

Table 1. Number of Assembly Errors due to Upstream Processes

Process	Error Description	Number of Errors
Cockpit Assembly	Stud Bolt Placement	258
Door Removal	Hinge Jammed	114
Door Installation	Dimensional Accuracy Hinge	39
Installation of Trailer Hitch	No Overlap	33
Door Installation	Hinge Jammed	14

The table shows the cases for approximately 239.000 vehicles built and is sorted according to the frequency of occurrence. It is noticeable that most of the coupled errors occur during the assembly of larger assembly modules with several joints, e.g. during the assembly of the cockpit module, the doors or the trailer hitch. All errors relate to a deviation in the position tolerance of screws, bolts or threads. The most common error in this evaluation, with 258 cases, is the incorrect position of the stud bolt, which is required for mounting the cockpit to the car body. During the assembly of the cockpit, the influences of intermediate process steps from upstream processes can be limited, as the screw-on points are used exclusively for the assembly of the cockpit. The following is an example of an error text that was assigned to this error pattern: *"Stud bolt cockpit mount does not fit > Joining not possible > Assembly line stop".*

In a next step, another vehicle model was identified at a different production site, where the automated assembly process of the cockpit is carried out using the same automation technology and a significantly different mounting concept is used in the cockpit area. This was to enable a detailed comparison of the two component variants regarding a singular possible error in the automated assembly process. Once again, a longer period of production errors was evaluated and the procedure was repeated in the same way as the first evaluation. In this case, only four errors of misplaced stud bolts were identified during automated cockpit assembly. This leads to the assumption that the other body concept in the cockpit mounting area has a lower error potential and higher production robustness. In both cases considered, these are assembly lines that already produce vehicles in a steady state, so ramp-up effects can be neglected. The following table compares the error cases and the data of the two product concepts (Table 2).

Table 2. Comparison of the Body Concepts and their Faults

Product Concept	Concept A: Mounted Angle Elements	Concept B: Attached Mounting Pot
Production Site	Plant A	Plant B
Period Under Review	2019–2021	2020–2022
Vehicles Produced	239.000	330.000
Evaluated Fault Messages	1.400.000	537.000
Occurred Faults "Stud bolt misalignment"	258	4
Error Rate per Vehicle Produced	0,1079%	0,0012%

The data supports the assumption and shows that Concept B leads to the described error of a misplaced stud bolt in automated cockpit assembly 100 times less frequently. The next step is to check whether the new simulation approach also reflects this statement.

In order to check this, a simulation model was created for each concept based on the planned component handling processes, clamping, and joining operations to evaluate the coherence between the modelling approach of the FEA coupled tolerance simulations and the frequency of faulty assembly. The effects of geometric deviations are then propagated into the connection area, which is relevant to the problem under consideration. The validity of the applied method is confirmed by a simulation result that provides a coherent explanation with the evaluated production data. The next section describes the structure and execution of the simulation. Two things need to be checked here: firstly, whether the simulation can make a trend statement on the suitability of a product concept for this automation process and, secondly, whether it is possible to illustrate the order of magnitude of the superiority of one design concept.

4   Theoretical Background

Tolerance analyses and simulations using finite element analysis are widely used in automotive engineering. Monte Carlo algorithms in connection with tolerance analysis of car body components offer the potential to obtain stochastic statements about the distribution of certain characteristics through a large number of random experiments. By manipulating the distribution of certain input parameters, valid insights into the actual behavior of systems are obtained [15–18].

4.1   The Cockpit Assembly Process

The process of automated assembly of vehicle cockpits is described in this section. A robot system picks up the pre-assembled and sequenced cockpit from a conveyor belt and waits for the corresponding vehicle body to arrive. As soon as this arrives in the work area, the seventh axis synchronizes with the conveyor speed of the assembly belt and moves the cockpit through the door cut-out on the driver's side into the vehicle interior. The cockpit is threaded and screwed onto two threaded bolts, known as stud bolts, during a joining process towards the front of the vehicle. One stud bolt is located on the left and one on the right side of the vehicle in the area of the A-pillar. The cockpit was previously mounted on an extruded profile, the module crossbar, which contains the holes required for threading the stud bolts. Once the cockpit has been installed, this module carrier is no longer visible to the customer. The schematic diagram below (Fig. 1) shows the threading on the stud bolts.

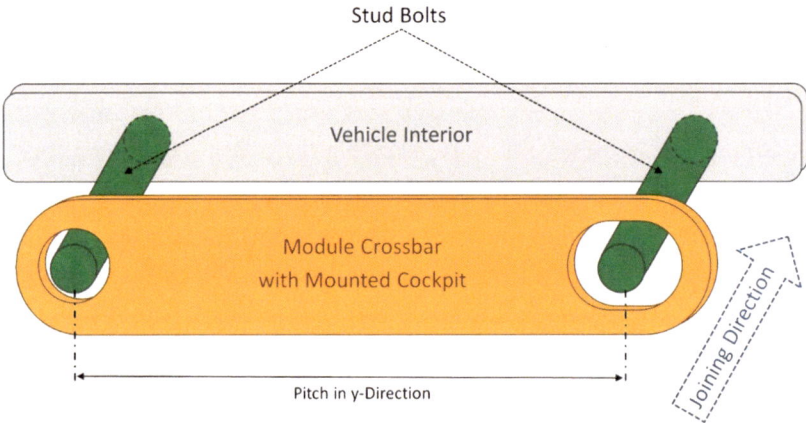

Fig. 1. Illustration: Module Crossbar Mounting on Stud Bolts.

If the robot cannot place the cockpit on the two stud bolts, a second attempt is made. If the stud bolts are not in the correct position, this attempt also fails. This causes the system to malfunction, which stops the entire assembly line in this section and also blocks the upstream and downstream assembly cycles. These were the cases identified in the production data. It is therefore essential to avoid such faults.

The dimension between the two mounting points, the stud bolts, is crucial for dimensional accuracy and may be ± 6.7 mm. A slotted hole in the module crossbar can be used to compensate for fluctuations. The original design parameters of the components allow tolerances of a maximum of ± 3.9 mm. In theory, this results in an automated threading process if the tolerances are observed. The important measured variable for the pitch is the distance between the right and left connection points, i.e. the pitch in the y-direction. The occurrence of cockpits that cannot be joined due to incorrect positioning of the stud bolts indicates a tolerance problem in car body construction. There are different concepts that allow a more frequent misalignment or, on the other hand, allow a less frequent misalignment.

4.2 Sensitivity Analysis

The linearity coefficient obtained through sensitivity analysis can be used to determine the dependency between analysis points in a linear system by making a specific change within a certain parameter range, determined through sensitivity analysis.

This method can provide information about the importance of components in a tolerance chain to a specific analysis criterion when investigating complex body structures. Thus, a sensitivity analysis can make clear statements about the contribution and the position of a feature in the tolerance chain.

The modeling strategy refers to the hierarchical structure of the product structure. It starts with each individual part of the welding group under consideration, and the simulation model includes welding points that determine the geometry, as well as component handling and clamping points that are critical for successful results. All process factors defined as boundary conditions were included as potential contributors to the sensitivity analysis of the random dimension in the combined product structure of the platform versions considered. It is important to point out that one parameter is changed in each case and the influence on the sample size is evaluated. Within the simulation process, every specified process factor is assessed for its sensitivity to the benchmark. The determining factor for both platform concepts is calculated in the manner described. Real measurement data is also considered within the simulation model for a precise evaluation. The mean displacement and measurement range are transferred to the main width beams by determining linearity coefficients [19–21].

5   Results

5.1 Simulation Model

For the specific application in this context, an FEA-coupled tolerance analysis model was developed based on existing planning data. This includes the positioning of the individual parts according to the 3-2-1 principle using defined reference points as well as the definition of planned clamping and joining points. It should also be noted that only the geometry-determining weld points are integrated into the model. The meshes relevant for the calculation of the deformation were created on the basis of the nominal design data. In order to make full use of the potential of the Monte Carlo method and to consider the influencing variables in a stochastic way, only linear relationships are simulated. This means that deviations in the part shape are only considered in the reversible, linear-elastic case. Furthermore, a 1000-fold variation of the deviation at the relevant clamping and joining points is performed and the contribution to the functional dimension under consideration is determined for each defined boundary condition. The sensitivity analysis, which is a part of the investigations, is an objective determination of the most significant contributors to the functional dimension along the entire tolerance chain. In total, the top 20 contributors of clamping and mating points are responsible for 80% of the influence on the gauge block. These 20 contributors are the subject of the investigations. Through the application of sensitivity analysis, the following results were obtained by examining the described modelling strategy. The simulation model yielded in a primary finding. The sensitivity analysis contributors display greater interlinking degrees on Concept A (Fig. 2), whereas those for the Concept B (Fig. 3) exhibit a shorter effective tolerance chain. The product structure of both platform concepts is characterized by the fact that the relevant pitch is formed by the assembly of 4 weld groups. The following images illustrate the difference in terms of the geometry-determining connection areas of the weld groups to each other.

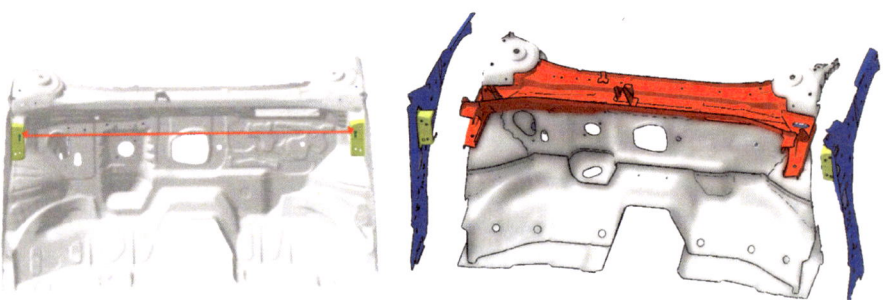

Fig. 2. Y-Pitch (Left) and Weld Groups (Right) of Construction Concept A

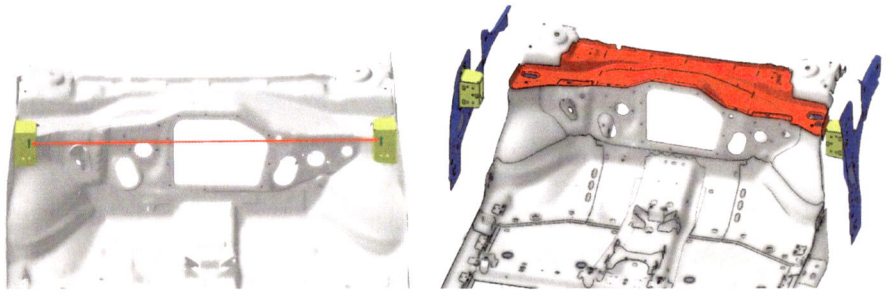

Fig. 3. Y-Pitch (Left) and Weld Groups (Right) of Construction Concept B

In addition to the y-aligned connecting flanges in the area of the wheel housing, product Concept A shows further welding contact in the immediate area of the stud bolt. Product Concept B is characterized by the fact that the geometry is determined only by the connection of the welding flanges in Y-alignment around the wheelhouse area. Both platforms show an area marked in red that differs in complexity. This area is part of a weld sub-group with many individual components. The large number of interconnections leads to a systematic transfer of deviations, making it difficult to compensate for deviations in the functional area.

5.2 Sensitivity Analysis

As part of the sensitivity analysis, a normally distributed random variable within the 2 mm range was assigned to each clamping and welding point. After running simulations for both platform concepts, the top 20 contributors to the functional dimension were identified and analysed. A comparison of the major contributors is shown below (Fig. 4).

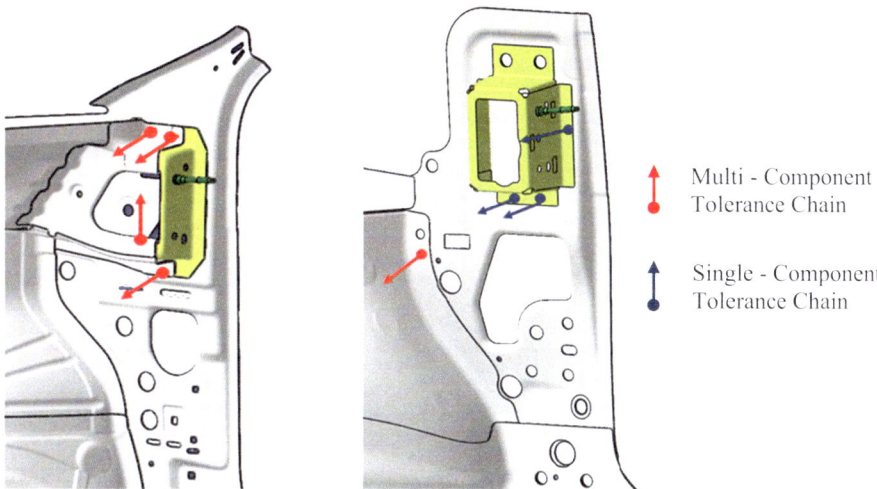

Multi - Component
Tolerance Chain

Single - Component
Tolerance Chain

Fig. 4. Comparison of Major Contributors Focusing Concept A (Left) and Concept B (Right)

The simulation results for Concept B reveal that the welding group, comprising three individual parts, exerts 90% of the impact on the gauge block. Consequently, the individual links in the tolerance chain can be indirectly controlled more effectively. The blue arrows indicate the contribution providers with a larger share of influencing points and their corresponding direction of effect. The red arrow indicates a contributor in the tolerance chain of the complex welding group with many individual parts. Even if this point is part of the interlinked structure, the influence on the tolerances can be controlled indirectly due to the y-alignment of the flange. After conducting the simulation, it is evident that the distribution of contributors in product Concept A is different. Notably, 80% of the contribution to the gauge block is on the side of the interconnected weld group, which is inherently complex. The left-hand figure shows the location and direction of the clamping and welding points with the greatest influence. The analysis shows that the most significant influence on the gauge block in this platform design comes from the deviations of the components located closest to the joining points. Additionally, the direct connection in the functional area, combined with the higher number of links in the tolerance chain with different vector directions, complicates tolerance compensation (Table 3).

Table 3. Tolerance Chain Relationships

Product Concept	Distribution of Proportion on Top 20 Contributors for y-Pitch Measurement	
Concept A	80 %	Multi - Component Tolerance Chain
	20 %	Single - Component Tolerance Chain
Concept B	10 %	Multi - Component Tolerance Chain
	90 %	Single - Component Tolerance Chain

The table provides a summary of the top 20 contributors' ratio concerning tolerances on the y-pitch in the relevant welded groups. The overview illustrates the proportion of the top 20 contributors that are a part of a tolerance chain with several components. Additionally, the table indicates the proportion of contributors that belong to a tolerance chain with a small number of links, which can be better controlled. The color coding allows for a rapid evaluation of the relationships between tolerance chains within the contributor chain. The bar chart summarizes the impact of contributors in the tolerance chain. Overall, the two concepts differ in terms of the proportion of top 20 contributors in the tolerance chain. In Concept B, the proportion of contributors in the complex welding group is significantly lower than Concept A. However, Concept B has a higher proportion of contributors in the shorter tolerance chain of the flanged welding group.

6 Conclusion

In conclusion, this study reviewed and quantified the failure rates of the different platform concepts, with a particular focus on module assembly. It was observed that Platform Concept A had a significantly higher incidence of errors. The study utilized a simulation approach to investigate the impact of design-geometric and process-related variables on assembly errors. A detailed simulation model was constructed based on existing planning data to develop a comprehensive plan for clamping and joining points. The use of both Monte Carlo simulation and linear finite element analysis enabled a detailed assessment of stochastic frequency distributions, leading to the accurate identification of critical factors within the tolerance chain.

During this analysis, we focused on the y-gauge block, which is located among the stud bolts on the installation side, as a crucial variable. Simulations, which were limited by predetermined planning variables, revealed that different elements within the tolerance chain had varying degrees of impact on the positioning of the gauge block. A comprehensive sensitivity analysis, along with direct comparisons, highlighted significant differences in the tolerance chains of the two platform concepts being reviewed. This distinction was made by examining two fundamentally divergent characteristics of the tolerance chain, which are shaped by the intrinsic structure of the product. The study identified two types of tolerance chains: a single component chain, which is associated with the intended outer welding group and directly transmits changes in individual components to the overall component dimension, and a more complex multi-component tolerance chain that describes the inner welded assembly. The second is characterized by a layered and intricate systematic propagation of deviations, which presents challenges in interpretability.

After conducting a detailed simulation that objectively ranked the top 20 contributors for both platform concepts, it was concluded that Platform Concept B is the more robust design. This conclusion was reached through an indirect evaluation that highlighted the beneficial impact of systematically introduced measures on the functional dimension. This finding highlights that Concept B is more compatible with automated assembly in a multifaceted production process with numerous interconnected steps. Thereby it was demonstrated that the simulation is capable of making trend statements regarding the suitability of a product concept for an automated production process. Additionally, the successful validation of the simulation methodology not only supports its use for future design concept verification without the need for preliminary error frequency classification via data mining but also demonstrates its effectiveness in improving the precision and efficiency of manufacturing processes.

7   Critical Evaluation and Outlook

The study of interrelated process dynamics across diverse fields utilizing stochastic simulation models and the linking of the linear finite element analysis has shown notable advancements in exploring fundamental possibilities. However, it is crucial to critically observe the level of development of simulation techniques. It is worth mentioning that the modeled influences adhere to a strictly linear model behavior, and the simulated effects are solely mechanical in nature. Weld distortions and non-linearities are not simulated. Using the Pareto principle, it can be concluded that the chosen methodology yields valuable results and facilitates a comprehensive evaluation of car body structures in terms of buildability, tolerance chain investigations, and stochastic tolerance influences. The approach described also enables more sophisticated investigations in the future. In this case, the simulation was unable to demonstrate the superiority of a design concept in terms of magnitude. However, this feature is desirable for future simulations. Assuming linear dependencies in the simulation process and the potential for parallelizing calculations can yield useful insights on complete body structures in shorter time frames. Moreover, identifying actual errors in the final assembly and the conducted data mining poses a certain degree of uncertainty. Undocumented or misclassified errors may have occurred due to the semi-automated classification process and the large quantity of production data collected and analyzed.

Acknowledgements. This research has received no external funding. The results, opinions and conclusions expressed in this work are not necessarily those of Volkswagen Aktiengesellschaft.

References

1. S. von Praun, "Toleranzanalyse nachgiebiger Baugruppen im Produktentstehungsprozess," Dissertation, Technische Universität München, München, 2002.
2. C. Germer, "Interdisziplinäres Toleranzmanagement," Dissertation, Technische Universität Braunschweig, Braunschweig, 2004.
3. R. Leuschel, "Toleranzmanagement in der Produktentwicklung am Beispiel der Karosserie im Automobilbau," Dissertation, Technische Universität Bergakademie Freiberg, München, 2010.
4. K. Feldmann, V. Schöppner, and G. Spur, *Handbuch Fügen, Handhaben, Montieren: Edition Handbuch der Fertigungstechnik*. München: Hanser, 2014.
5. W. Eversheim, *Organisation in der Produktionstechnik*. Berlin, Heidelberg, New York: Springer, 1996.
6. C. Schwarz, "Verfahren zur Überwachung und Regelung eines Produktionsprozesses mittels multipler Regression," Patent DE102019201192A1, 30 Jul., 2020.
7. T. Bauernhansl, J. Krüger, and G. Schuh, "WGP-Standpunkt Industrie 4.0," Wissenschaftliche Gesellschaft für Produktionstechnik WGP e.V., 2016.
8. B. Lotter and H.-P. Wiendahl, *Montage in der industriellen Produktion*. Berlin, Heidelberg, New York: Springer, 2013.
9. G. Schuh, G. Reinhart, J.-P. Prote, F. Sauermann, J. Horsthofer, F. Oppolzer, and D. Knoll, "Data mining definitions and applications for the management of production complexity," *Procedia CIRP : 52nd CIRP Conference on Manufacturing Systems*, vol. 81, pp. 874–879, June, 2019.
10. U. Winkelhake, *Die digitale Transformation der Automobilindustrie: Treiber – Roadmap - Praxis*. Berlin, Heidelberg, New York: Springer, 2021.
11. R. Glebke, M. Henze, K. Wehrle, P. Niemietz, D. Trauth, P. Mattfeld, and T. Bergs, "A case for integrated data processing in large-scale cyber-physical systems," *Proceedings of the Annual Hawaii International Conference on System Sciences*, January, 2019.
12. A. Mockenhaupt, *Digitalisierung und künstliche Intelligenz in der Produktion: Grundlagen und Anwendung*. Berlin, Heidelberg, New York: Springer, 2021.
13. Y. Cao, T. Liu and J. Yang, "A comprehensive review of tolerance analysis models," *The international Journal of Advanced Manufacturing Technology*, vol. 97, pp. 1055–3085, May, 2018.
14. R. S. Tabar, H. Zheng, F. Litwa, K. Paetzhold-Byhain, L. Lindkvist, K. Wärmefjord and R. Söderberg, "Digital Twin-Based Clamping Sequence Analysis and Optimization for Improved Geometric Quality," *Applied Sciences*, vol. 17, January, 2024.
15. R.Söderberg, L. Lindkvist, K. Wärmefjord, and J. S. Carlson, "Virtual Geometry Assurance Process and Toolbox," *14th CIRP Conference on Computer Aided Tolerancing (CAT), Procedia CIRP 43 (2016)*, vol. 43, p. 3–12, May, 2016.
16. Y. Xia, L. Jiang, D. Yang, and Y. Zhou, "The Dimensional Tolerance Analysis in Different Assembly Procedures Based on 3DCS," *Proceedings of the 19th Asia Pacific Automotive Engineering Conference & SAE-China Congress*, pp. 77–90, October, 2017.
17. P. K. Singh and V. Gulati, "Tolerance analysis and yield estimation using Monte Carlo simulation– case study on linear and nonlinear mechanical systems," *Sadhana*, November, 2021.

18. Y. Gao, "Tolerance analysis and optimization based on 3DCS," *Journal of Physics: Conferences Series* 2137, October, 2021.
19. A. Deif, *Sensitivity Analysis in Linear Systems*. Berlin, Heidelberg, New York: Springer, 1986.
20. R. Söderberg, L. Lindkvist, and S. Dahlström, "Computer-aided robustness analysis for compliant assemblies," *Journal of Engineering Design*, vol. 17, no. 5, pp. 411–428, January, 2007.
21. Y. Liu, Z. Liu, H. Zhong, H. Qin, and C. Lv, "Gauge sensitivity analysis and optimization of the modular automotive body with different loadings," *Structural and Multidisciplinary Optimization*, vol. 60, pp. 363–374, January, 2019.

Tiered Approach for Prospective Sustainability Assessment of New Technologies to Assess Benefits and Challenges (TAPAS)

Martin Möller[1,2(✉)], Olga Speck[1,3], and Thomas Speck[1,3]

[1] Cluster of Excellence livMatS @ FIT – Freiburg Center for Interactive Materials and Bioinspired Technologies, University of Freiburg, Georges-Köhler-Allee 105, D-79110 Freiburg, Germany

martin.moeller@livmats.uni-freiburg.de, {olga.speck, thomas.speck}@biologie.uni-freiburg.de

[2] Oeko-Institut e.V., Merzhauser Str. 173, D-79100, Germany

[3] Plant Biomechanics Group @ Botanic Garden Freiburg, University of Freiburg, SchCSnzlestr. 1, D-79110 Freiburg, Germany

http://www.livmats.uni-freiburg.de

Abstract. When assessing the sustainability of novel technologies and materials systems, the question arises as to how a robust sustainability assessment for technology and material developments with a low readiness level is nevertheless possible and how the necessary data can be collected for this purpose. Based on this research question, TAPAS as a new, tiered methodological framework for a prospective assessment of the sustainability aspects of novel technologies and materials systems is presented.

Within the conceptual development of TAPAS, important groundwork for the new methodological framework was devoted to the methodological interface between biomimetics and sustainability research. We selected the plant growth form liana as an example; biological concepts were used to sharpen the sustainability strategies of efficiency, consistency, and sufficiency and to derive practical design principles for more sustainable products in the technosphere.

On this basis, we developed guiding principles for TAPAS and defined a total of five stages of analysis, allowing self-reflection to be integrated into the research and innovation process. In some cases, established instruments were adopted or refined, such as those used in the prospective screening of chemicals and in life cycle assessment studies. In addition, new methodological ground has been broken to meet the demand for an integrated and normatively based assessment that covers both benefit and risk aspects. This applies in particular to the domain of an in-depth benefit analysis and, for the first time, a set of 30 indicators has been developed that establish a direct link to the 2030 Agenda and its Sustainable Development Goals.

© The Author(s) 2025

K. Dröder and T. Vietor (Eds.): CD 2024, Proceedings, pp. 110–119, 2025.

https://doi.org/10.1007/978-3-658-45889-8_9

We conducted two case studies to test the practical applicability of TAPAS. They showed that the developed tools are in principle suitable for basic research projects. Concrete recommendations for more sustainable design options (e.g., substitution of toxicologically problematic solvents) were provided.

Keywords: Benefit analysis · *liv*MatS · Sustainability assessment · Sustainable Development Goals · TAPAS

1 Introduction and Background

Sustainability assessments of novel technologies and materials systems face the challenge of an inverted relationship between the data available for assessment and the ability to influence technology development. Early in the innovation process, design options are large, but data availability is often poor. Later, the amount and quality of data increase significantly, but the design options may be considerably limited by existing path dependencies. This phenomenon is known as the Collingridge dilemma [1].

Particularly in the case of basic research & development (R&D), new developments are characterized by a low technology readiness level (TRL), i.e., they are still relatively far from coming onto the market in the form of tangible products. A characteristic of technologies with a low TRL is that, on the one hand, little quantitative data is available concerning subsequent product specifications and potential environmental and sustainability impacts. On the other hand, addressing sustainability issues at such an early stage of the innovation process provides an excellent opportunity for identifying and avoiding potential undesirable consequences (e.g., the use of materials with a high carbon footprint, the inclusion of problematic substances, or the poor recyclability of the final product) [2].

Given the constraints of the Collingridge dilemma, an intermediate stage in the R&D process would be an ideal point at which to assess novel technologies against sustainability criteria. At this juncture, the ability to avoid sustainability risks is still relatively high, while the amount and quality of data required for a sustainability assessment has already increased significantly. However, sustainability assessment at the stage of basic materials research needs to occur well before this ideal moment.

Against this background, the question arises as to how a robust sustainability assessment for technology and material developments with a low readiness level is nevertheless possible and how the necessary data can be collected for this purpose [3]. Based on this research question, this paper presents a Tiered Approach for Prospective Assessment of Benefits and Challenges (TAPAS) as a new, tiered methodological framework for a prospective assessment of the sustainability aspects of novel technologies and materials systems.

The conceptual design of TAPAS is embedded in the research and development work of the DFG Cluster of Excellence "Living, Adaptive and Energy-autonomous Materials Systems" (*liv*MatS). This cluster envisions the development of novel, bio-inspired materials systems that can autonomously adapt to their environment and thereby exploit previously untapped energy sources in the sense of "energy harvesting". In addition, the materials systems should be damage resistant or capable of selfrepair. Questions concerning the sustainability of the developed materials systems play an important role and therefore form an integral part of the work from the very beginning [4].

In the following, the normative framework for sustainability research as well as biological concepts that have proven to be helpful in sharpening the normative framework are highlighted (Sect. 2). On this basis, the cornerstones of the TAPAS methodological framework are outlined and the concrete procedure for sustainability assessment with its individual instruments is described (Sect. 3). This is complemented by an insight into the practical application of the new methodology in two ongoing R&D projects within the *liv*MatS cluster (Sect. 4). Finally, the results and findings are discussed (Sect. 5) and summarized (Sect. 6).

2 Normative Framework and Biological Concepts

With regard to the normative framework for sustainability assessment, TAPAS is based on the 2030 Agenda, which was adopted in 2015 by all 193 member states of the United Nations.

With its 17 Sustainable Development Goals (SDGs), the associated 169 targets, and its 231 indicators, the 2030 Agenda creates much more clarity and accountability for the understanding of sustainable development. For the first time, a universal normative basis for sustainability assessment has been created that covers all three dimensions of sustainable development, i.e., environmental, economic and social aspects [5–8]. However, its target system is primarily aimed at states and local authorities. For this reason, the associated indicators are not fully suitable for an assessment of technologies and materials systems without prior adaptation. This paper therefore employs an elaborate procedure and clearly defined criteria to filter out or further develop the indicators that are suitable for the sustainability assessment of technologies and materials systems.

Furthermore, TAPAS also integrates three basic strategies aiming to achieve more sustainable production and consumption patterns. These are the strategies of efficiency, consistency and sufficiency. As many developments at *liv*MatS are inspired by biological models, the methodological interfaces between biomimetics and sustainability research are highlighted in this context. As Speck et al. [9] have pointed out, biological concepts can be assigned to the three sustainability strategies. This has been illustrated in an interdisciplinary study using lianas as an example. Lianas are climbing plants, i.e., a special plant growth form that has evolved independently many times over the course of biological evolution in different plant clades. Indeed, lianas and their interactions with host trees are prime examples of complex interactions and interdependence among diverse plant species. Despite this complexity, the evolutionary characteristics of lianas allow only a limited range of biological concepts to be extracted from them, making them particularly suitable for alignment with sustainability strategies. Within this context, several biological concepts were identified that can sharpen and further strengthen the understanding and management of sustainability strategies.

For example, their non-self-supporting growth form enables climbing lianas in dense forests to reach the canopy and thus an advantageous position for pollination, fruit, or seed dispersal and for receiving direct sun light for photosynthesis with little investment in their own stem material (concept of external support). The resource savings allow liana to use its metabolic energy for other tasks. Hence, this concept can be attributed to the sustainability strategy of efficiency [9].

During secondary growth, following the secure attachment of the stem to a host tree, the newly formed flexible wood increases in volume to such an extent that the closed ring of strengthening tissue is ruptured. These internal fissures are rapidly sealed by adjacent thin-walled sealing cells. In the subsequent self-healing phase, some of the thin-walled sealing cells develop into thick-walled strengthening cells, leading to a partial recovery of the original mechanical property. This concept of damage repair can be related to the sustainability strategy of consistency [9].

In the self-healing phase, further thin-walled cells are formed in the fissures, which lead to the disintegration of the formerly closed strengthening ring into individual parts connected by the thin-walled cells (concept of less is more). Although the thin-walled cells cannot completely restore the original mechanical properties, the self-repair is sufficient to maintain the function of the plant (good enough concept). The concept of less is more and the good enough concept can be attributed to the sustainability strategy of sufficiency. This leads not only to interesting theoretical considerations within the sufficiency debate, which often focuses on the aspect of renunciation, but also to practical design principles for more sustainable products in the technosphere [9].

3 Cornerstones, Workflow and Instruments of the TAPAS Framework

Building on the research into the normative framework and the refinement of the sustainability strategies, the cornerstones of the TAPAS framework were developed. Over the past five decades, a wide range of different methodological approaches and frameworks to technology assessment have been developed. TAPAS is based on the framework of Prospective Technology Assessment (ProTA) as described by Liebert and Schmidt [10], because it seeks to start as early as possible in the innovation process. ProTA aims at designing new technologies from the perspective of their expected consequences and realistic potentials. As a participatory approach, it also emphasizes the self-reflection of the actors involved. In contrast to observations from an external and ex-post perspective (as practiced in earlier concepts of technology assessment), ProTA should become part of a self-reflection and self-criticism of scientists and engineers in the R&D phase itself, which also includes the perspective of social and political actors. To achieve this goal,

ProTA requires a well-defined normative framework [10]. Reflecting the international consensus reached by the 2030 Agenda for universally applicable sustainability goals, the TAPAS assessment framework is consistently aligned with the normative requirements of the 17 SDGs and the 169 SDG targets. These normative principles, goals and targets are considered and implemented in "pragmatic places" [11], i.e., places relevant to the ongoing innovation process.

As an evolution of the ProTA approach, TAPAS envisages a tiered assessment in which, depending on the maturity of the analyzed object and the available data, different low-threshold instruments are applied that are as robust as possible and enable self-reflection to be integrated into the development process [12]. Against the background of the methodological cornerstones described above, a total of five analysis stages have been defined (see Fig. 1).

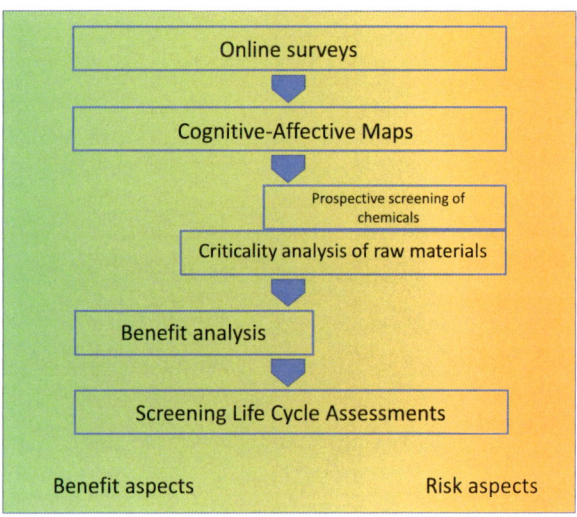

Fig. 1. Tiered workflow and associated instruments of the TAPAS framework, covering both benefit and risk aspects. Adapted from [12].

In the first stage, qualitative online surveys are used to obtain an understanding of fundamental features of the analyzed object, such as its functions and the materials used. In the second stage, "Cognitive-Affective Maps" [12, 14], a method from social sciences, are used to analyze social acceptance. In the third stage, a prospective screening of chemicals with "REACH-Radar" [15] and an analysis of the criticality of raw materials are carried out by means of "ÖkoRess" [16, 17]. The fourth stage is a detailed benefit analysis [18] with a set of indicators derived from the 2030 Agenda (see above). In the case of more advanced technologies or materials systems, life cycle oriented approaches, e.g., in the form of a life cycle assessment [19, 20], are conducted in a fifth and final stage.

As a sustainability assessment with a systemic claim, TAPAS is not limited to the early identification and minimization of existing weaknesses and risk aspects. It also provides a strategic "radar" for making optimal use of existing strengths and opportunities, and for strengthening and shaping them at an early stage in line with the normative goals (see Sect. 2). Therefore, in addition to an "early warning system", an "early encouragement system" is an integral part of the TAPAS assessment architecture. This underlying principle is also reflected in the instruments employed (see Fig. 1). For example, the online surveys cover both benefit and risk aspects, while the detailed benefit analysis focuses on the benefit aspects and the prospective screening of critical chemicals with "REACH Radar" centers on the risk aspects [12]. Furthermore, the design of TAPAS allows its users to analyze the potential impacts of their development work in an autonomous and robust way, even in the earliest stages of the innovation process. For example, online surveys promote early internal awareness and scoping of relevant sustainability aspects as a starting point for in-depth analysis. Cognitive-affective maps enable early stakeholder engagement and external verification of sustainability claims. Prospective chemical screening and criticality analysis of raw materials help to identify potential lock-in effects and path dependencies at an early stage. Through the self-reflection of the scientists involved in the innovation process, TAPAS ensures that the best possible data basis is available for the prospective sustainability assessment [12].

4   Results of Initial Case Studies

In order to evaluate the practical applicability of TAPAS, two case studies have been carried out in the *liv*MatS context in which selected instruments of TAPAS are applied and tested in cooperation with the *liv*MatS projects "SolStore" and "ThermoBatS." Both projects pursue the goal of supplying the *liv*MatS materials systems with energy: "SolStore" involves integrated energy storage systems based on perovskite solar cells [12], whereas "ThermoBatS" uses thermoelectric micro-generators to produce electrical energy [12].

The case studies show that the developed tools are generally suitable for research and development projects in the field of basic research. Furthermore, their application also provides specific suggestions for the substitution of solvents of toxicological concern. In addition, the "SolStore" project conducted a detailed benefit analysis based on the newly developed set of indicators. The results show potential positive contributions to the 2030 Agenda in eleven of the 30 benefit indicators. In this context, some concrete fields of application for the newly developed materials system could already be outlined, especially in the field of off-grid sensors, see Table 1.

Table 1. Results of the benefit analysis for the SolStore materials system (B=Benefit). Adapted from [12].

#	SDG target	Indicator	Relevance for SolStore materials system / suggestions for application fields
B1	2.1, 2.2	Reduction of hunger and malnutrition	Energy detection in agriculture and nutrition; sensor on food containers to assess shelf life
B2	2.3	Increasing incomes of small-scale food producers	Currently no contribution recognizable
B3	2.4, 2.5	Strengthening sustainable food production systems	Sensors for monitoring agricultural systems
B4	3.1, 3.2, 3.3, 3.4	Reducing mortality	Automatic drug delivery; UV light warning sensor
B5	3.5	Strengthening the prevention and treatment of substance abuse	Currently no contribution recognizable
B6	3.6	Reducing deaths / injuries from road traffic accidents	Warning sensors (e.g. for blind persons)
B7	3.9	Reducing deaths / injuries from hazardous chemicals and air, water and soil pollution and contamination	Sensors for the detection of water and air pollution
B8	4.4, 4.7	Strengthening knowledge and skills related to sustainability issues	Currently no contribution recognizable
B9	6.1, 6.2	Improving the access to safe drinking water, sanitation and hygiene	Currently no contribution recognizable
B10	6.3	Improving water quality by reducing the release of hazardous chemicals and materials	Currently no contribution recognizable
B11	6.4	Increasing water-use efficiency and strengthening sustainable supply of freshwater	Currently no contribution recognizable
B12	7.2	Enabling / increasing the production of renewable energy	On-grid and off-grid energy harvesting
B13	7.3	Enabling / increasing energy efficiency	Currently no contribution recognizable
B14	8.5, 8.6	Creation of well-paid jobs / reducing youth unemployment	Currently no contribution recognizable
B15	8.8	Strengthening secure working conditions	Warning sensors (e.g. for handling goods)
B16	8.9	Strengthening sustainable tourism	Currently no contribution recognizable
B17	8.10	Expanding the access to banking, insurance and financial services	Energy source for bank cards
B18	9.4	Fostering decarbonization and resource efficiency of industries	Currently no contribution recognizable
B19	11.5	Reducing deaths / people affected by disasters	Avalanche warning system
B20	11.6	Improving urban air quality	Sensors for the detection of air pollution
B21	12.2	Strengthening sustainable management and efficient use of natural resources	Currently no contribution recognizable
B22	12.3	Reducing food losses and food waste	Sensors for assessing shelf life
B23	12.4	Reducing the release of chemicals / hazardous waste into air, water and soil	Currently no contribution recognizable
B24	12.5	Reducing waste generation through waste prevention, recyclability and reusability	Currently no contribution recognizable
B25	13.2	Significant contribution to GHG emission reductions	Currently no contribution recognizable
B26	14.1	Reducing marine pollution / marine littering	Currently no contribution recognizable
B27	14.7	Strengthening the sustainable use of marine resources (fisheries, aquaculture and tourism)	Currently no contribution recognizable
B28	15.1	Fostering the conservation and sustainable use of ecosystems / biodiversity	Currently no contribution recognizable
B29	16.10	Strengthen public access to information	Currently no contribution recognizable
B30	1.3, 3.8, …, 11.2	Strengthening the availability of affordable and sustainable products / services	Currently no contribution recognizable

5 Discussion

Until recently, the criteria and indicators for sustainability assessments in science and practice had to be determined subjectively or intersubjectively through time-consuming stakeholder panels, because there was no universal, normative reference framework.

In contrast to this rather unsatisfactory practice of the past, TAPAS is consistently aligned with the globally accepted SDGs of the 2030 Agenda, creating an evaluation framework with much greater clarity and accountability. In an integrated approach, both benefits and risks are the subject of the analysis, in the sense of an early warning system for the early detection of risks, but also as a strategic radar for the proactive positive shaping of the normative goals of the SDGs. In this context, new methodological ground has been broken in the area of detailed benefit analysis, as for the first time a set of 30 indicators has been developed that establish a direct link to the 2030 Agenda.

Moreover, one of the core strengths of TAPAS is the interdisciplinary interaction between biology, biomimetics and sustainability research during its development phase. We have carefully investigated the methodological interface between biomimetics and sustainability research and have carried out an in-depth analysis of fundamental biological concepts in order to specify the central sustainability strategies of efficiency, consistency and sufficiency with respect to their application in the context of TAPAS. This finally formed an important basis for the development of analytical instruments suitable for the evaluation of the bio-inspired materials systems of the *liv*MatS cluster. In addition, the design of the methodological framework as a multi-stage approach and the implementation of the sustainability assessment in the context of self-reflection by the actors of the innovation process proved to be particularly beneficial and expedient for the successful implementation of TAPAS.

However, the consequent use of TAPAS and its instruments can cause additional workload for the users and may initially slow down the development of novel technologies. This is particularly the case if the data required for the analysis are not available or can only be obtained with great effort. On the other hand, the integration of sustainability assessment into the innovation process and the approach of self-reflection can also significantly enhance the ability of science to respond to the challenges of ever faster development and more consequential influence of technologies [21, 22]. Consciously pausing at important milestones in research and development can therefore shorten society's overall reaction time if existing benefits and risks are recognized as early as possible and lead to concrete, targeted and pragmatic adjustments in ongoing innovation processes [3].

6 Conclusions

In summary, TAPAS provides a field-proven methodology for an early sustainability assessment of novel technologies and materials systems. In order to enable sustainability assessments depending on the level of technological maturity and existing data, a stepwise and iterative approach combines low-threshold instruments even for the earliest stages of the innovation process, e.g., in the field of basic research. Consistently aligned with the 2030 Agenda and the Sustainable Development Goals, TAPAS combines a high level of commitment regarding its indicators with the pragmatic approach of self-reflection. For example, a set of 30 indicators has been derived from the 169 SDG targets of the 2030 Agenda for the purpose of a detailed benefit assessment that allows TAPAS users to analyze not only the risks, but also the benefits of novel technologies and materials systems. By placing the analysis and assessment steps in the hands of the actors of the innovation process, the best possible database can be provided. Further advantages arise in terms of "capacity building", i.e., strengthening the methodological knowledge and skills for sustainability assessment among the users.

Acknowledgements. Funded by the Deutsche Forschungsgemeinschaft (DFG, German Research Foundation) under Germany's Excellence Strategy – EXC-2193/1 – 390951807.

References

1. Collingridge, D.: The social control of technology. Frances Pinter, London (1980).
2. Köhler, A. R., Som, C.: Risk preventative innovation strategies for emerging technologies the cases of nano-textiles and smart textiles. Technovation **34**, 420–430 (2014). https://doi.org/10.1016/j.technovation.2013.07.002.
3. Möller, M., Höfele, P., Reuter, L., Tauber, F. J., Grießhammer, R.: How to assess technological developments in basic research? Enabling formative interventions regarding sustainability, ethics and consumer issues at an early stage. TATuP – Journal for Technology Assessment in Theory and Practice **30**(1), 56–62 (2021). https://doi.org/10.14512/tatup.30.1.56.
4. *livMatS* Homepage, https://www.livmats.uni-freiburg.de/en, last accessed 2024/03/12.
5. United Nations Homepage, https://www.un.org/en/development/desa/population/migration/generalassembly/docs/globalcompact/A_RES_70_1_E.pdf, last accessed 2024/01/17.
6. BMUV Homepage, https://www.bmuv.de/en/topics/sustainability/overview-sustainability/2030-agenda, last accessed 2024/01/17.
7. SDGF Homepage, https://www.sdgfund.org/sites/default/files/Report-Universality-and-the-SDGs.pdf, last accessed 2024/01/17.
8. United Nations Homepage, https://unsdg.un.org/2030-agenda/universal-values/leave-no-one-behind, last accessed 2024/01/17.
9. Speck, O., Möller, M., Grießhammer, R., Speck, T.: Biological concepts as a source of inspiration for efficiency, consistency, and sufficiency. Sustainability **14**(14), 8892 (2022). https://doi.org/10.3390/su14148892.
10. Liebert, W., Schmidt, J. C.: Towards a prospective technology assessment: Challenges and requirements for technology assessment in the age of technoscience. Poiesis and Praxis **7**, 99–116 (2010).
11. Grunwald, A.: Ethische Grenzen der Technik? Reflexionen zum Verhältnis von Ethik und Praxis. In: Stephan Saupe and Armin Grunwald (eds.): Ethik in der Technikgestaltung. Praktische Relevanz und Legitimation, pp. 221–252. Springer, Berlin (1999). https://doi.org/10.1007/978-3-642-60033-3_11.
12. Möller, M.: Entwicklung eines mehrstufigen Methodenrahmens für eine prospektive Nachhaltigkeitsbewertung von neuartigen Technologien und Materialsystemen (TAPAS). Dissertation, University of Freiburg (2022). https://doi.org/10.6094/UNIFR/235292
13. Höfele, P., Reuter, L., Estadieu, L., Livanec, S., Stumpf, M., Kiesel, A.: Connecting the methods of psychology and philosophy: Applying Cognitive-Affective Maps (CAMs) to identify ethical principles underlying the evaluation of bioinspired technologies. Philosophical Psychology 1–24 (2022). https://doi.org/10.1080/09515089.2022.2113770.

14. Thagard, P.: EMPATHICA: A computer support system with visual representations for cognitive-affective mapping. In: K. McGregor (Ed.), Proceedings of the workshop on visual reasoning and representation, pp. 79–81. AAAI Press (2010).
15. Öko-Institut Homepage, https://www.oeko.de/reach-radar, last accessed 2024/01/17.
16. Dehoust, G., Manhart, A., Möck, A., Kießling, L., Vogt, R., Kämper, C., Giegrich, J., Auberger, A., Priester, M., Rechlin, A., Dolega, P.: Erörterung ökologischer Grenzen der Primärrohstoffgewinnung und Entwicklung einer Methode zur Bewertung der ökologischen Rohstoffverfügbarkeit zur Weiterentwicklung des Kritikalitätskonzeptes (ÖkoRess I). UBA-Texte 87/2017. Oeko-Institut (2017), https://www.umweltbundesamt.de/sites/default/files/medien/1410/publikationen/2017-09-28_texte_87-2017_oekoress_konzeptband_2.pdf, last accessed 2024/01/17.
17. Dehoust, G., Manhart, A., Dolega, P., Vogt, R., Auberger, A., Kämper, C., von Ackern, P., Rüttinger, L., Rechlin, A., Priester, M.: Weiterentwicklung von Handlungsoptionen einer ökologischen Rohstoffpolitik (ÖkoRess II). UBA-Texte 79/2020. Oeko-Institut (2020), https://www.umweltbundesamt.de/sites/default/files/medien/1410/publikationen/2020-06-17_texte_79-2020_oekoressii_abschlussbericht.pdf.
18. Möller, M., Grießhammer, R.: Product-related benefit analysis based on the Sustainable Development Goals – integrating the voice of society into Life Cycle Sustainability Assessment. Journal of Industrial Ecology 28, 397–409 (2024). https://doi.org/10.1111/jiec.13464.
19. ISO 14040: Environmental management—life cycle assessment—principles and framework ISO 14040:2006. Beuth, Berlin (2006).
20. ISO 14044: Environmental management—life cycle assessment—requirements and guidelines ISO 14044:2006. Beuth, Berlin (2006).
21. Steffen, W., Grinevald, J., Crutzen, P., McNeill, J.: The Anthropocene: conceptual and historical perspectives. Philosophical Transactions of the Royal Society A 369(1938), 842–867 (2011). https://doi.org/10.1098/rsta.2010.0327.
22. Steffen, W., Broadgate, W., Deutsch, L., Gaffney, O., Ludwig, C.: The trajectory of the Anthropocene: The Great Acceleration. Anthropocene Review 2(1), 81–98 (2015). https://doi.org/10.1177/2053019614564785.

Analysis of Model Specific Features for the Development of a Non-Contact Reference Point System on a Flexible Gripper

Julia-Christina Sattler[1(✉)], Marcel Todtermuschke[2], Alexander Voigt[1], and Steffen Ihlenfeldt[2,3]

[1] Volkswagen AG, Berliner Ring 2, 38440 Wolfsburg, Germany
Julia-Christina.Sattler1@volkswagen.de
[2] Fraunhofer Institute for Machine Tools and Forming Technology IWU, Reichenhainer Straße 88, 09126 Chemnitz, Germany
[3] Institute for Mechatronic Engineering, TU Dresden, Helmholtzstraße 7a, 01069 Dresden, Germany

Abstract. The increasing diversity of models and variants in the automotive industry is leading to new challenges in the flexibilization of the body shop production process, especially the gripper systems. One way of making the component handling process more flexible by using type-variable grippers is to replace the mechanical reference point pins with virtual sensor technology, which ensures positionally and repeatedly accurate handling per component.

One major challenge in using sensory technology for type-variable components is determining the necessary measuring range, which depends on the varying position and size of the reference points per model. In order to design a sustainable solution, the definition of a measurement range for both current and future model types is essential as well.

According to the current state of the art, no optimal sensor parameters are known for the use case described. Additionally there are no existing studies that provide a guideline for the recommended position of reference points depending on the sensor technology used.

This paper presents the approach of investigating characteristics of various reference points with regard to their position and geometry in order to define a measurement range for the usage of sensor technology. To optimize the determined measuring ranges, various scenarios of component alignment are examined and their influence on the sensor parameters identified. Here the example of interior door parts of different vehicle segments is applied as a use case.

In the result, a system framework for positioning reference points is defined based on the investigated use case and a recommendation is made for the parameters of the sensor technology to be used. Furthermore the systematic alignment of components leads to a reduction of sensor parameters. These results can be used for positioning reference points in future models.

Keywords: Automotive · Flexibility · Reference point system · Position determination · Body-in-white

© The Author(s) 2025
K. Dröder and T. Vietor (Eds.): CD 2024, Proceedings, pp. 120–135, 2025.
https://doi.org/10.1007/978-3-658-45889-8_10

1   Introduction

The flexibilization of production processes has become unavoidable in the course of increasing model and variant diversity in the automotive industry. In addition type-specific devices with mechanical fixtures for centering objects are also used in other branches of industry. Examples can be found in the component clamping in milling and turning machines or the automated tire assembly [1, 2]. In all areas, reducing the complexity of the handling system results in increased flexibility and improved process efficiency of the overall system [3].

With a degree of automation of around 95%, flexibilization is a major challenge, particularly in body shop processes, in order to counteract a greater variety of models while maintaining quality and keeping costs low [3]. In a conventional body shop production plant, 2–3 vehicle models are usually manufactured on one production line. Each model requires type-specific grippers and fixtures with customized mechanical fixture parts and contour pieces. In particular, the individual design of the type-specific reference point system (RPS) requires additional planning effort and increased investment costs [4]. The adaptation of the conventional RPS is essential when making handling systems more flexible and is therefore considered in more detail below.

The mechanical RPS consists of pins on the gripper and matching holes on the component, thus ensuring that the component is spatially fixed in all six degrees of freedom. The positioning of the holes and pins varies per each component and gripper. A type-variable design of the mechanical RPS using movable pin units is not possible due to the limited installation space of the gripping systems or devices and adherence with cycle time specifications [5]. Due to the model-specificity of the mechanical RPS, reusability of the grippers for subsequent models is also not ensured [4]. A non-contact measuring system is therefore required for handling several model types, which substitutes the function of a mechanical RPS and ensures flexible handling for both current and subsequent models.

The concept of a non-contact RPS is based on the principle of determining the position of the component by recording RPS position data via measuring sensors and the calculation of the actual position and orientation via a calculation unit. The final position correction of the component replaces the mechanical positioning with the pin-hole principle of conventional RPS. The first step in the development of a non-contact RPS is therefore to analyze the image acquisition process and the relevant sensor and component parameters.

For this purpose it is necessary to analyze a use case of position data of model-specific RPS holes and their relation to geometric parameters of the components. The aim of this paper is to use the determined data to identify the required sensor parameters and to define a limitation of the non-contact RPS for flexible further utilization for subsequent models.

In the first part, the relevant component and sensor parameters and their dependencies and interrelationships for the image acquisition process are identified for the applied use case. For this purpose an overall system consisting of a flexible gripper, components (here: interior door parts) and measurement sensors (2D camera) is used. Based on the results a RPS framework is defined and a suitable sensor system is recommended. Finally, the optimization of the system is verified by systematically arranging the components to reduce the necessary sensory parameters.

2 Approach

There are different approaches for measuring component geometries and features using sensor technology. A distinction is made between non-contact and tactile measurement principles. A major advantage of non-contact measurement technology is the measurement speed and the possibility of measuring very large components, such as those used in shipbuilding or aircraft construction [6]. Tactile measuring systems are mainly used via coordinate measuring machines with touch probes for the high-precision measurement of components, e.g. in engine or transmission construction, and are not discussed further in this paper [7]. Non-contact measuring principles can be divided into optical, electromagnetic, acoustic and capacitive measuring methods in terms of their functionality [8]. Optical systems are often used for measuring geometries, position determination and alignment correction, but also for quality control and part recognition [9]. The other measuring principles are mainly used for distance measurement of large or small distances and for position control [9, 10].

The following section focuses on the non-contact, optical measurement of component geometry and alignment using two-dimensional cameras. Area scan cameras enable the position of the component to be determined via optical detection of RPS features and subsequent data conversion into coordinates by a calculation unit. The fundamental challenge of sensor-based measurement of type-variable components lies in a variable required measuring range with a fixed measuring distance between the measuring system and the object. The appropriate selection of technical parameters as well as the optimum positioning and alignment of the 2D-cameras to the measurement object is decisive.

There are a few publications in the literature on the optimum positioning of sensors depending on the measurement object. However, these are specific to the respective application and do not consider variable measurement parameter (e.g. [11, 12]). Furthermore, there are some approaches for optimizing the camera position via the development of image processing algorithms (e.g. [13–15]), but these are not considered further in this paper.

Overall, there is still no study that analyzes the optimal measuring range of cameras in the flexible gripping system with type-variable components. Furthermore, there are no publications on the positioning of known and (as yet) unknown RPS features for a sustainable use of 2D cameras on a flexible gripping system. In order to consider these focal points, the first step is to identify the required features of the non-contact RPS.

Precise data measurement requires the determination of the external and internal technical parameters of the camera system [16]. Calibration of the camera and subsequent image data processing are further topics that are not covered in detail here. Examples for analyzing the influence of data processing and calibration parameters are given in or in the analyses of [16, 17].

The question of **positioning** and the suitable **number of cameras** as external parameters are decisive for the optimal detection of all features [18]. These not only affect the image recognition and post-processing time of the data, but also the measurement accuracy and the covered image area of the object surface [15]. To determine the positioning and number of cameras, it is first necessary to define the **required field of view (FOV)**. For precise feature detection, the FOV must cover the entire area of all necessary RPS features (in this case: holes) across all components. In the case of flexible model compositions, the positions of the RPS features and therefore the required size of the FOV per model vary. If the hole positions vary on a large scale, the selected FOV for a component may be too small and may not cover the RPS features of all components. The FOV can be maximized, for example, by using several cameras at different positions or a moving camera [17]. However, moving the cameras is not feasible in this use case due to the small installation space and the inline measurement (compliance with cycle time). Furthermore, the complexity of the overall system must be kept low, as the computing effort and the calculation time of the overall system increase as the number of cameras increases [15]. Another way of increasing the FOV is to increase the **working distance** *WD* to the object. A flexible gripping system is characterized by its compact design, which enables the handling of components with different geometries. For this reason, the working distance to the object should be kept as small as possible.

Furthermore, the RPS features must not only be in the FOV, but also in the **DOF** (depth of field) **range,** as this range ensures the optimum focus for detecting the features. The DOF is influenced by the aperture opening. With a constant aperture setting, an increase in the FOV can be achieved by increasing the **focal length** with a smaller depth of field.

In addition to the variable parameters mentioned above, there are also constant internal, technical parameters that are fixed when the sensor is selected. These include the sensor size, the resolution (including pixel size and number of pixels) and the magnification, which have an influence on the measuring accuracy. All relevant parameters for the alignment of sensors on the flexible gripping system are shown in the figure below (Fig. 1).

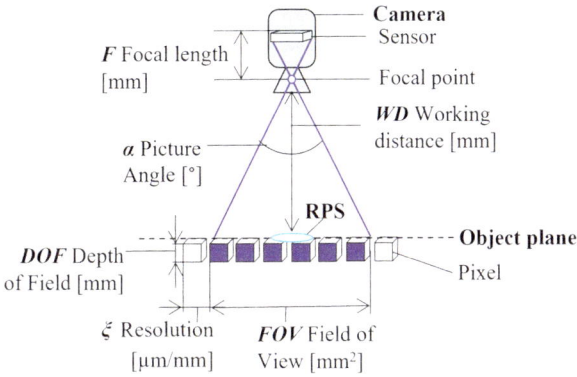

Fig. 1. Schematic representation of the necessary parameters for the alignment of sensors on the flexible gripping system

In summary, the challenge of determining the necessary measuring range of cameras in the flexible gripping system lies in maximizing the FOV and depth of field to cover all current and future features while minimizing the working distance and number of cameras. In addition, sufficient measurement accuracy must be ensured via the technical parameters.

3   Method for Part Analysis

For the analysis of possible positioning criteria and patterns of RPS features depending on the component geometry, 15 interior door parts (front left door) of vehicle segments A, B, C, D, M, J (designation according to EU Commission) were used. In order to cover a variable sample size, interior door parts of various sizes and widths were used in the component selection (see table below) (Table 1).

Table 1. Naming and definition of components examined for analysis

No	Vehicle segm	Designation	Production Start	Feature
1,2	A	Microcar	2012	2-, 4-door
3,4	B	Small car	2018	2-, 4-door
5	J	Sport utility & off-road vehicle	2019	
6,7	C	Middle class	2020, 2021	ICE-, E-vehicle
8,9	J	Sport utility & off-road vehicle	2021,2018	E-, ICE-vehicle
10	M	Multivan	2007	
11, 12	J	Sport utility & off-road vehicle	2017, 2018	
13,14	D	Upper middle class	2011, 2018	
15	J	Sport utility & off-road vehicle	2019	

Two of the 15 interior door parts are from electric vehicles (no. 7, 8), two others are from 2-door vehicles (no. 1, 3). The production start of the components is between 2007 and 2021. For data acquisition, the components were measured manually using a CAD program, aligned in the vehicle coordinate system. A round hole (*RPS1*) and a slotted hole (*RPS2*) are used as RPS features in this use case. For the analysis, geometric input variables of the component and the RPS feature are considered, which have an influence on the positioning. The recorded input variables are analyzed in three investigation steps I, II, III and are listed below with regard to the designation, measurement unit and the measured coordinates (Table 2):

Table 2. Definition of input variables and naming of variables

No	Parameter	Description	Unit/vehicle coordinate
Analysis part 1			
1	B_T	Wide door	[mm] in X
2	H_T	Door height	[mm] in Z
3	d_{RPS1}	Diameter RPS1	[mm]
4	d_{RPS2}	Diameter longest side RPS2	[mm]
Analysis Part II			
5	$A_{h.RPS1-2}$, $A_{v.RPS1-2}$	Distance RPS1–2 hor. Resp. Vert	[mm] in X or Z
6	$A_{O.RPS1}$, $A_{O.RPS2}$	Distance top edge to RPS feature	[mm] in Z
7	$A_{L.RPS1}$, $A_{R.RPS2}$	Distance left - RPS1 or right - RPS2	[mm] in X
Analysis Part III			
8	W_{RPS1}, W_{RPS2}	Size viewing window RPS feature	Width X * Height Z * Depth Y [mm]
9	x,y,z_{RPS1}; x,y,z_{RPS2}	Position RPS feature in x,y,z	X,Y,Z

The parameters are measured using a predefined scheme that can be used for components with a rectangular shape. In principle, the component geometry is determined by the average value of the maximum and minimum height and width of the component. Using the example of the inner door part, the height is defined as the length between the window well and sill edge (Z value) and the width as the length between the edge of the A and B pillars (X value).

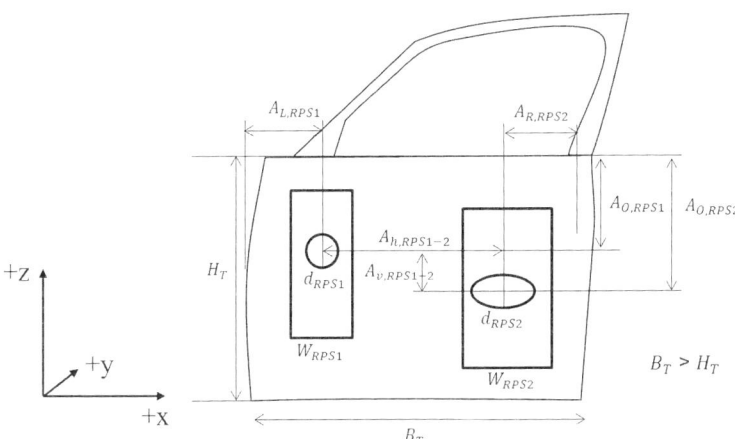

Fig. 2. Schematic representation of the measurement method applied to the use case

The diameter of the RPS holes $d_{RPS1, RPS2}$ is described via the horizontal, longest side. The measurement of the positions x,y,z_{RPS1} and x,y,z_{RPS2} is carried out via the collar hole center point in maximum Y-value (Fig. 2).

4   Results

4.1   Analysis Part I: General Influencing Variables, System Delimitation

To restrict the system space, the geometric dimensions of the component and RPS feature are first analyzed. The measured average component heights in Z-direction and widths in X-direction are shown below (Fig. 3):

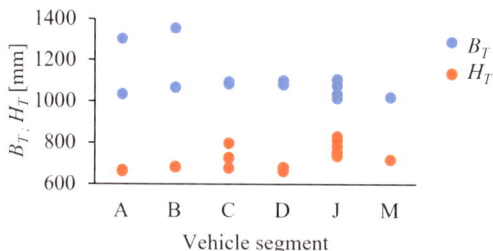

Fig. 3. Heights and widths of the examined sample parts depending on the vehicle class

The tested samples have a **height** of 663.6 to 831 mm, measured from the window shaft to the sill. This results in an average height of 724.54 mm. The values of the samples from vehicle segments A, B, C, D and M are in the range between 663.6 and 729.9 mm. Samples 5, 8, 9, 11, 12, 15 (vehicle segment J), on the other hand, have door heights in the range between 739.65 and 831 mm. The **door width** of the sample parts of all vehicle segments considered is between 1017 and 1108.6 mm. Exceptions are samples 1 and 3 (2-door) with a width of 1306 and 1356.6 mm. Overall, the average width of all 15 sample parts is 1102.81 mm.

In summary, the geometry of the doors under consideration varies more in height (Δ 167.4 mm) than in width (Δ 91.6 mm without samples 1 and 3, Δ 339.6 mm including samples 1 and 3:

$$B_T > H_T \tag{1}$$

For this reason, a greater scattering of the RPS features in the Z direction is to be expected. The results of the analysis of the RPS hole diameters d_{RPS1} and d_{RPS2} are shown graphically below (Fig. 4):

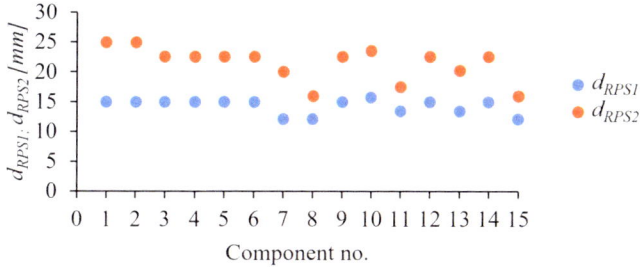

Fig. 4. Diameter of the RPS holes of the examined components

The values of d_{RPS1} are between 13.5 and 15.7 mm. The most frequent value is 15 mm (9 samples). Samples 7, 8 (electric vehicles) and 15 have an RPS1 diameter of 13.5 mm. Sample 10 (Production Start 2007) differs from the other samples with a value of 15.7 mm.

The vertical diameter of RPS1 corresponds to the vertical diameter of RPS2 per component (not shown here). This implies that the same RPS pin diameter is used per component. Exceptions are samples 1 and 2 with a vertical RPS diameter of 20 mm compared to 15 mm (RPS1). As the horizontal diameter of RPS2 of all samples varies more, these values are used for the analysis of d_{RPS2}. Values between 16 and 25 mm are measured here.

To summarize, the horizontal diameter of RPS2 within the sample set under consideration is larger than the diameter of RPS1. It can therefore be assumed that a larger FOV is required for RPS2. Therefore it can be concluded:

$$d_{RPS1} < d_{RPS2} \tag{2}$$

4.2 Analysis Part Ii: Positioning of RPS Features Depending on the Component Geometry

In the second part, the distances of the RPS features to each other and to the edge of the component are examined in order to draw conclusions about positioning patterns. The distances between RPS1 and RPS2 are measured parallel to the X-axis (horizontal) or Z-axis (vertical) (Fig. 5).

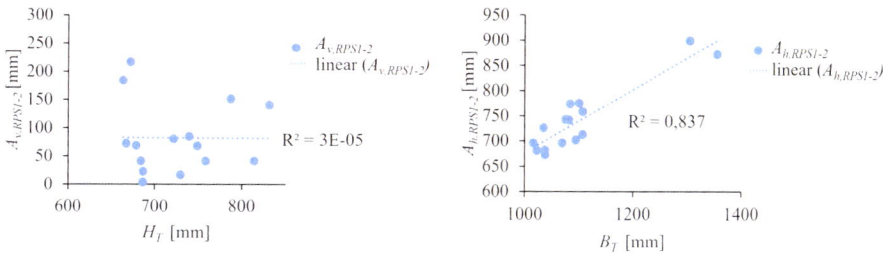

Fig. 5. Distances between RPS1 and 2 as a function of component height/width

The measured distances $A_{h.RPS1-2}$ are between 671.5 and 899 mm (figure on the right). Strong deviations can be seen in samples 1 and 3 with values between 899 and 871 mm. The remaining sample values are between 680 and 773.5 mm. A linear relationship between increasing component width and increasing distance from RPS1–2 in the vehicle X direction can be seen for the majority of the samples examined:

$$A_{h.RPS1-2} \sim B_T \tag{3}$$

This implies that the distance between the features and the edge of the component remains the same or becomes smaller as the component width increases. This can be explained by the need to ensure component stability during the handling process. Maximum stability during the handling process is ensured by the greatest possible distance between the RPS features in direction of the longest component side.

When examining the distance in the vehicle-Z direction ($A_{v.RPS1-2}$), no comparable correlation is recognizable (figure on the left). The smallest value is 3 mm, the maximum distances are 217 mm, 184 mm and 151 mm (sample 1, 2, 12). These findings correlate with the results from Analysis I: Due to the high scattering of the component heights, the distances between the RPS features vary in the Z-direction with a similar degree of severity. As the values of $A_{v.RPS1-2}$ does not increase linearly, no uniform pattern can be recognized here. This may be due to necessary, component-specific cut-outs that prevent the positioning of RPS features for space reasons. It can therefore be concluded that.

$$A_{h.RPS1-2} < A_{v.RPS1-2} \tag{4}$$

The next step is to analyze the distances to the nearest component edge in vehicle Z direction (upper edge) and vehicle X direction (left and right edge) (Fig. 6):

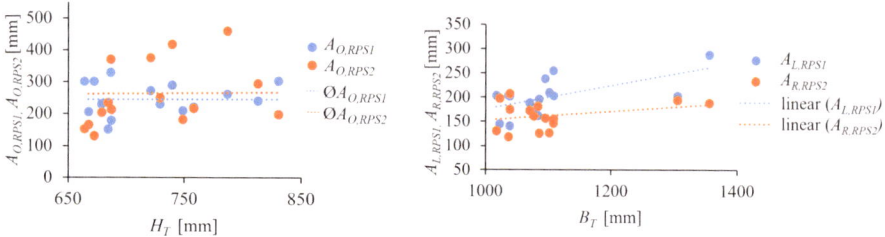

Fig. 6. Distances from RPS1–2 to the upper or left/right edge as a function of the component height or width

The values of $A_{O.RPS1}$ are between 151.1 mm and 329.6 mm (figure on the left), whereas the values of $A_{O.RPS2}$ are between 130.6 mm and 460.3 mm. On average, the distance from the upper edge to RPS1 is smaller compared to RPS2 (248.7 mm vs. 257.64 mm):

$$A_{O.RPS1} < A_{O.RPS2} \tag{5}$$

This deduces that RPS1 features are more frequently positioned in a certain area near the upper edge. Furthermore, the values of RPS2 scatter more than the values of RPS1 in the vehicle Z-direction (standard deviation $\pm$ 50 mm vs. $\pm$ 102 mm). This supports the hypothesis that a larger FOV in Z-direction is necessary to capture all RPS2 and RPS1 features. In addition, there is no linear correlation between the component height and the positioning of RPS features in Z-direction.

The analysis of the distances to the left and right edge ($A_{L,RPS1}$, $A_{R,RPS2}$) results in an average lower value for RPS2 with 162.2 mm compared to RPS1 with 200.3 mm. This may be related to the specificity of the examined component (inner door part). The rebate edge on the left side of the com-ponent is larger, which implies a greater distance requirement for the placement of the RPS1 feature:

$$A_{L,RPS1} > A_{R,RPS2} \tag{6}$$

Overall, a linear relationship between increasing component width and increasing distance of both RPS features to the left or right edge can be seen in a larger number of samples examined:

$$A_{L,RPS1};A_{R,RPS1} \sim B_T \tag{7}$$

4.3 Analysis Part III: Theoretical Window Size in Vehicle Coordinate System

To investigate the theoretical viewing window size in all three spatial directions, the position coordinates of the RPS features x,y,z_{RPS1} and x,y,z_{RPS2} in their absolute position in the vehicle coordinate system are compared with each other (Fig. 7):

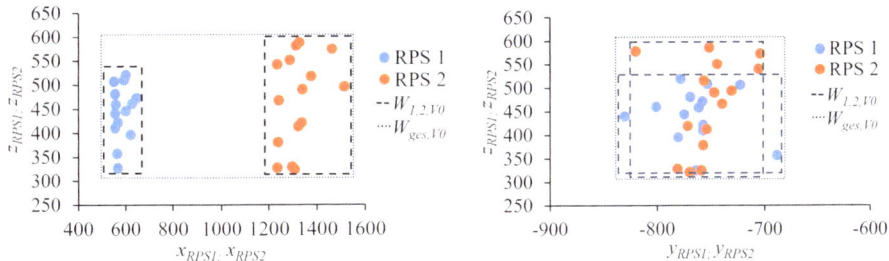

Fig. 7. Positions RPS1, RPS2 in XZ and YZ planes and theoretical window sizes for scenario V0

Here, the width and height of the theoretical FOV is analyzed via the positions of the RPS holes in the X and Z directions (figure on the left). Overall, the necessary FOV in the XZ plane is smaller for RPS1 than for RPS2. In addition, the standard deviation for RPS1 is smaller in both the Z and X directions ($\pm$59.3 mm and $\pm$ 30.4 mm) compared to the standard deviation of the RPS2 coordinates in the Z and X directions ($\pm$95.4 mm and $\pm$ 80.3 mm). This implies the necessity of a larger FOV as well as the consideration of a larger tolerance in the edge area for RPS2 features.

$$z_{RPS1},x_{RPS1} < z_{RPS2},x_{RPS2} \tag{8}$$

The necessary depth of field DOF of the theoretical viewing window is defined by the distance in the Y direction (figure on the right). For all component samples, y_{RPS1} is between -830.6 and -689. Consequently, the range of the necessary depth of field $y_{RPS1,max} - y_{RPS1,min}$ of 141 mm is greater than the distance $y_{RPS2,max} - y_{RPS2,min}$ of 114 mm.

$$\left| y_{RPS1,max} - y_{RPS1,min} \right| > \left| y_{RPS2,max} - y_{RPS2,min} \right| \tag{9}$$

With a value of ± 37.0 mm, the standard deviation of the values for y_{RPS1} is also higher than the standard deviation of the values for all y_{RPS2} (± 28.4 mm). The reason for the greater variation of the values in the depth for RPS1 may lie in the necessary "bulges" for loudspeakers, for example.

In summary, the findings from Analysis III can be summarized as follows:

$$W_{RPS1} < W_{RPS2} \tag{10}$$

In addition, the viewing windows W_{RPS1} and W_{RPS2} can be combined into an overall theoretical viewing window W_{ges}. In this window, a large area with a length of 500 mm in the X direction can be identified in which no RPS features are placed. If two sensors are used, this area could be saved in the form of the FOV. The same applies to the different heights of the viewing windows in Z direction, where the FOV for RPS1 could also be reduced by a range of approx. 100 mm.

The results of this analysis describe the theoretical viewing window size in relation to the absolute position of the RPS features in the vehicle coordinate system (scenario V0). The systematic alignment of the components is necessary to further reduce the viewing window size.

4.4 Analysis Part IV: Systematic Alignment of Components to Reduce the Viewing Window

The systematic alignment of the components to reduce the viewing windows is considered in three scenarios V1, V2, V3. The components are moved parallel to the vehicle coordinate system.

One way of reducing W_{ges} is to minimize one viewing window (here: W_{RPS1}) to the position of a coordinate point (Scenario V1). Scenario V2 describes the alignment of the components based on the center of the distance between RPS1 and RPS2 in all three spatial directions (typical constructive use case). For scenario V3, the components are positioned that W_{RPS1} and W_{RPS2} are minimal and of the same size (verification of the use of two sensor systems with the same parameters).

The following figure shows the results of scenarios V1, V2 and V3 for the XZ plane as an example. In addition, the determined viewing window of scenario V0 is shown (Fig. 8, Fig. 9).

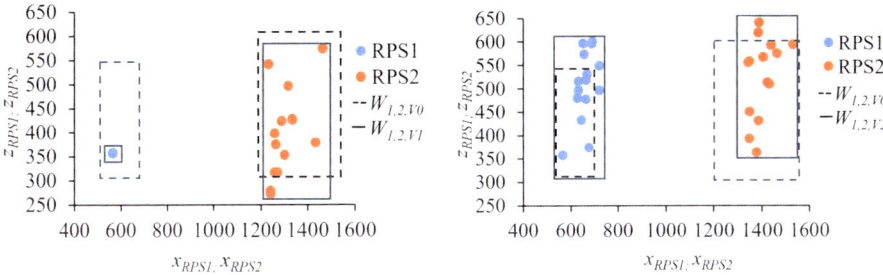

Fig. 8. Positions RPS1, RPS2 in the XZ planes and theoretical window sizes for scenario V1–V2

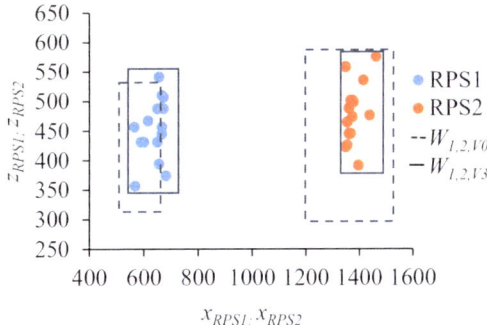

Fig. 9. Positions RPS1, RPS2 in the XZ planes and theoretical window sizes for scenario V3

As a result, the maximum reduction of the FOV to one coordinate point due to geometric dependencies of the RPS features leads to an enlargement of W_{RPS2} (V1). The maximum window size is 227 x 368 x 46.8 mm. The alignment of the components in their center of gravity position (V2) leads to a reduction of W_{RPS2} and an increase of W_{RPS1}. The reason for this is the initial position of the components in V0: the original alignment of the components along the A-pillar of the vehicle results in a reduced window width W_{RPS1} whereas a centered alignment along the X-axis results in a similar window width of 155 mm (RPS1) and 183 mm (RPS2). Placing the components with the aim of minimizing the viewing windows W_{RPS1} und W_{RPS2} to the same size (V3) results in a reduction of both viewing windows compared to scenario V0. This results in a minimum FOV of 113.75 x 184 x 23.4 mm.

5 Conclusion

To narrow down the examined system, the considered use case can be represented using the following geometric values (Table 3):

Table 3. Determined system limits and parameters of the examined overall system of the 15 sample parts

No	Parameter	Average value	System limit min	System limit max
1	B_T	1102,8	1017	1356,2
2	H_T	724,5	663,6	831
3	d_{RPS1}	14,2	12	15,7
4	d_{RPS2}	21,4	16	25
5	$A_{h,RPS1-2}$	741,1	671,5	899
	$A_{v,RPS1-2}$	82	3	217
6	$A_{O,RPS1,}$ $A_{O,RPS2}$	248,7 257,6	151,1 130,6	329,6 460,3
7	$A_{L,RPS1,}$ $A_{R,RPS2}$	200,3 162,2	140,9 118,2	288 207,3

The alignment of the components and the geometric dependencies of the RPS features on each other have an influence on the three theoretical viewing windows $W_{ges,theor}$, $W_{1,theor}$ and $W_{2,theor}$. Taking into account the maximum RPS point diameter $d_{RPS1,max}$ and $d_{RPS2,max}$ as well as a tolerance of 0.1 mm for a possible circle of confusion in the edge area of the sensor system, the real viewing windows ($W_{ges,real}$; $W_{1,real}$; $W_{2,real}$) could be calculated. The following table lists the theoretical and calculated real viewing window sizes exemplary for V0 (Table 4).

Table 4. Calculated viewing windows (theoretical and real) for the examined sample parts V0

Viewing window	Theoretical [mm] (width x * height z * depth y)	Real [mm] (width x * height z * depth y)	Volume [mm^3]
W_{ges}	966*264,5*141,6	986,5*284,7*141,8	39.827.477,3
W_1	95*194*141,6	110,9*209,9*141,8	3.300.807,64
W_2	278,5*264,5*115,2	303,7*284,7*115,4	9.977.875,2

To determine a suitable sensor system for the described use case, the required accuracy for the process must be taken into account. With a required accuracy of 0.2 mm for joining processes, a 5-fold higher accuracy is specified for the sensor technology in order to compensate for possible tolerances due to external and internal influences. Accordingly, a pixel pitch of 0.4 mm/px is required. The necessary resolutions for all scenarios V0–V3 are shown below (Table 5).

Table 5. Necessary resolution of the camera systems for the examined scenarios V0–V3

	Total resolution [MP] and percentage deviation compared to V0 [%]							
	V0	V1		V2		V3	
W_{ges}	175,544	223,106	+27,09%	200,264	+14,08%	136,323	-22,34%
W_1	14,549	0,158	-99,91%	27,480	+65,03%	16,198	-72,26%
W_2	54,04	61,311	+47,49%	41,568	-16,95%	17,733	-71,08%

To summarize, the systematic alignment of the components enables a reduction in the viewing windows and, based on this, a reduction in the sensor parameters. The maximum reduction of the viewing window W_{ges} is achieved by the uniform alignment of the RPS features on equally sized, minimal viewing windows (V3). The parameter $d_{RPS1,2}$ and the geometric relationships between the RPS features *(e.g. $A_{h.RPS1-2}$, $A_{v.RPS1-2}$)* have a significant influence on the viewing window size and the required resolution of the sensor system. In this use case a sensor system with a resolution of at least 136 MP is required to detect all RPS features in the specified range (V3). This high resolution is due to the widely varying positions of the RPS features. Nevertheless approximately 80,5% of the area covered by the viewing window contains no RPS features. For this reason, it is recommended to use one sensor per RPS feature. This allows a lower resolution per sensor to be selected. In this case, a resolution of 16 MP is required for RPS1 features and a resolution of approx. 17 MP for RPS2 features due to the higher RPS diameter. As camera systems in the range between 5 and 21 MP are usually used in industry, scenario V3 is recommended as a suitable variant for reducing the viewing windows and sensor parameters A reduction in the sensor parameters (e.g. resolution, depth of field) is achieved by selecting similar geometric positioning parameters for RPS1 and RPS2 across all models. The aim is to achieve similarly large distances between RPS1 and 2 ($A_{h.RPS1-2}$, $A_{v.RPS1-2}$), and to minimize the system limits min and max of d_{RPS1} und d_{RPS2} respectively. The reduction of the viewing window is proportionally related to the required resolution.

These results can be used as a basis for further investigation of the systematic alignment of components for the reduction of RPS viewing windows and sensor parameters. However, the third system component must be taken into account for a holistic view of the application of a virtual RPS within a flexible gripping system. The inclusion of this third system component and the effects on the viewing window has to be checked in the next step.

References

1. N. Cottone, D. Kotzor, T. Albrecht, "Verfahren und Vorrichtung zum automatisierten Stapeln von Reifen auf einem Träger," DE 10 2005 053 296 A1, Deutschland.
2. W. E. Boyes and R. Bakerjian, *Handbook of jig and fixture design,* 2nd ed. Dearborn, Mich.: Society of Manufacturing Engineers, 1989.
3. B. Ebel and M. B. Hofer, *Automotive Management: Strategie und Marketing in der Automobilwirtschaft,* 2nd ed. Berlin: Springer, 2014.
4. R. Fritzsche, A. Richter, and M. Putz, "Product Flexible Car Body Fixtures with Position-dependent Load Balancing Based on Finite Element Method in Combination with Methods of

Artificial Intelligence," *Procedia CIRP*, vol. 67, pp. 452–457, 2018, https://doi.org/10.1016/j.procir.2017.12.241.

5. R. Fritzsche, E. Voigt, R. Schaffrath, M. Todtermuschke, and M. Röber, "Automated design of product-flexible car body fixtures with software-supported part alignment using particle swarm optimization," *Procedia CIRP*, vol. 88, pp. 157–162, 2020, https://doi.org/10.1016/j.procir.2020.05.028.

6. C. P. Keferstein and M. Marxer, *Fertigungsmesstechnik*. Wiesbaden: Springer Fachmedien Wiesbaden, 2015.

7. V. Böhm *et al.*, "Mess- und Prüftechnik," *Prozesskette Präzisionsschmieden*, pp. 311–430, 2014, https://doi.org/10.1007/978-3-642-34664-4_6.

8. G. Schnell, *Sensoren in der Automatisierungstechnik*, 2nd ed. Wiesbaden, s.l.: Vieweg+Teubner Verlag, 1993.

9. M. Schuth and W. Buerakov, *Handbuch optische Messtechnik: Praktische Anwendungen für Entwicklung, Versuch, Fertigung und Qualitätssicherung*. München: Hanser, 2017.

10. S. J. Maier, *Inline-Qualitätsprüfung im Presswerk durch intelligente Nachfolgewerkzeuge*, 1st ed. München: TUM.University Press, 2018.

11. P. Puerto, I. Leizea, I. Herrera, and A. Barrios, "Analyses of Key Variables to Industrialize a Multi-Camera System to Guide Robotic Arms," *Robotics*, vol. 12, no. 1, p. 10, 2023, https://doi.org/10.3390/robotics12010010.

12. E.-C. Lovasz, I. Maniu, I. Doroftei, M. Ivanescu, and C.-M. Gruescu, Eds., *New Advances in Mechanisms, Mechanical Transmissions and Robotics*. Cham: Springer International Publishing, 2021.

13. L. Roveda *et al.*, "Robot End-Effector Mounted Camera Pose Optimization in Object Detection-Based Tasks," *J Intell Robot Syst*, vol. 104, no. 1, p. 239, 2022, https://doi.org/10.1007/s10846-021-01558-0.

14. A. G. Poyraz and M. Kaçmaz, "Edge Detection Based Autofocus Algorithm to Detect Accurate Camera Working Distance," *acperpro*, vol. 5, no. 3, pp. 406–416, 2022, https://doi.org/10.33793/acperpro.05.03.7422.

15. H. Zhang, J. Eastwood, M. Isa, D. Sims-Waterhouse, R. Leach, and S. Piano, "Optimisation of camera positions for optical coordinate measurement based on visible point analysis," *Precision Engineering*, vol. 67, pp. 178–188, 2021, https://doi.org/10.1016/j.precisioneng.2020.09.016.

16. J. Sun, H. He, and D. Zeng, "Global Calibration of Multiple Cameras Based on Sphere Targets," *Sensors (Basel, Switzerland)*, vol. 16, no. 1, 2016, https://doi.org/10.3390/s16010077.

17. S. Gai, F. Da, and M. Tang, "A flexible multi-view calibration and 3D measurement method based on digital fringe projection," *Meas. Sci. Technol.*, vol. 30, no. 2, p. 25203, 2019, https://doi.org/10.1088/1361-6501/aaf5bd.

18. G. Olague and R. Mohr, "Optimal camera placement for accurate reconstruction," *Pattern Recognition*, vol. 35, no. 4, pp. 927–944, 2002, https://doi.org/10.1016/S0031-3203(01)00076-0.

A Novel 3D Parametrization Approach for Topology Optimization of Rollbonded Cooling Plates

Frederik Schewe[1](✉), Niklas Klinke[2], and Ali Elham[3]

[1] Mubea Rollbonding Products, Mubea Tailor Rolled Blanks GmbH, Mubea-Platz 1, 57439 Attendorn, Germany
frederik.schewe@mubea.com
[2] Foxbyte, csi Verwaltungs GmbH, Robert-Mayer-Str. 10, 74172 Neckarsulm, Germany
niklas.klinke@csi-online.de
[3] University of Southampton, Boldrewood Innovation Campus, Southampton, SO16 7QF, United Kingdom
a.elham@soton.ac.uk

Abstract. Electric mobility depends on batteries, which typically require a thermal management system. Those systems are often realized as sheet metal plates with integrated channel structures. Rollbonding technology is one of the most promising technologies to produce such cooling plates due to the competitive cost structure in combination with its unmatched design freedom. However, the design of a cooling plate manufactured using rollbonding is challenging. Such a design requires thermal and hydraulic targets to be considered while manufacturing constraints must be satisfied. Due to the given design freedom and the conflicting targets manual design of cooling plates is challenging and requires significant development time and effort. Topology optimization is a popular method for automated and optimal design of components.

In this paper, the authors present a strategy to design rollbonded cooling plates by topology optimization, taking thermal, hydraulic, and manufacturing requirements into account. State-of-the-art methods usually consider channel shapes with rectangular cross-sections, ignoring the effects of realistically curved channel cross-sections as they result from the manufacturing technology. The authors present a novel parametrization strategy, considering the 3D channel shape in a realistic manner. The mathematical formulation of the novel modelling approach is shown and validated based on a small, simplified test case. The methodology is implemented into a custom, solver-agnostic framework and coupled with commercial Finite Element software.

Keywords: Topology Optimization · Battery Cooling Plate · Rollbonding · Foced Convection · Flow · Thermal

K. Dröder and T. Vietor (Eds.): CD 2024, Proceedings, pp. 136–149, 2025.
https://doi.org/10.1007/978-3-658-45889-8_11

1   Introduction

Thermal management of batteries for automotive applications is one of the key challenges of electric mobility, since it influences charging speeds, efficiency in operation and prevents premature ageing effects. Heat is transferred by forced convection of a fluid that passes through a channel structure inside sheet metal cooling plates. One of the most promising technologies for the production of such cooling plates is rollbonding due to its unmatched design freedom and its competitive cost structure for high and low-volume applications. Nevertheless, designing optimal channel structures is challenging due to conflicting thermal and hydraulic targets. The maximum temperature over the plate shall be low while the pressure loss from inlet to outlet shall be small. Topology optimization is a well-known method to generate optimal structures starting from a given design space. [1] One of the first approaches for heat transfer problems was presented by [2]. It can also be used to design optimal channel patterns for heat transfer by liquid cooling as shown in [3–5]. An application with dedicated constraints for rollbonded cooling plates is shown in [6]. This approach lacks the consideration of a realistically curved channel cross-sections though, which leads to a mismatch for the pressure and temperature prediction and therefore makes the optimization for defined target values difficult. Therefore, this paper presents a novel parametrization method to consider feasible channel cross-sections as they result from the rollbonding process in thermo-fluid topology optimization.

2   Governing Equations

Heat exchanger design can be classified as a conjugate heat transfer (CHT) problem. The following section describes the governing equations for modelling the physical properties of the problem. Furthermore, the discretization using the Finite Element Method (FEM) is described. The turbulent flow problem can be modelled with the Reynold-Averaged Navier Stokes (RANS) equations. However, integrating with topology optimization, RANS is computationally expensive. Therefore, the simplified linear Darcy approach is used for optimization purposes due to its superior computational efficiency.

2.1   Daray Flow Analysis

The incompressible Navier-Stokes equations describe the flow problem as follows:

$$\rho u \cdot \nabla u = -\nabla p + \mu \nabla^2 u - \rho b \tag{1}$$

$$\nabla \cdot u = 0 \tag{2}$$

ρ describes the physical density, **u** the velocity field. **p** is the pressure field, μ denotes the dynamic viscosity, and **b** describes the body force vector per mass unit. The idea of the Darcy approach is to describe flow through porous media by a specific solution of the Navier-Stokes equations. In this approach, the flow is described as potential flow as given in Eq. 3.

$$u = -\frac{\kappa}{\mu}\nabla p \tag{3}$$

κ denotes the material permeability of the porous media. The material permeability is the main parameter that must be selected reasonably to be able to consider friction effects implicitly. The problem can be transferred to a linear finite-element problem as

$$K_p p = f_p \tag{4}$$

where K_p is the permeability matrix, p denotes the pressure vector and f_p the pressure load vector. For a detailed description refer to [7].

2.2 Thermal Analysis

The thermal analysis is based on the convection-diffusion model given in Eq. 5.

$$\rho c_p u \cdot \nabla T = k \nabla^2 T + Q \tag{5}$$

The equation is described by the temperature field T and the volumetric heat source Q using ρ as density, c_p as heat capacity and k as conductivity. The finite-element problem states as

$$[K_t(x) + C(u, x) + D(u, x)]t = f_t \tag{6}$$

where K_t is the conductivity matrix, $C(u,x)$ the conduction matrix, $D(x,u)$ a discontinuity capturing term for optimization purposes, f_t the thermal load vector and t the nodal temperature vector. The details of the approach are described in [3].

3   The Rollbonding Process

Rollbonding classifies as solid-state joining process, which connects two sheet metal parts under high pressure by a rolling process [8]. In the context of heat exchangers there are applications in the automotive industry as well as in the solar and refrigeration industries.

3.1 Manufacturing Process

Figure 1 illustrates the manufacturing process. Two coils of aluminium are decoiled while a separator agent is printed on one of the coils. The printed area marks the later channel position. The two coils are rolled onto each other after heat treatment, leading to a solid-state joining in all areas without an applied separator print.

The coils are separated into smaller parts and the channels are formed by an inner pressure within an inflation press. Using this technology, it is possible to produce different channel patterns without the need for a dedicated tool for every single pattern, as the same flat toolplate can be used for different kinds of channel patterns. Therefore, the rollbonding process offers unmatched flexibility and design freedom for the production of cooling plates.

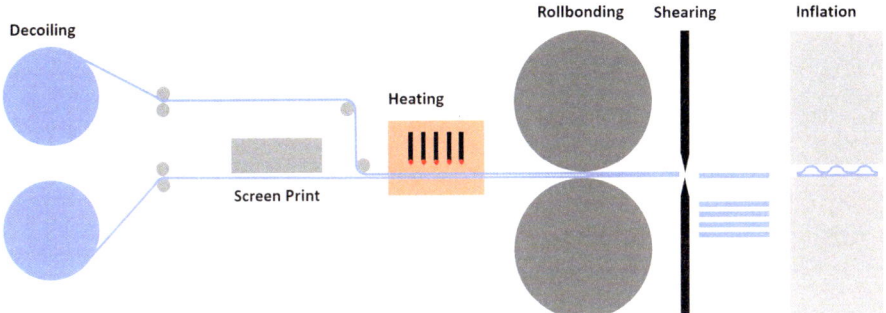

Fig. 1. Schematic representation of the rollbonding process

3.2 Geometric Description of Channel Shapes

There are multiple relevant parameters to describe the channel shape of rollbonded cooling plates. In the plane of the screen print, as shown in Fig. 2 there are length scale constraints limiting the minimum and maximum channel width $l_{w_{min}}$ and $l_{w_{max}}$ as well as the minimum bonded width between channels $l_{b_{min}}$.

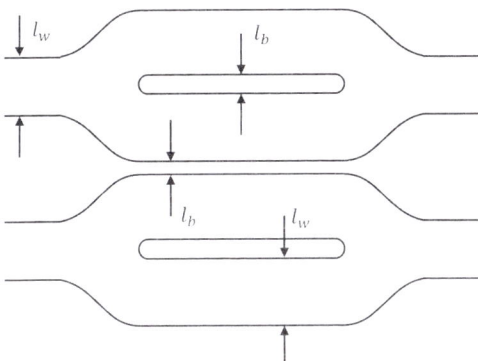

Fig. 2. Schematic representation of length scale measures in the screen print plane with bonding width l_b and channel width l_w

   The cross-section of the channels can be described by the total inner channel height l_h, the chamfer angle α, the inner radius r and the sheet thickness of the channel side t_{cs} as shown in Fig. 3.

   This paper presents a novel parametrization of the channel cross-section. The constraint formulations used to enforce the length scales in the screen print plane as shown in Fig. 2, within the optimization are described in [6].

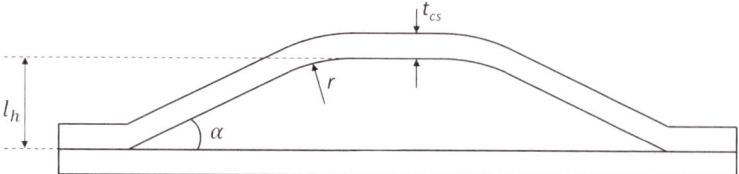

Fig. 3. Schematic representation of the channel cross-section with parameters for the total inner channel height l_h, chamfer angle α, inner radius r, and sheet thickness of the channel side t_{cs}

4 Benchmark Cooling Plate

The method of operation of the parametrization methodology presented in the paper will be shown based on a small benchmark cooling plate, with a rectangular shape and two equally sized heat input zones. Each heat input zone is cooled by a symmetric channel pattern with four channels. The channel patterns are connected by a single channel on the opposite side of the inlet and outlet spigots. A heat flux of $h_f = 4000\,\tilde{W}/m^2$ is applied at the heat input zones, while a constant flow rate of $\dot{V} = 4^l/min$ and an inlet temperature $t_{in} = 25°C$ are defined as boundary conditions at the inlet. The model is shown in Fig. 4.

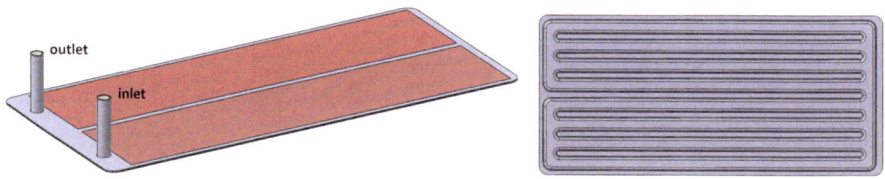

Fig. 4. Benchmark cooling plate for demonstration of the parametrization strategy in iso (left) and bottom view (right) with heat input zones (red)

4.1 Conjugate Heat Transfer Analysis

All analysis and optimizations shown in the paper are performed using the simplified Darcy flow model. This model comes with the benefit of linearity. Therefore, the calculation times are small compared to high-fidelity RANS models. A drawback is a loss of accuracy. For the fast conceptional design of rollbonded cooling plates the approach is feasible as shown in [7] since it is still able to predict the correct temperature hotspot locations. Therefore, the qualitative temperature distribution can be optimized.

The thermal and hydraulic performance is evaluated using the Darcy model using the solver OptiStruct. Figure 5 shows the resulting temperature, total pressure and velocity fields.

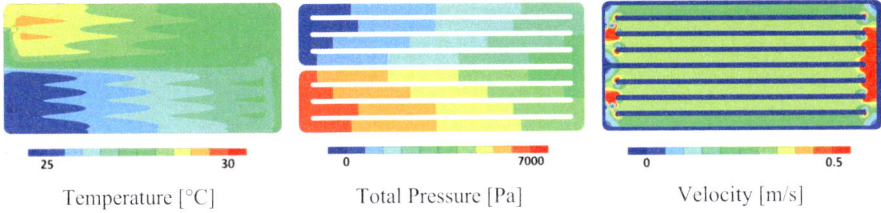

| Temperature [°C] | Total Pressure [Pa] | Velocity [m/s] |

Fig. 5. Benchmark cooling plate evaluated with Darcy solver and body-fitted mesh

The material permeability is selected as $\kappa = 3.18 \cdot 10^5 mm^2$ to reproduce the same pressure drop as given from a comparable RANS analysis. Qualitatively, the resulting temperature distribution is quite homogeneous. The velocity field shows a homogeneous flow distribution, while the pressure drop is in a reasonable range. Desired target values highly depend on the vehicle and cooling concept though. Therefore, this paper uses the performance of this benchmark plate for a fair comparison between optimization and manual design.

5   3d Parametrization Strategy

Existing topology optimization strategies generate channel patterns in 2D as described in [6]. This leads to a mismatch for the pressure drop and for the thermal performance. Therefore, these methods can be used to generate qualitatively reasonable trade-offs between thermal and hydraulic performance, while optimization for defined thermal or hydraulic target values is difficult. Strategies for 3D topology optimization exist for manufacturing technologies like extrusion or casting [9]. To the authors knowledge, there is no parametrization strategy described in the literature that describes channel shapes as they result from the rollbonding process properly. Strategies without feasible parametrization will probably result in channel structures with rectangular channel shapes due to hydraulics benefits of this shape. This shape is not realistic for sheet metal cooling plates though. Therefore, the authors developed a parametrization strategy for topology optimization that represents the channel shape resulting from the rollbonding process in a realistic manner. This way another step towards realistic modelling of cooling plates in topology optimization is taken enabling to generate optimal channel patterns considering realistic target values for the thermal and hydraulic performance.

5.1   Mesh Design

The basis of the parametrization strategy is a modelling that representing the rollbonded cooling plate as a mesh with a suitable structure, as shown in Fig. 6. The top-layer (grey) represents the flat side of the plate, while the channel side is split into two parts. First, the design variable layer, is a thin layer with a single element depth (yellow). The element artificial densities of this layer are the input parameters for the optimization. The second part (blue) consists of multiple element layers, where the total height is the total inner channel height plus the sheet thickness. All elements of this second part are directly dependent on the design variables in the first part. That means, that their artificial densities will result from the underlying design variable layer by a logic described in the later section of this paper. This collection of elements will later form the channel cross-sections and therefore, the channel domain.

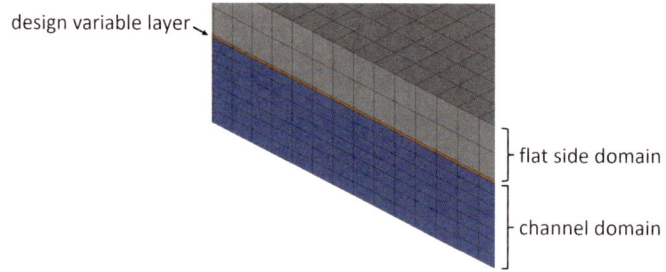

design variable layer

flat side domain

channel domain

Fig. 6. Mesh structure for 3D parametrization of the channel shape divided into layer sections for the flat side of the plate (grey), a design variable layer (yellow) and a section to represent the channel cross sections (blue)

5.2 Parametrization Steps

Figure 7 describes the steps of the parametrization. For visualisation a shrunk version of the benchmark cooling plate is shown. The first steps are performed in the design variable layer, described by the design variable vector x. Within this layer a 2D parametrization strategy is used to control length scales and to ensure mesh independence and convergence. The strategy utilizes filtering, resulting in the filtered design variable field $\tilde{x}$ and projection, resulting in $\bar{x}$. For details on the approach refer to [6]. The projected density field $\bar{x}$ is the input for the 3D parametrization which is illustrated in Fig. 8 and described in the following.

In the first step the design domain is divided into a passive (dark grey) and an active (light grey) domain. The artificial densities of the passive areas are defined as $x_i = 1$ to represent a bonded domain of the plate. They will not change during the optimization. The artificial densities of the active domain will change during the optimization, resulting in a channel pattern with $x_i \in [0, 1]$. All variables with $x_i = 1$ represent a bonded area of the plate, while variables with $x_i = 0$ represent a channel area.

The projected design variable field $\bar{x}$ will not be perfectly discrete during the optimization. Therefore, a second, binary projection according to Eq. 7 is performed in the second step resulting in the binary field $\bar{\bar{x}}$.

$$\bar{\bar{x}}_i = \begin{cases} 0 \ if \ \bar{x}_i < \vartheta \\ 1 \ otherwise \end{cases} \tag{7}$$

where the projection threshold ϑ is selected as $\vartheta = 0.9$. As an example, a density field representing the channel structure of the benchmark design is assigned.

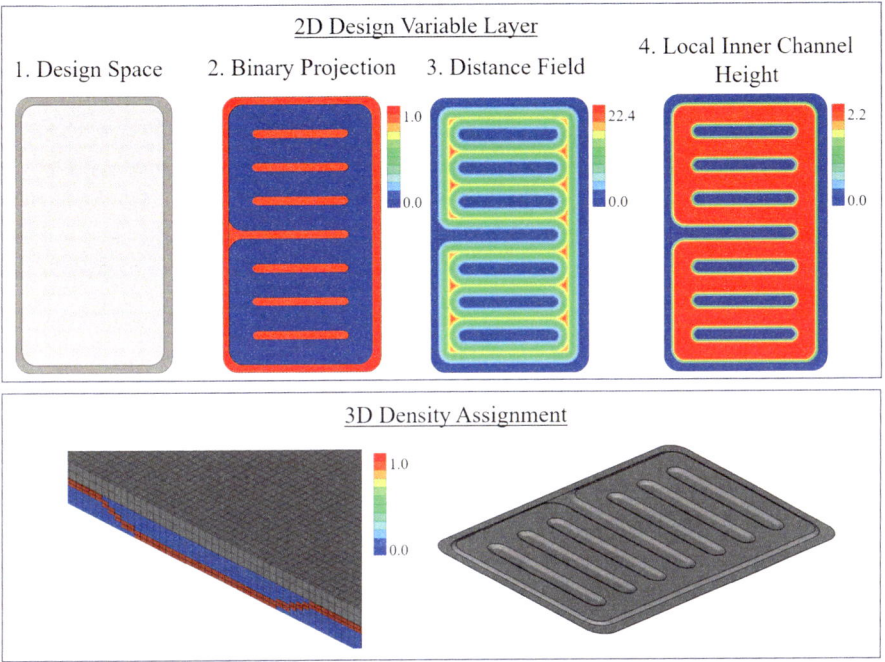

Fig. 7. Steps of the 3D parametrization strategy for rollbonded cooling plates. Steps 1.–4. Show the design variable layer. Based on the design variable layer the artificial densities in the third dimension are assigned

Benefit of the binary field is the minimum distance of every cell in the design variable layer to a bonded domain can now be calculated efficiently by Eq. 8 and Eq. 9.

$$D_{b_{i,j}} = D_{i,j} \cdot \left(I_{i,j} \bar{\bar{x}}_i \right) \tag{8}$$

$$d_{b_i} = \min_{i=1}^{n} \{ D_{b_{i,j}} \mid D_{b_{i,j}} \neq 0 \, for \, all \, j \} \tag{9}$$

D donates a matrix containing the distance of every cell in the design variable layer to each other, while I donates the identity matrix. Consequently D_b describes the distance of every cell to every cell in the bonded domain. d_{b_i} Gives the minimum distance of a cell i in the layer to the bonded domain. n donates the number of cells in the design variable layer.

The calculation of this field is the third step of the parametrization. The resulting field is the input of a function, that describes the local inner channel height depending on the minimum distance to the bonded domain. This function is illustrated at a channel cross-section in Fig. 8.

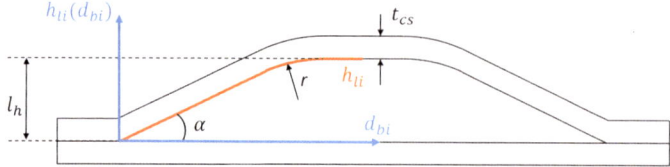

Fig. 8. Illustration of the function describing the local inner channel height dependent on the distance to the bonded domain

The function of the local inner channel height can be describe as defined in Eq. 10–12.

$$
h_{li} = \begin{cases} d_{bi} \cdot \tan(a) & \text{if } d_{bi} \leq c_1 \\ \sqrt{r^2 - (d_b - c_1 = r \cdot \cos(\pi/2 - a))^2} = l_h - r & \text{if } c_1 < d_{bi} \leq c_2 \\ l_h & \text{otherwise} \end{cases} \tag{10}
$$

$$
c_1 = \frac{l_h + \sin(\pi/2 - \alpha - 1)}{\tan(\alpha)} \tag{11}
$$

$$
c_2 = \frac{l_h + r(\sin(\pi/2 - \alpha - 2))}{\tan(\alpha)} + r \cdot \cos(\pi/2 - \alpha) \tag{12}
$$

The function depends on the main geometric parameters, total inner channel height h, inner radius r and chamfer angle α. It describes the inner channel height at every element dependent on the minimum distance to the next bonded area d_{bi}. Applied to every element of the design variable layer the function gives a field that describes the border distance between the inner of the channel (fluid) and the channel material (solid). Based in this 2D field the artificial densities in the 3D field can be assigned by measuring the distance of every cell in the 3D domain to its corresponding base cell in the design variable layer in z-direction. This distance is stated as d_{zi} in the following. If the distance is lower than the local inner channel height of the base cell the cell belongs to the domain within the channel and the artificial density of the base cell is assigned. If it is higher than the local inner channel height, but lower than the sum of local inner channel height and sheet thickness the cell belongs to the channel solid and an artificial density of 1 is assigned. For elements with distance above this threshold an artificial density of 0 is assigned. The logic is summarized in Eq. 13.

$$
x_{3Di} = \begin{cases} \bar{x}_i & \text{if } d_{zi} \leq h_{li} \\ 1 & \text{if } h_{li} < d_{zi} < h_{li} + t_{cs} \\ 0 & \text{otherwise} \end{cases} \tag{13}
$$

The result is shown in Fig. 7 (lower left). A realistically curved channel shape according to the channel function based on a cartesian mesh is formed and can be used for topology optimization.

5.3 Verification of the Parametrization

Due to the nature of the mesh used in topology optimization which is cartesian and not body-fitted, the results might differ dependent on the chosen mesh resolution. With high mesh resolution the resulting temperature, total pressure and velocity fields of the cartesian mesh will converge against the body-fitted results. Figure 9 shows the analysis results of the model with cartesian mesh with application of the proposed parametrization strategy.

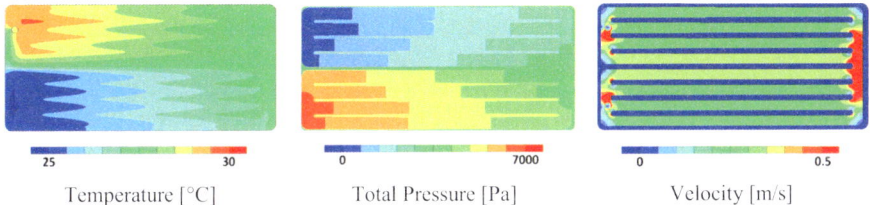

| Temperature [°C] | Total Pressure [Pa] | Velocity [m/s] |

Fig. 9. Benchmark cooling plate with applied 3D parametrization using a cartesian mesh, evaluated with Darcy solver

The results of the body-fitted mesh and the proposed parametrization approach match very well. The maximum temperature differs by 0.12 °C while the difference in the pressure drop is 1.4 mbar. Qualitatively, the fields match very well. Therefore, it is shown that the proposed parametrization strategy reproduces the results of the body-fitted reference with reasonable accuracy.

5.4 Sensitivity Analysis

For optimization purposes the sensitivities of every design variable regarding the objective function df/dx are needed. The sensitivity of every cell regarding the objective function $\partial f/\partial x_{3D}$ is exported by the solver. Due to the parametrization, there is a dependency between multiple cells, as a change a variable in the design variable layer leads to a change in the artificial densities in the overlying layers. Therefore, the sensitivity exported by the solver cannot be used directly. The dependency between the design variables and the cells in the channel domain must be considered by the partial derivatives given in Eq. 14.

$$\frac{df}{dx} = \frac{\partial \tilde{x}}{\partial x} \frac{\partial \overline{x}}{\partial \tilde{x}} \frac{\partial x_{3D}}{\partial \overline{x}} \frac{\partial f}{x_{3D}} \tag{14}$$

As the artificial density of every cell in the channel domain only depends on the artificial density of the underlying variable in the design variable layer the calculation of the partial derivative can be done efficiently by finite-differencing. The partial derivatives of the filtering and projection steps $\partial \tilde{x}/\partial x$ and $\partial \tilde{x}/\partial \overline{x}$ can be derived analytically. To evaluate the feasibility of the implementation the resulting sensitivity df/dx resulting from Eq. 14 is compared to the sensitivity evaluated by a finite differencing approach

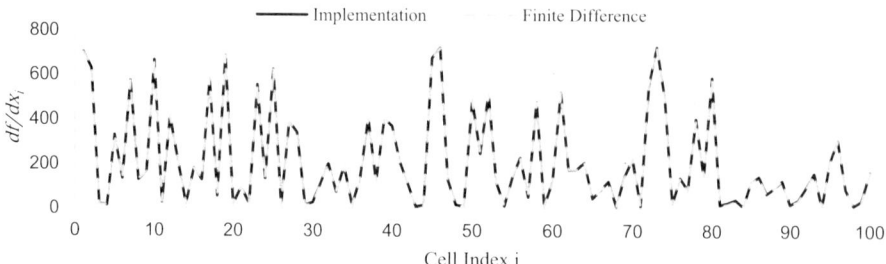

Fig. 10. Verification of the used implementation by comparison of sensitivities generated by the implementation compared to those generated by finite differencing

over all steps of the parametrization. The comparison is performed for 100 randomly selected cells as shown in Fig. 10.

The sensitivities of both approaches match. Therefore, it is shown that the presented implementation calculates the sensitivities as expected.

5.5 Optimization

As a next step, an optimization is defined and performed based on the benchmark plate. The objective is to minimize pressure drop from inlet to outlet Δp while ensuring the thermal compliance T_c to be lower to a given threshold $T_{c_{max}}$. $T_{c_{max}}$ is defined as the thermal compliance of the benchmark design to ensure an equal or better thermal performance. The pressure drop is given in Eq. 15, while the thermal compliance is defined in Eq. 16.

$$\Delta p = p_{in} - p_{out} \tag{15}$$

$$T_c = \frac{1}{2} t^T f_t \tag{16}$$

The optimization problem is given in Eq. 17.

$$\begin{aligned} min \quad & \Delta p \\ s.t. \quad & T_c - T_{c_{max}} \leq 0 \\ & g_{max}^f \leq 0 \\ & g_{min}^s \leq 0 \\ & g_{min}^f \leq 0 \end{aligned} \tag{17}$$

Furthermore, minimum and maximum length scales on the fluid domain g_{max}^f, g_{min}^f and a minimum length scale constraint for the solid domain g_{min}^s are considered. The resulting topology is shown in Fig. 11 while results for temperature, total pressure and velocity are presented in Fig. 12.

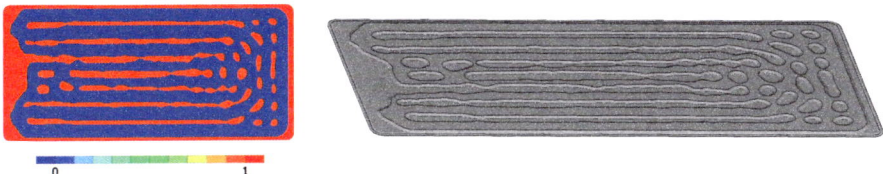

Fig. 11. Topology of the optimized plate. Artificial densities in the design variable layer (left) and 3D topology view on top of plate (right)

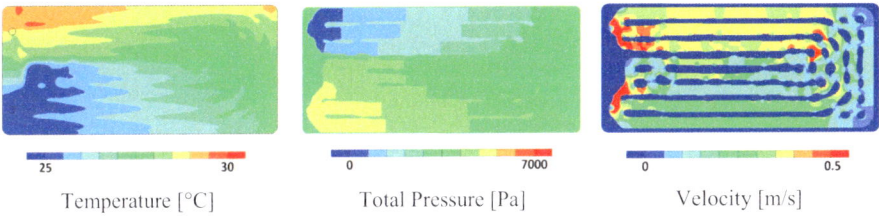

Temperature [°C]   Total Pressure [Pa]   Velocity [m/s]

Fig. 12. Analysis results of optimized cooling plate

Comparing the benchmark plate with cartesian mesh shown in Fig. 9 and the optimized plate presented in Fig. 12. Shows that the temperature field are qualitatively similar, as enforced by the constraint of thermal compliance. The maximum temperature is almost equal with a reduction of -0.01 °C. The same accounts for the thermal compliance with a small reduction of 4.7 WK, while the pressure drop is optimized by 8.95 mbar which equals a reduction of 13.92%. All relevant performance indicators are summarised in Tab. 1. In comparison to the benchmark design, the optimized plate shows locally narrow channels in the middle section, restricting the flow in this area to guide it into the corners of the plate instead of guiding all the flow to the right section of the plate by a long middle separation. This way the optimized plate shows a comparable thermal performance, while the pressure drop is reduced.

Table 1. Performance indicators of benchmark cooling plate and optimized plate

	Benchmark Plate	Optimized Plate	Difference
Max. Temperature	29.53 °C	29.52 °C	0%
Thermal Compliance	1052.8 WK	1048.1 WK	-0.45%
Pressure Drop	67.77 mbar	58.82 mbar	-13.92%

The results show that the presented approach generates topologies that optimize the pressure drop while ensuring a defined thermal performance. The temperature targets are fulfilled and even slightly optimized while the pressure drop is reduced.

6 Conclusion

A novel strategy for parametrization of realistically curved 3D channels as they result from the rollbonding process was presented. The shown parametrization is a significant advance from known methods which usually use rectangular or completely unconstrained channel cross sections. This way the influence of the channel shape to the temperature and pressure fields can be considered. Therefore, another step towards realistic modelling of channel structures in topology optimization is taken, enabling to consider thermal and hydraulic target values in a more realistic manner. The sensitivity calculation is validated and the approach is applied to a benchmark cooling plate. Based on the simplified Darcy flow model a topology is generated that fulfills the thermal requirements while optimizing the pressure drop. This way it is shown that the presented strategy can be used for fast conceptional generation of channel patterns for battery cooling plates and especially for those manufactured by rollbonding. By application on large scale problems the approach can help engineers to find optimal designs that use the full design freedom offered by the rollbonding technology. The presented approach is independent from the used FEM solver and flow model. Consequently, it can be coupled with high-fidelity RANS solvers which will lead to another increase of accuracy in future studies.

References

1. Bendsoe MP, Sigmund O.: Topology optimization: theory methods and applications. Springer, Berlin, (2004)
2. Bruns, T.: Topology optimization of convection-dominated, steady-state heat transfer problems. Int J Heat Mass Transf 50(15), (2014)
3. Ghasemi, A., Elham, A.: A novel topology optimization approach for flow power loss minimization across fin arrays. Energies 13:1–20, (2020)
4. Dienemann, R., Schewe, F., Elham A.: Industrial application of topology optimization for forced convection based on darcy flow. Structural and Multidisciplinary Optimization 65(9), (2022).
5. Dilgen S, Dilgen C, Fuhrman D: Density based topology optimization of turbulent flow heat transfer systems. Struct Multidisc Optim 57(5), (2018)
6. Schewe, F., Fleischer, D., Elham, A.: Industrial application of thermofluid topology optimization to rollbonding cold plates with dedicated manufacturing constraints. Structural and Multidisciplinary Optimization (2023).
7. Zhao X, Zhou M, Sigmund O.: A "poor man's approach" to topology optimization of cooling channels based on a darcy fow model. Int J Heat Mass Transf 116:1108–1123, (2018)
8. Khan HA, Asim K., Akram F.: Rollbonding processes: state-of-the-art and future perspectives. Metals, (2022)
9. Liu Jikai, Ma Yongsheng: A survey of manufacturing oriented topology optimization methods, Advances in Engineering Software 100, (2016)

Framework for Modeling and Evaluating Information Flows in Flexible Manufacturing Systems

Kira Welzel, Anna-Sophia Wilde$^{(\boxtimes)}$, Yasin Bulut, Max Juraschek,
and Christoph Herrmann

Institute of Machine Tools and Production Technology, Chair of Sustainable Manufacturing and
Life Cycle Engineering, Technische Universität Braunschweig, Langer Kamp 19B, 38106
Braunschweig, Germany
a.wilde@tu-braunschweig.de

Abstract. In the manufacturing industry, flexible manufacturing systems (FMS) are gaining importance in meeting the demand for product customization while maintaining high manufacturing efficiency. The digitization of physical manufacturing additionally expands the relevance of industrial communication networks, resulting in complex interactions of material, energy and information flows that pose challenges for the manufacturing system design. Existing decision support frameworks for the design of FMS mainly incorporate the evaluation of the material flow. A holistic view of FMS requires the consideration of the interdependencies of the material, energy and information flow. Particularly, the performance of the material flow of a manufacturing system increasingly depends on the information flow of the connected communication network. Against this background, the paper presents a simulation-based framework that enables the integrated quantitative evaluation of material, energy and information flows in FMS as decision support in manufacturing system design. The framework contains a modular reference model of a generic FMS and its communication network to reduce modeling efforts. It further provides a procedure and an evaluation model that enable a multi-criteria comparison of different FMS configurations. The functionality and benefits of the developed framework are examined with the help of an exemplary case study involving an FMS for producing and recycling polymer parts.

Keywords: Information flow · Flexible manufacturing systems · Cyber-physical manufacturing systems · Simulation · Manufacturing system design

1 Introduction

Manufacturing companies use product customization strategies to create product differentiation and market competitiveness. This results in a trade-off between product variety, production volume, and productivity losses. In addition, there is an increasing focus on energy efficiency due to high energy costs [1–5]. Integrating flexibility as well as automation and digitalization measures in manufacturing systems are two

K. Dröder and T. Vietor (Eds.): CD 2024, Proceedings, pp. 150–162, 2025.
https://doi.org/10.1007/978-3-658-45889-8_12

crucial synergetic strategies to address the resulting challenges [1, 6]. As a result, flexible manufacturing systems (FMS) are evolving into cyber-physical production systems (CPPS), including high-performing communication networks for data transmission [2]. FMS show increasing interactions between incorporated material, energy, and especially information flows compared to traditional non-CPPS manufacturing systems [7]. The digitalization and the associated increase in complexity of FMS can significantly impact the complexity of manufacturing system design [7, 8]. Using computational simulation as decision support allows an extensive interdisciplinary understanding of the FMS despite the increased complexity [9]. Today, many simulation-based approaches that serve different decision-making contexts in manufacturing system design can be found in the literature [5]. For a holistic simulation-based examination of FMS, the interdependencies between material, energy and information flows must be considered. The influence of the information flow gains in significance due to the rising connectivity, as the performance of the physical material flow in an FMS increasingly depends on the performance of the information flow (data transmission) in the linked communication network, which additionally requires energy-intensive hardware (e.g., servers) [10]. A peak load or malfunction in data transmission, for example, is expected to be critical for the overall system's performance. Hence, the necessity to examine the modeling and evaluation of information flows in FMS arises.

Existing approaches mainly incorporate the evaluation of material flows [11]. Therefore, a simulation-based framework that enables the integrated quantitative evaluation of material, energy and information flows in FMS as a decision support for manufacturing system design was developed, according to the Design Science Research Methodology [12]. The framework consists of a generic procedure model, a reference model for a generic FMS and an evaluation model enabling a multi-criterial assessment of different FMS configurations. For a target-oriented evaluation, technical as well as environmental criteria are taken into account. The framework aims to enable a model-based assessment of the material, energy, and, in addition, the information flows to gain knowledge about the relations of all these flows in FMS and, thus, a sophisticated selection of the best-fitted FMS configuration. In addition, an efficient application of the framework in the manufacturing system design process is aimed at by reducing the modeling effort with the help of the reference model and the structured procedure. The paper is structured as follows. At first, the technical background of FMS and communication networks is explained (Sect. 2). Afterward, the state of research is presented, and the research gap is identified (Sect. 3). Based on that, the framework and its sub-models are described in Sect. 4. In Sect. 5, the framework is applied to a case of an exemplary FMS for manufacturing and recycling polymer parts. Finally, a conclusion and a short outlook are given.

2   Technical Background

2.1   Flexible Manufacturing Systems

The trade-off between a high product variety and a high production volume is challenging for traditional manufacturing systems such as mass production, assembly lines or batch production [1]. Pursuing each goal isolated leads to idle or blocked workstations, which have negative impacts like lower capacity utilization [13]. As a solution, FMS are designed to produce medium volumes of medium variety products to overcome these challenges [14]. Therefore, an FMS aims to produce several product types of one product family on a single system with low set-up times and at the highest possible capacity utilization [15]. The similarity within a product family provides the basis for built-in flexibility measures [14]. This includes implementing individual cycle times for each workstation to achieve high equipment utilization. In addition, using redundant workstations per process reduces congestion, and designing individual workstations for multiple processes prevents idle time. Routing flexibility enables flexibility with regard to the material flows of each product through the FMS.

FMS also differ from traditional manufacturing systems concerning the system elements and the layout. Essential system elements are workstations, processes carried out at the workstations, material handling systems (e.g., Automated Guided Vehicle (AGV) fleets) and the different product types. A possible layout of FMS is the matrix production [16]. Smart FMS are also considered CPPS [2]. The main characteristics of CPPS are intelligence, connectivity and responsiveness [17].

2.2   Industrial Communication Networks

In CPPS, the performance of the physical manufacturing system depends on the information flow in the linked communication network [10]. An information flow is defined as the transmission of data between a spatially separated source and sink using a digital transmission system [18]. Single data transmissions can be aggregated to form the total information flow at the FMS level [7]. Industrial communication networks include all necessary resources to support business processes by providing information or automating business activities [19]. Communication networks based on the automation pyramid have a strictly hierarchical structure, as the connections of one level are limited to the neighboring levels, and each component is assigned to a specific purpose. The interface between the physical manufacturing system and the communication network is defined at the field level by sensors and actuators [20]. This level is followed by the control level, where Programmable Logic Controllers (PLCs) cyclically process sensor signals and respond to the actuators [21]. The PLCs, in turn, communicate with a Supervisory Control and Data Acquisition (SCADA) server at the higher supervisory level. Manufacturing Execution Systems and Enterprise Resource Planning Systems at the planning and management level are not considered in this paper as the developed framework is limited to modeling and evaluating data transmission between the control and supervisory levels [20].

The digital transmission system on which data transmission between two components of a communication network is based does not in itself ensure faultless communication. This requires standardized communication protocols [22]. The basic structure of a communication protocol following the Open System Interconnection (OSI) Basic Reference Model distinguishes between the payload, i.e., the volume of actual information, and the overhead, which contains the data volume required for data transmission [7, 23]. Popular communication protocols are the OPC Unified Architecture (OPC UA) and the Message Queuing Telemetry Transport (MQTT) protocol [24].

3 State of Research

To identify the current state of research, a systematic literature research was conducted according to Durach et al. [25]. The search of common databases within the scope of the addressed topic and an initial screening identified ten relevant approaches, which can be found in.

Table 1. The detailed analysis of existing approaches has revealed the potential of modeling and simulating FMS as decision support in manufacturing system design. There is a heterogeneous range of methods, reference models and simulation studies dedicated to the quantitative modeling and evaluation of FMS. Mourtzis et al. point out that existing approaches focus on material flow analysis [11]. Florescu and Barabas present an approach based on mathematical modeling, using transformation functions and coupling matrices [2]. The approaches of Evans et al. and Rabe et al. both combine static Value Stream Analysis with Discrete Event Simulation (DES) to incorporate dynamic properties of system elements and a flexible material flow [10, 26]. Berger et al. apply Petri nets to model the state transition of a workstation during the operation of an FMS by places and transitions [27]. Most of the approaches are based on Agent-Based Simulation (ABS) [3, 4, 7, 28, 29]. In ABS, the elements of an FMS are modeled using heterogeneous agents that interact in an adaptive environment [30]. Feldkamp et al. found that the ABS is generally closer to the nature of FMS than the DES [31]. The core of these approaches are modular reference and process models for simulating FMS that reduce the modeling effort for specific use cases. Decision-making in manufacturing system design is additionally supported by evaluating the performance of an FMS using key performance indicators (KPIs) [3, 4, 7, 28, 29]. The existing approaches mainly select KPIs evaluating the material flow. Only a few approaches also include a quantitative evaluation of the energy flow. Information flows are merely modeled qualitatively to ensure the functionality of the simulation model.

Table 1 Identified approaches of the systematic literature research

Approach	Material flow	Energy flow	Information flow
[2] Florescu und Barabas 2020	●	●	◑
[3] Greinacher et al. 2020	●	●	◑
[28, 32] Voit et al. 2021	●	○	◑
[4] Schönemann et al. 2019	●	●	◑
[33] Saez-Mas et al. 2018	●	○	◑
[10] Evans et al. 2021	●	○	◑
[26] Rabe et al. 2020	●	○	◑
[27] Berger et al. 2020	●	○	◑
[29] Komoto et al. 2019	●	○	◑
[7] Thiede et al. 2019	●	○	●

Modeling and evaluation: ○ = none, ◑ = qualitative, ● = quantitative

Saez-Mas et al. identify the need to explicitly model the information flow to analyze both physical and logical decision processes in manufacturing operations. The decomposition of the physical and the information system of an FMS into different layers of their simulation architecture promises a better manufacturing system design [33]. Berger et al. also note that manufacturing systems are increasingly connected to communication networks. Therefore, are not only the physical components (workstations) of an FMS modeled as a Petri net, as described above, but also the components of the associated communication network (PLCs) and their interfaces [27]. Only the approach of Thiede et al. allows for a quantitative modeling of information flows in FMS. Based on a hierarchical communication network, Thiede et al., therefore, developed mathematical functions that calculate the data volume of bidirectional cyclic or event-triggered information exchange between two PLCs using OPC UA or MQTT as the communication protocol [7].

The analysis of existing approaches has shown that complex interdependencies between the material, energy, and information flows must be considered in manufacturing system design since the deliberate design of the communication network is seen as a potential way to maintain competitiveness [34]. Existing decision support frameworks, however, focus on evaluating material and energy flows. Consequently, there is a need for a framework that additionally integrates the quantitative modeling and evaluation of the information flow into the performance evaluation of different FMS configurations in the manufacturing system design process.

4   Framework

Based on the identified research gap, a framework for modeling and evaluating information flows in the manufacturing system design process was developed for a defined decision context. The framework aims to provide a decision support system focusing on FMS. One application potential lies in the quantitative assessment of the current state of a specific FMS. Furthermore, the insights gained about the current state can be compared with a desired state or alternative FMS configurations, leading to the development of the decision support system to improve the examined FMS. The framework can be used to compare and evaluate different configurations of an FMS regarding, e.g., the layout, the resource availability or material, energy and information flows. The focus of utilizing the framework for this paper lies on different configurations of material, energy, and information flows, with a specific emphasis on modeling information flows varying due to different data collection and data processing strategies.

In developing the framework, illustrated in Fig. 1, the following requirements and assumptions are considered. The reference model is based on a generic FMS, as described before. Elements of the physical system of the FMS are, therefore, the workstations, industrial robots that can perform either operational or transport process steps, an AGV fleet, processes, products, employees and the flow of materials and energy. The system boundaries are defined as gate-to-gate. The material flow enters and leaves the system via the incoming and outgoing warehouse. The energy flow in the physical system is limited to the workstations, industrial robots and the AGV fleet. Both the source and the sink of the energy flow are not considered. The individual battery management of the AGVs is also not considered.

A generic architecture of a communication network is defined, which is linked to the physical system of the FMS via sensors and actuators at the field level. A hierarchical system is assumed to be the generic communication network. Workstations, industrial robots and AGVs have a PLC unit. The PLCs are linked to a server, which represents the SCADA system. Communication is limited to the control and supervisory levels. The consideration of the information flow is limited to the FMS and its system boundaries, so there is no communication with the external environment. The information flow modeling is based on the approach developed by Thiede et al. [7]. It is assumed that the connection between two communication components is permanent and transmits defined information with a constant payload. The potential damage caused by disruptions and interferences is not considered. The framework does not envisage any intervention by employees in the production process via a human-machine interface (HMI). The HMI is only used to process and visualize information for the employees.

As shown in Fig. 1, the framework consists of one superordinated model and several sub-models. The superordinated model has a defined input, which consists of the specifications and possible configurations of the FMS, and a defined output, which is the decision support for selecting the best-fitted configuration for the FMS. The first sub-model is the procedure model, which is based on the model of the VDI 3633 [35]. The first step of the procedure model is the definition of objectives for the FMS, which is based on a KPI system, also serving as a foundation for the evaluation in the end. The second step is the model formalization. In this step, the reference model of the FMS with all necessary sub-models is developed. The sub-models of the reference model, i.e., the layout model, the coupling model, the process chain model and the system element models, are hierarchically structured according to an increasing level of detail. Afterward, the implementation phase, which is the development of the simulation model using a specific simulation tool, is performed.

In the end, the results of the simulation study are evaluated. Therefore, the Decision Support Model aggregates the relevant KPIs measuring the performance of the physical system and the communication network of each alternative FMS configuration to enable a multi-criterial comparison of the FMS configurations, visualized in a portfolio diagram. The throughput time, reject rate and equipment utilization can be exemplarily enumerated as KPIs for the assessment of the technical performance of physical systems, while, e.g., the average data volume, the maximal data load and the payload/overhead-ratio assess the technical performance of the communication network [7, 36]. Energy consumption and power demand are analyzed to evaluate the environmental performance of both systems. Additional detailed analysis can be executed to evolve a deeper understanding of the nature and interdependencies of material, energy, and information flows. Based on the process described in the VDI 3633, continuous verification and validation are performed [35].

Since the model formalization is a crucial step, the development of the reference model with all sub-models is of high importance. The reference model is coherent with the principles of ABS, describes the assumed generic FMS in its entirety, and is effortlessly applicable to a specific use case due to its modularity. The first sub-model is the layout model. It aims to provide a spatial description of all system elements, such as the location of the machines, transport routes of AGVs, and the shop floor layout

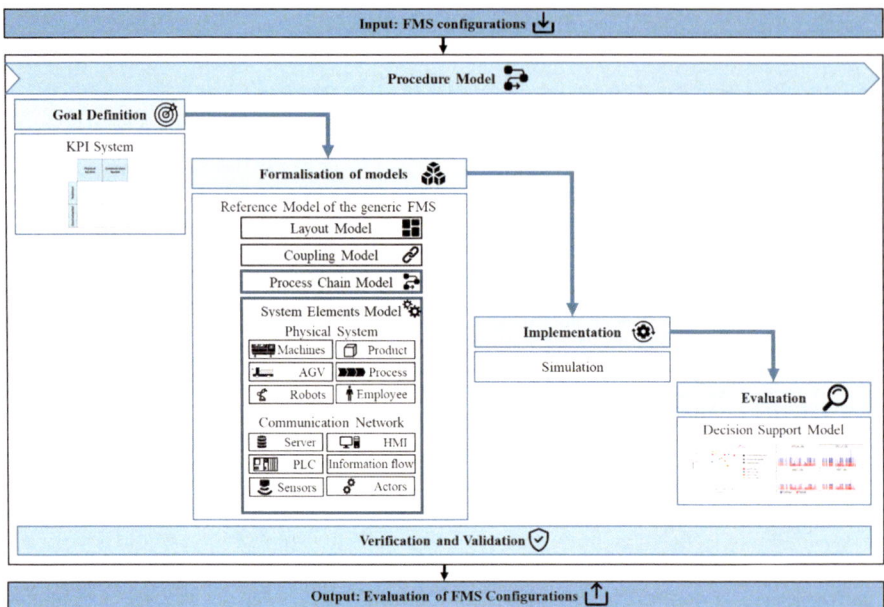

Fig. 1 Framework for modeling and evaluating material, energy, and information flows in FMS according to VDI 3633 [35]

of the factory building. Here, defining the system boundaries and the area considered is essential. The coupling model is inspired by the approach of Schönemann et al. and visualizes all system elements from the physical system and the communication network, as well as their properties and interconnections [4]. The next sub-model is the process chain model, which has to be individually developed for each product type (see Fig. 2). The process chain model visualizes the whole process chain with all interactions and all permutations caused by the built-in routing flexibility, i.e., the flexible material flow. The modeling language used is the Business Model and Notation (BPMN) language [37]. As visualized in Fig. 2, the process chain model consists of different processes and their interconnections. Each individual process is based on the generic process model, one of the developed element models and consists of an operation module and a transport module, similar to the approach of Voit et al. [28]. For each module, activities and sub-processes of the physical system and the communication network are defined in detail. An exemplary extract of the transport module is shown in Fig. 2 on the right side. The detailed modeling of the operation and transport module enables allocating time, volume, source and sink of a data transmission along the material flow. Since single data transmissions can be aggregated to the total information flow of an FMS, the quantitative modeling, simulation and evaluation of information flows are enabled by the reference model. Other element models formalize, e.g., workstations, AGVs or PLCs, based on their operating states (off, idle, processing, etc.). Therefore, the energy flow of the FMS can also be modeled and quantified, assuming a state-dependent average power demand of the system elements.

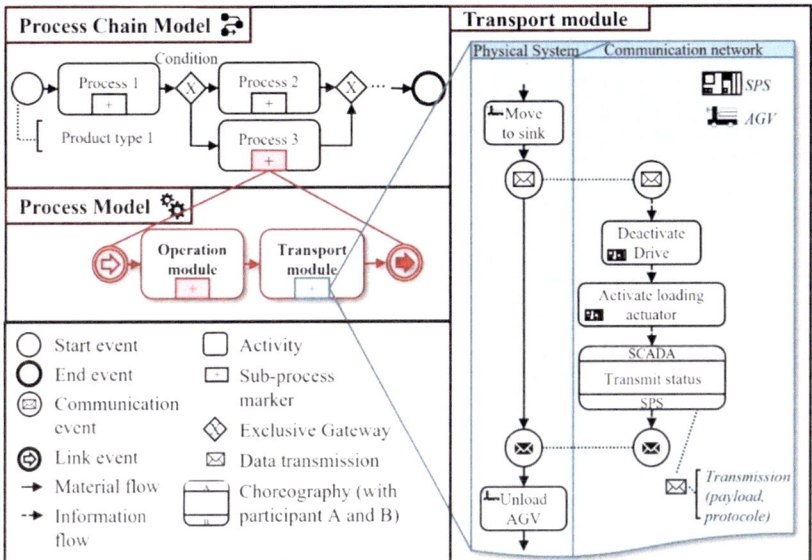

Fig. 2 Exemplary extracts of the reference model based on a generic FMS

5 Case Study

The developed framework is tested on a use case of an exemplary FMS producing three different product types of a polymer part (see Fig. 3) to demonstrate its applicability. Four different configurations of the presented FMS are generated as input of the framework application by varying the data acquisition and processing strategies using two variation parameters. The first variation parameter is the choice between implementing the MQTT or the OPC UA communication protocol for all data transmission in the communication network. Secondly, the communication strategy of the AGV fleet is varied. In the first case, the AGVs only communicate bi-directionally with the SCADA server. In the second case, the AGVs also communicate with each other at the start of each transport order. As a first step of the procedure model, an objective is developed. The objective is to maximize overall equipment efficiency (OEE) and minimize the environmental impact. In addition, the aim is to minimize data volume and its volatility to reduce the risk of malfunctions caused by overload or overload peaks. Goal-related KPIs are selected from the framework's KPI system to assess the physical system performance, such as throughput time, reject rate, system and utilization, and the communication network performance, such as average data volume, maximal data load and payload/overhead-ratio. For evaluating the environmental performance of both systems, the energy consumption and the power profile are analyzed. The reference model is used to create a formal model of the FMS, incorporating the set of variation parameters to model all FMS configurations. The system boundaries are defined as gate-to-gate. The workstations are set up according to the reference model. An additional agent type is created for the assembly and disassembly stations, the industrial robot, and the AGVs. AGV malfunctions are not considered in this case study. A control strategy is implemented as a function for routing the product unit. The workstations whose input buffers are available are given priority.

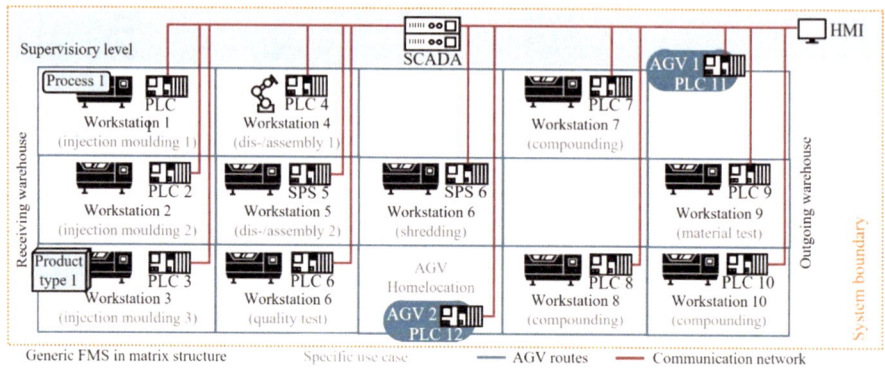

Fig. 3 Adaption of the generic FMS to the use case

The formal model is prototypically implemented in the software AnyLogic. The process chain model, in particular, is implemented in AnyLogic with function blocks. Function calls of a data transfer function model the event-triggered information flows along the process chain. Similarly, the routing function is called in each transport function block to control the movement of a product unit through the FMS and to adjust the individual processing time depending on the process and the workstation. The resources, e.g., the workstations, are modeled as agent types in accordance with the system element models. An agent is instantiated for each resource of the FMS and provided with individual parameter values.

The PLC of the resources, a server for the SCADA system and a PC as HMI are also implemented as instances of the agent types modeled according to the formal model. Variables assigned to the PLCs are used to aggregate the occurring data volume of event-triggered and cyclical communication, modeled by function calls in equidistant time intervals. The KPIs are calculated accordingly.

Applying the Decision Support Model in the evaluation phase results in a portfolio, visualizing the technical and environmental performance of the FMS configurations and, in particular, their physical systems and communication networks in relation to an ideal FMS configuration (see Fig. 4). Technical and environmental performance is derived by aggregating and normalizing the selected KPIs, describing the physical system and the communication network. Statistical significance is ensured by averaging the relevant KPI values from several simulation runs. The average values are then normalized to the occurring minimum and maximum values. The normalized KPIs are combined with equal weighting to form the technical and environmental performance index. The ideal FMS configuration is defined by summarizing the best values of all KPIs occurring among the FMS configurations in one configuration. This allows the best-fitted FMS configuration to be selected based on the FMS configurations' Euclidean distance from the ideal configuration [38]. Since the variation parameters only affect the communication network, differences in the performance of the physical systems seem to be caused by stochastic influences distorted by normalization. The evaluation of the communication network's performance leads to the recommendation to select the configuration with MQTT as the communication protocol and the second communication strategy of the AGV fleet. Additional detailed analysis can be executed to validate the decision-making and evolve a deeper understanding of the nature and interdependencies of material, energy and information flows. In this case study, the material flow influences the information flow through discrete events that trigger data transmission, resulting in an increase in data volume and volatility of data load. The energy flow is influenced by the material and information flow in that the resources of the physical FMS and its communication network cause energy consumption and volatility of power load according to their operational states. Dependencies of the material flow on the energy and information flow can further be taken into account by, e.g., modeling individual malfunctions of energy resources or data transmission. The information flow can be affected by the energy flow if an energy management system is considered in the simulation model.

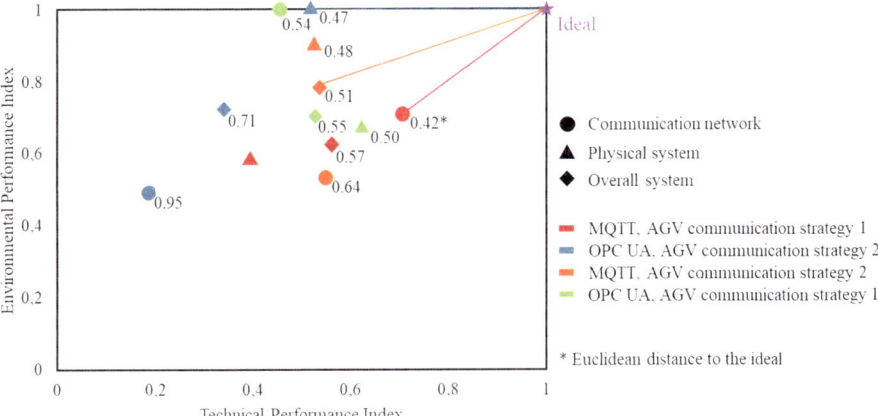

Fig. 4 Portfolio analysis of the FMS configurations presented in the use case

6 Conclusion and Outlook

The first simulation study based on the use case presented highlights the necessity of a target-driven design of not only the physical system but also the communication network, and thus the information flow, of FMS as a part of the manufacturing system design process. For a holistic assessment of FMS, interdependencies between material, energy and information flows must be considered. Since existing decision support frameworks for the planning and design of FMS incorporate the evaluation of the material and energy flow, the newly developed simulation-based framework also integrates the quantitative modeling and evaluation of the information flow into the performance evaluation and comparison of different FMS configurations. While the included procedure model structures the decision-making process in manufacturing system design, the modular reference model of the physical system and the communication network of a generic FMS reduces the modeling effort for a specific use case. The modular structure of the reference model ensures good scalability of the simulation model. Its hierarchical structure, with an increasing level of detail, not only allows the entire FMS under consideration to be easily captured but also provides an in-depth understanding of the system. The agent-based modeling of the physical system and particularly the communication network of FMS allows both to be analyzed sophistically. The evaluation model enables a multi-criterial comparison of different FMS configurations using technical and environmental KPIs. Applying the developed framework results in a simple portfolio representation of the decision alternatives. Further detailed analysis of the generated simulation results also enables a deeper understanding of the interactions between the material, energy, and information flows of the FMS.

As a next step, the new framework must be further validated with various more complex use cases. Future work could focus on integrating an economic evaluation of the decision alternatives into the evaluation model. To further explore the modeling of information flows, the reference model of the generic FMS could be adapted to modern Internet of Things architectures, which are expected to replace hierarchical communication networks [20].

Acknowledgements. The research and development project *RePro*, on which this publication is based, is funded by the Federal Ministry of Education and Research (BMBF) under the funding code 16KISR015K. The authors are responsible for the content of this publication.

References

1. Y. Yin, K. E. Stecke, and D. Li, "The evolution of production systems from Industry 2.0 through Industry 4.0," *International Journal of Production Research*, vol. 56, 1–2, pp. 848–861, 2018, https://doi.org/10.1080/00207543.2017.1403664.
2. A. Florescu and S. A. Barabas, "Modeling and Simulation of a Flexible Manufacturing System—A Basic Component of Industry 4.0," *Applied Sciences*, vol. 10, no. 22, p. 8300, 2020, https://doi.org/10.3390/app10228300.
3. S. Greinacher, L. Overbeck, A. Kuhnle, C. Krahe, and G. Lanza, "Multi-objective optimization of lean and resource efficient manufacturing systems," *Production Engineering*, vol. 14, no. 2, pp. 165–176, 2020, https://doi.org/10.1007/s11740-019-00945-9.
4. M. Schönemann, H. Bockholt, S. Thiede, A. Kwade, and C. Herrmann, "Multiscale simulation approach for production systems," *The International Journal of Advanced Manufacturing Technology*, vol. 102, 5–8, pp. 1373–1390, 2019, https://doi.org/10.1007/s00170-018-3054-y.
5. M. Münnich, S. Ihlenfeldt, and S. Thiede, "Simulation platform for energetic considerations in matrix production systems," (in de), 2023, https://doi.org/10.22032/DBT.57790.
6. A. Hellmich *et al.*, "Umsetzung von cyber-physischen Matrixproduktionssystemen: Expertise des Forschungsbeirats der Plattform Industrie 4.0," Jun. 2022.
7. S. Thiede, M.-A. Filz, B. Thiede, N. L. Martin, J. Zietsch, and C. Herrmann, "Integrative simulation of information flows in manufacturing systems," *Procedia CIRP*, vol. 81, pp. 647–652, 2019, https://doi.org/10.1016/j.procir.2019.03.170.
8. S. Schumacher, A. Bildstein, and T. Bauernhansl, "The Impact of the Digital Transformation on Lean Production Systems," *Procedia CIRP*, vol. 93, pp. 783–788, 2020, https://doi.org/10.1016/j.procir.2020.03.066.

9. Y. Jeong, A. Singh, M. Zafarzadeh, M. Wiktorsson, and J. Baalsrud Hauge, "Data-Driven Manufacturing Simulation: Towards a CPS-Based Approach," in *Advances in Transdisciplinary Engineering, SPS2020: Proceedings of the swedish production symposium, october 7–8, 2020*, K. Säfsten and F. Elgh, Eds.: IOS Press, 2020.

10. M. Evans, A. Yarbrough, G. Harris, and G. T. Purdy, "A simulation testbed for the evaluation of product and information flows in a manufacturing system," in *IISE Annual Conference and Expo 2021: Online, 22–25 May 2021*, A. Ghate, Ed., Norcross, GA, USA: Institute of Industrial & Systems Engineers (IISE), 2021, pp. 560–565.

11. D. Mourtzis, "Simulation in the design and operation of manufacturing systems: state of the art and new trends," *International Journal of Production Research*, vol. 58, no. 7, pp. 1927–1949, 2020, https://doi.org/10.1080/00207543.2019.1636321.

12. K. Peffers, T. Tuunanen, M. A. Rothenberger, and S. Chatterjee, "A Design Science Research Methodology for Information Systems Research," *Journal of Management Information Systems*, vol. 24, no. 3, pp. 45–77, 2007, https://doi.org/10.2753/MIS0742-1222240302.

13. P. Greschke, M. Schönemann, S. Thiede, and C. Herrmann, "Matrix Structures for High Volumes and Flexibility in Production Systems," *Procedia CIRP*, vol. 17, pp. 160–165, 2014, https://doi.org/10.1016/j.procir.2014.02.040.

14. H. A. ElMaraghy, "Flexible and reconfigurable manufacturing systems paradigms," *International Journal of Flexible Manufacturing Systems*, vol. 17, no. 4, pp. 261–276, 2005, https://doi.org/10.1007/s10696-006-9028-7.

15. H. A. ElMaraghy, "Reconfigurable Process Plans For Responsive Manufacturing Systems," in *Digital Enterprise Technology: Perspectives and Future Challenges*, P. F. Cunha and P. G. Maropoulos, Eds., Boston, MA: Scholars Portal, 2007, pp. 35–44.

16. M. Schönemann, C. Herrmann, P. Greschke, and S. Thiede, "Simulation of matrix-structured manufacturing systems," *Journal of Manufacturing Systems*, no. 37, pp. 104–112, 2015, https://doi.org/10.1016/j.jmsy.2015.09.002.

17. L. Monostori *et al.*, "Cyber-physical systems in manufacturing," *CIRP Annals*, vol. 65, no. 2, pp. 621–641, 2016, https://doi.org/10.1016/j.cirp.2016.06.005.

18. M. Meyer, *Kommunikationstechnik: Konzepte der modernen Nachrichtenübertragung*, 6th ed. Wiesbaden: Springer Vieweg, 2019.

19. O.-E. Heiserich, K. Helbig, and W. Ullmann, "Informations- und Kommunikationssysteme der Logistik," in *Logistik*, Heiserich, Ed., Wiesbaden: Gabler Verlag, 2011, pp. 337–382.

20. D. Siepmann and N. Graef, "Industrie 4.0—Grundlagen und Gesamtzusammenhang," in *Einführung und Umsetzung von Industrie 4.0*, A. Roth, Ed., Berlin, Heidelberg: Springer Berlin Heidelberg, 2016, pp. 17–82.

21. P. Linke, "Speicherprogrammierbare Steuerungen," in *Handbuch Maschinenbau: Grundlagen und Anwendungen der Maschinenbautechnik*, A. Böge and W. Böge, Eds., Wiesbaden: Springer Vieweg, 2017, pp. 1591–1609.

22. H. Merz, T. Hansemann, and C. Hübner, *Building Automation: Communication Systems with EIB/KNX, LON and BACnet*: Springer Cham, 2018.

23. G. Schnell and B. Wiedemann, *Bussysteme in der Automatisierungs- und Prozesstechnik: Grundlagen, Systeme und Anwendungen der industriellen Kommunikation*, 9th ed. Wiesbaden: Springer Vieweg, 2019.

24. D. Silva, L. I. Carvalho, J. Soares, and R. C. Sofia, "A Performance Analysis of Internet of Things Networking Protocols: Evaluating MQTT, CoAP, OPC UA," *Applied Sciences*, vol. 11, no. 11, p. 4879, 2021, https://doi.org/10.3390/app11114879.

25. C. F. Durach, J. Kembro, and A. Wieland, "A New Paradigm for Systematic Literature Reviews in Supply Chain Management," *J Supply Chain Manag*, vol. 53, no. 4, pp. 67–85, 2017, https://doi.org/10.1111/jscm.12145.

26. M. Rabe, W. Wincheringer, and T. Sohny, "Referenzmodell basierend auf Wertstromsimulation zur Bewertung von Produktionssystemen in der Angebotsphase," in *Proceedomgs ASIM SST*, 2020, pp. 373–380.
27. S. Berger, M. Borgenreuther, B. Häckel, and O. Niesel, "Modelling Availability Risks of IT Threats in Smart Factory Networks—A Modular Petri Net Approach," in *Proceedings of the 27th European Conference on Information Systems (ECIS)*, Stockholm und Uppsala, 2019.
28. P. Voit, M. Jira, M. Horn, and C. Seidel, "Agentenbasierte Simulation für CPPS," *Zeitschrift für wirtschaftlichen Fabrikbetrieb*, vol. 116, no. 11, pp. 841–846, 2021, https://doi.org/10.1515/zwf-2021-0191.
29. H. Komoto, S. Kondoh, Y. Furukawa, and H. Sawada, "A simulation framework to analyze information flows in a smart factory with focus on run-time adaptability of machine tools," *Procedia CIRP*, vol. 81, pp. 334–339, 2019, https://doi.org/10.1016/j.procir.2019.03.058.
30. S. Andrae and P. Pobuda, *Agenbasierte Modellierung: Eine interdisziplinäre Einführung*. Wiesbaden: Springer Gabler, 2021.
31. N. Feldkamp, S. Bergmann, and S. Straßburger, "Modellierung und Simulation von modularen Produktionssystemen," in *18. ASIM-Fachtagung Simulation in Produktion und Logistik*, Chemnitz, 2019, pp. 391–401.
32. P. Voit, M. Beller, and G. Reinhart, *A methodology for simulation production systems considering the degree of autonomy*. Hannover: publish-Ing, 2020.
33. A. Saez-Mas, J. P. Garcia-Sabater, and J. Morant-Llorca, "Using 4-Layer Architecture to Simulate Product and Information Flows in Manufacturing Systems," *International Journal of Simulation Modelling*, vol. 17, no. 1, pp. 30–41, 2018, https://doi.org/10.2507/IJSIMM 17(1)408.
34. G. Harris, C. Peters, A. Yarbrough, C. Estes, and Aberanthy D., "Industry Readinesss for Digital Manufacturing May Not Be as We Thought: Preliminary Findings of MxD*Project 17-ß1-ß1," in *Proceedings of the 10th Model-Based Enterprise Summit (MBE 2019)*, Gaithersburg, Maryland, pp. 110–116.
35. *Simulation von Logistik-, Materialfluss- und Produktionssystemen - Grundlagen*, VDI 3633–1:2014, VDI, 2014–12–00.
36. R. Joppen, S. von Enzberg, J. Gundlach, A. Kühn, and R. Dumitrescu, "Key performance indicators in the production of the future," *Procedia CIRP*, vol. 81, pp. 759–764, 2019, https://doi.org/10.1016/j.procir.2019.03.190.
37. OMG, *Business Process Model and Notation (BPMN): Version 2.0.2*. [Online]. Available: https://www.omg.org/spec/BPMN/2.02/PDF
38. M. Tavana, "Euclid: Strategic alternative assessment matrix," *Multi Criteria Decision Anal*, vol. 11, no. 2, pp. 75–96, 2002, https://doi.org/10.1002/mcda.318.

Sustainable Materials & Surfaces

3d-Shaped High-Strength Parts from Partially Delignified and Densified Wood—Introduction of the Project HolzF³

Thomas Grosse[1(✉)], Ulrich Müller[2], Matthias Jakob[3], Florian Feist[4], Mathias H. Luxner[5], and Wolfgang Knoebl[6]

[1] Volkswagen AG, Berliner Ring 2, 38440 Wolfsburg, Germany
thomas.grosse5@volkswagen.de

[2] Department of Material Science and Process Engineering, Institute of Wood Technology and Renewable Materials, Vienna, University of Natural Resources and Life Sciences, Konrad Lorenz-Straße 24, 3430 Tulln an Der Donau, Austria
ulrich.mueller@boku.ac.at

[3] Department of Material Science and Process Engineering, Institute of Wood Technology and Renewable Materials, University of Natural Resources and Life Sciences, Konrad Lorenz-Straße 24, 3430 Vienna, Tulln an Der Donau, Austria
matthias.jakob@boku.ac.at

[4] Vehicle Safety Institute, Graz University of Technology, Inffeldgasse 23/I, 8010 Graz, Austria
florian.feist@tugraz.at

[5] LUXNER Engineering ZT GmbH, Franz-Xaver-Renn-Straße 4 / 18, 6460 Imst, Austria
m.luxner@luxner-zt.com

[6] Weitzer Wood Solutions GmbH, Klammstraße 24C, 8160 Weiz, Austria
Wolfgang.Knoebl@weitzer-woodsolutions.com

https://www.volkswagen-group.com/de,
https://boku.ac.at/map/holztechnologie,
https://www.tugraz.at/institute/vsi/home/,
http://www.luxner-zt.com/,
https://www.weitzer-woodsolutions.com/

Abstract. This study explores the potential of wood-based materials in sustainable lightweight automotive applications. The research focuses on overcoming the challenge of balancing complex three-dimensional design with component stiffness and strength in crash-relevant vehicle structures. The study investigates a novel approach involving partial delignification and subsequent densification of wood, previously shown to significantly enhance mechanical properties in solid wood. The HolzF³ project applies this technique to chipboard materials (strands) and small-diameter birch and beech wood, which are typically used for energy production only. By utilizing smaller semi-finished products like strands, complex three-dimensional component shapes become feasible, with densification compensating for potential loss in mechanical performance. Initial results demonstrate the effectiveness of a two-step densification process in producing high-strength wood fractions from veneers and strands derived from small-diameter birch and beech roundwood. This process increases raw material density to 0.92-1.19 g/cmB3, while significantly improving tensile strength and stiffness. Notably,

© The Author(s) 2025
K. Dröder and T. Vietor (Eds.): CD 2024, Proceedings, pp. 165–178, 2025.
https://doi.org/10.1007/978-3-658-45889-8_13

densified birch veneers achieved tensile strength up to 400 MPa and stiffness up to 40 GPa, comparable to industrially available veneers. The study concludes that this approach enables the utilization of previously undervalued wood fractions in high-quality material applications, particularly in the automotive industry. This innovation not only improves product sustainability and reduces COb footprint but also creates new opportunies for sustainable value-added processes in the mobility sector.

Keywords: delignification · densification · lightweight design · sustainability · wood

1 Introduction

Tackling man-made climate change, which is probably the greatest global challenge of our time, requires all industrial sectors to innovate in terms of sustainable product design [1]. Especially in the field of mobility, the use of ecologically sustainable and lightweight materials is an effective lever for reducing the CO_2 footprint in both, the production and the use phase [2, 3].

Within the group of biomaterials, wood is a material with particularly good density-specific mechanical properties. The high strength resulting from the naturally grown structure of wood as a kind of "anisotropic fiber composite foam" ensures high mechanical performance, especially in bending and bulging load cases (Fig. 1). [4].

In addition, especially birch and beech wood are cost-effective raw materials that is available in large quantities, in Europe. With regard to the global warming potential, it can be said that wood itself is initially a CO_2-negative material, as the tree removes CO_2 from the atmosphere as it grows. Even in the case of burning wood as an end-of-life scenario of cascaded use, only as much CO_2 is emitted into the atmosphere as the tree has previously bound.

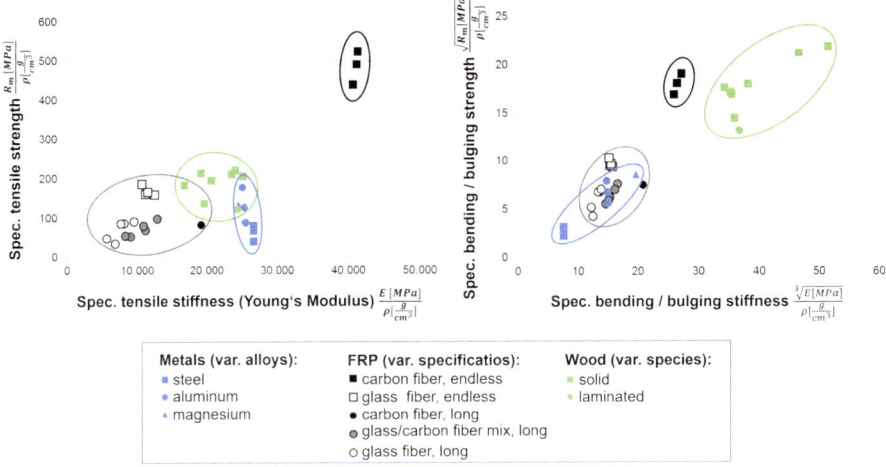

Fig. 1. Density-specific mechanical properties of exemplary woods and conventionally used materials for mobility applications according to Grosse et al. [4]

2   Current Research on Wood in Vehicle Body Applications

Even though wood is a material that was widely used, especially in the early days of the automobile, it was virtually forgotten with the use of ever-improving metal alloys and plastics, with the exception of purely decorative elements. It is only in recent years that more and more research activities have been carried out on the use of wood in structural and also crash-relevant vehicle structures using material-hybrid lightweight construction approaches.

In the "HAMMER" research project, for example, a door impact beam with a multi-layer veneer structure and further reinforcement layers was realized, which can be integrated into the standard process chains of automotive body-in-white production through the use of metallic connection points using conventional joining technologies [5].

The research project "WoodCAR" [6] focused on the development of suitable models and methods for the mapping of solid and laminated veneer lumber in crash load cases using the finite element method (FEM). An according demonstrator amongst others was a vehicle subfloor including a tunnel.

In the project "For(s)tschritt" [7], a focus was placed on functional integration in addition to FEM simulation. Using a car door as an example, the door impact beam, the hinge reinforcements and the acoustic insulation were integrated into a complexly shaped wood-based component. Further demonstrators for rail vehicles were implemented in the form of a train side wall and a sliding door.

In the field of rail vehicles, Siemens Mobility recently presented an aerodynamic chassis fairing as a wood-layer composite and tested it in real use under winter conditions at speeds of up to 360 km/h [8].

In the above-mentioned projects and the demonstrator components implemented in them, wood was essentially used analogous to the classic fiber composite construction method in the form of a multi-layer structure of individual wood veneers with defined fiber orientations (Fig. 2).

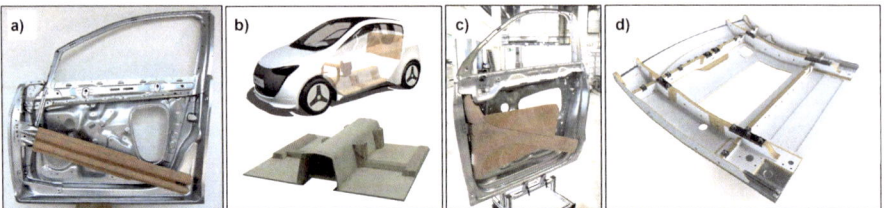

Fig. 2. Demonstrator parts of exemplary research projects on wood in structural carbody applications: a) door impact beam made of laminated beech wood veneers from the project HAMMER [5], b) vehicle subfloor incl. Tunnel made of plywood and solid wood from the project WoodCAR [6], c) function-integrated door inner part made of laminated beech wood veneers from the project For(s)tschritt [7], d) train chassis fairing made of plywood from Siemens Mobility [8]

In contrast to textile fiber composite semi-finished products, however, the formability of veneers is severely limited by their brittleness. In the current HyEndWood research project, approaches to the production of structural components from wood strands analogous to fiber composite technologies such as Sheet Molding Compound (SMC) and Bulk Molding Compound (BMC) are being investigated [9, 10] in order to be able to produce more complex three-dimensional geometries.

However, as the fiber length decreases, so does the strength and stiffness of the composite. A major limitation for the use of wood in structural vehicle applications is therefore currently the trade-off between 3D formability and mechanical properties, since unlike in construction, for example, complex three-dimensional structures are almost always required. This results from the boundary conditions of a small installation space, functional or design-related requirements. A further challenge arises from the production process of the automotive body shop at the end of which the cathodic dip painting with the furnace drying at approx. 200 °C stands for approx. 30 min. As a result, structural wood components already introduced in car body construction also have to go through the drying process, which limits the availability of suitable adhesives. Further challenges arise from the operating conditions of the vehicle, which are usually characterized by large temperature and humidity differences or corrosive media (e.g. road salt in winter). The wood-based materials used must therefore have a low moisture absorption in order to minimize effects such as swelling and shrinkage or delamination. The above-mentioned challenges of three-dimensional component geometries, high mechanical demands with low moisture absorption are addressed by the approach of delignification and compaction pursued in the HolzF^3 project.

3   Approach of the Project HolzF^3

In the review by Jakob et al. [11] various wood densification processes that have been established over the last 100 years, were compared with each other in terms of their potential and challenges. Two-step densification processes, in which the wood is first chemically pretreated (partial extraction of amorphous wood polymers) and then densified, proved to be very promising with regard to high-end applications, as it was shown that the mechanical properties can be increased beyond pure scaling in densification. Thereby, two different concepts are usually pursued, either the complete delignification of the wood material as carried out by Frey et al. [12], or the concept of partial delignification as carried by Song et al. [12]. Although both concepts have their individual advantages, partial extraction of lignin seems to have a greater potential in future. Three arguments for this are the astonishing results of Song et al. [12], with improvement factors of over 10 in tensile properties, the easier handling of the raw material after extraction and the presumed potential to transfer alkaline processes to the existing industry.

Therefore, the approach of the HolzF^3 project is now to apply partial delignification and densification to veneers and in particular wood strands made from small-diameter wood, that would usually be used energetically, to enable the production of complex three-dimensional and high-strength wood components.

Figure 3 shows the schematic model of the process used in HolzF^3 to improve the mechanical properties of wood as a raw material for structural components. As can be seen, different raw materials are used in HolzF^3 depending on the desired end use of the material. Both peeled hardwood veneers (Fig. 3a) and hardwood strands (Fig. 3b) are used, well knowing that both raw materials have their own potential for HolzF^3. On the one hand, veneers are already highly oriented, which makes them attractive for use in products where excellent mechanical properties along the grain direction are necessary and desired. On the other hand, HolzF^3 focuses on hardwood strands, which is obviously a fairly new topic. When comparing veneers with strands, veneers are clearly superior to strands in terms of mechanical properties. Strands, however, are believed to offer the possibility of shaping wood in 3D. It is also known that the size reduction of wood is associated with a reduction in mechanical properties. At the same time, however, the degree of homogenization increases, which can ultimately lead to an improvement in the characteristic values of the final wood product.

In addition, HolzF^3 focuses on the use of domestic hardwoods with small diameters as a valuable "new" resource. The definition of "small diameter" was limited to a maximum diameter of 20 cm at breast height. Up to now, this roundwood has mostly been burned to produce energy, as it is not classified as suitable for sawing. The first main part of the HolzF^3 project is to demonstrate the potential of small diameter untreated wood compared to industrial available wood. Subsequently, the possible improvements through modification and densification will be evaluated in order to obtain sufficient data for the simulation of densified wood.

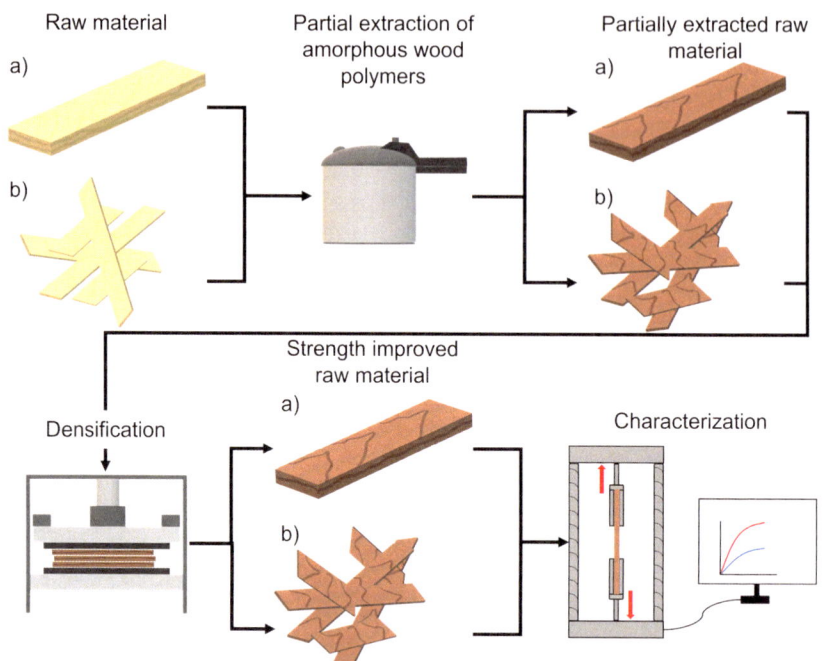

Fig. 3. Schematic model of the strength improving process of the raw material in the project HolzF[3]: a) hardwood veneers; b) hardwood strands.

4   First Experimental Results

4.1   Material and Methods

Extraction Protocol

One of the central objectives of the HolzF[3] project is the conversion of non-sawable hardwood (small diameter hardwood), which has mainly been used as an energy source to date, into construction elements. Until now, first experiments were done on industrial available veneers (I) and small diameter roundwood veneers (SD) from birch and beech wood. Thereby, industrial veneers served as a reference for the veneers currently available on the market. 30 oven dried (103 °C for 24 h) veneers with the dimensions of 100 × 50 × 1.5 mm (axial × angential × radial) and a total weight of ~ 120 g were first immersed in 3 L of an alkaline solution of 0.825 mol L^{-1} sodium hydroxide (NaOH). The applied extraction parameters are based on previous publications [12, 13] and were later adapted by means of pre-tests with the aim of reducing chemicals and energy. To ensure sufficient impregnation, vacuum was applied and veneers were left in the solution over night at room temperature. Cooking of the veneers was done in a commercial stainless steel pressure cooker at 100 °C for 2 and 4 h. To prevent overlapping and thus ensure full contact between wood and solution, veneers were separated with stainless steel meshes. After partial extraction of the amorphous wood polymers, veneers were washed in deionized water until the wash water was neutral (pH 7).

To evaluate the influence of sodium sulfite (Na_2SO_3) on the improvements of the wood, small diameter roundwood veneers were additional cooked in an alkaline solution of 0.825 mol L^{-1} sodium hydroxide (NaOH) and 0.13 mol L^{-1} sodium sulfite (Na_2SO_3). Thereby, the treatment conditions were the same as before.

In addition to veneers, tests were also carried out with strands of small diameter birch and beech roundwood. Similar to veneers, the strands were first impregnated with an alkaline solution of 0.825 mol L^{-1} sodium hydroxide (NaOH) and left over night under vacuum. Thereby, care was taken to ensure that the ratio of dry wood material to alkaline liquid was the same as in the tests with the veneers. Different to veneers, strands were boiled at 100 °C for only 4 h for the first trials.

Densification Protocol
After partial extraction of amorphous wood polymers, 10 out of the 30 washed and fully water-saturated veneers were densified in the radial direction in a hydraulic hot press (Langzauner Gesellschaft m.b.H., Lambrecht, Austria). The densification protocol was based on previous experiments [13] with slight adaptions regarding the maximum pressure. Therefore, densification temperature was set to 120 °C, and a maximum pressure of 11.67 MPa was built-up stepwise during 30 min. After cooling under pressure to 60 °C, specimens were left in a climate chamber operated at 20 °C and 65% humidity until further use.

Partially extracted strands were compacted according to a similar densification protocol. In contrast to veneers, it was not possible to determine an exact densification pressure for strands. However, care was taken to adjust the applied force depending on the spreading area of the strands so that a pressure of approx. 12 MPa could be built-up.

Tensile Testing
For the tensile tests, conditioned reference (Ref) and densified veneers (NaOH—NaOH; NaOH + Na_2SO_3—NaOHS) were first cut with a paper cutter to a length (axial) of 100 mm and a width (tangential) of 10 mm. For the testing of strands, small specimens were cut out of the untreated and densified strands with a razor blade. Care was taken to ensure that the samples were prepared along the fiber direction and did not exhibit any defects. As far as the strands allowed, samples were produced in the size of 100×10 mm (lengths × width). To prevent damages by clamping, the clamping areas were additional reinforced with birch veneers, before longitudinal tensile tests were performed using a universal testing machine (ZwickRoell 100 kN, ZwickRoell, Ulm, Germany) equipped with a 5 kN load cell (ZwickRoell, Ulm, Germany). The clamping distance was set to 40 mm and testing was performed at a speed of 1 mm min^{-1}. While testing, the displacement was tracked by means of a macro mechanical extensometer (MakroXtens II, ZwickRoell, Ulm, Germany). For each variant 7 specimens were prepared and tested according to this protocol.

4.2 Results

To improve the mechanical properties of wood without destroying its favorable structural orientation, industrial available veneers (I) and small diameter roundwood veneers (SD) of birch and beech were first chemically treated in a mild alkaline solution containing 0.825 mol L^{-1} NaOH or an alkaline solution containing 0.825 mol L^{-1} NaOH and

$0.13 \ \text{mol} \ L^{-1} \ Na_2SO_3$ at 100 °C for 2 and 4 h, respectively. The density was then increased by a combination of temperature, humidity and pressure. During this treatment, some of the amorphous wood polymers, mainly hemicelluloses and lignin, were removed from the cell walls, resulting in a distinct loss of dry mass, as shown in Fig. 4a and Fig. 4b. Comparing the two wood species, it is noticeable that birch wood has a slightly higher mean dry mass loss compared to beech wood when treated according to the same pretreatment process. Interestingly, while there are almost no differences between the average dry mass loss within beech wood samples, regardless of raw material, industrial available birch veneer has a clearly lower average dry mass loss compared to veneers and strands made from small diameter roundwood. Furthermore, it was shown that the addition of $0.13 \ \text{mol} \ L^{-1} \ Na_2SO_3$ to the pretreatment solution did not result in any distinct differences in dry mass loss for small diameter birch or beech roundwood veneers.

After densification and standard climate conditioning (20 °C, 65% rel. Humidity) the density of the specimens was measured. The comparison with the reference values shows that both the veneers and the strands were effectively densified after pretreatment, as the average density increased for all samples and lies between 0.92 and $1.19 \ \text{g cm}^{-3}$ (Fig. 4c and Fig. 4d). However, similar to the average dry mass loss, the average density of the

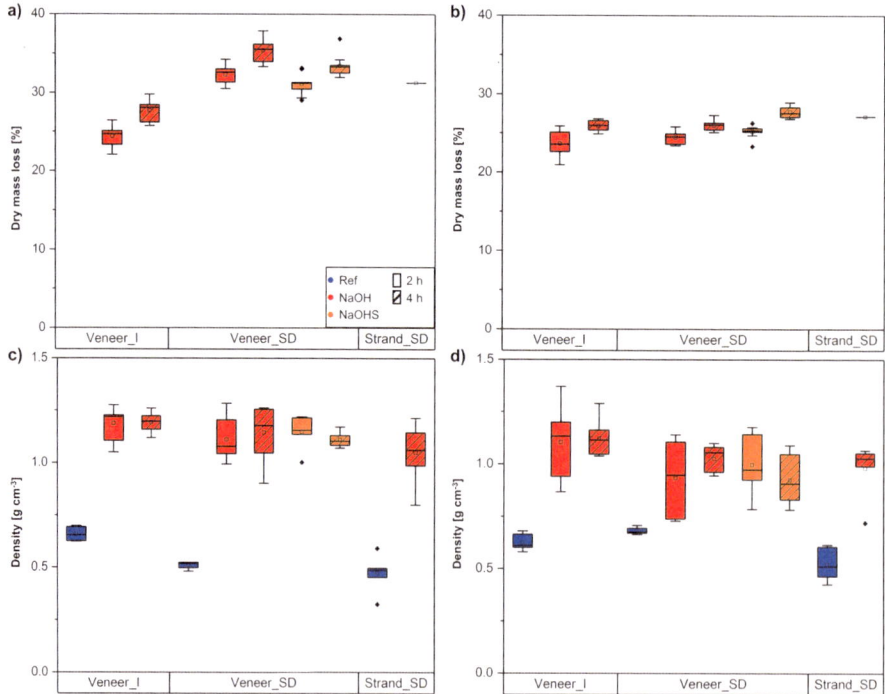

Fig. 4. Comparison of dry mass loss (a & b) and density (c & d) of untreated references (Ref), sodium hydroxide and densified (NaOH), and sodium hydroxide + sodium sulfite and densified (NaOHS) industrial (I) veneers, small diameter roundwood (SD) veneers and small diameter roundwood (SD) strands prepared from birch (a & c) and beech (b & d) wood. (n = 10 for each variant measuring the dry mass loss; n = 7 for each variant measuring the density)

pretreated and densified veneers and strands also differed slightly between the two wood species. Here, densified birch wood specimens tend to have higher average densities than the corresponding beech wood samples. This could be due to the anatomical differences between birch and beech wood. At this point, however, it has to be mentioned that not all pores were completely closed for both wood species and all pretreatment processes used, as the maximum possible density of wood is around 1.5 g cm^{-3}. There is therefore still potential to improve the densification process and thus the mechanical properties of the studied wood species.

To emphasize this right at the beginning: The densification of the tested raw material led to a clear improvement in the mean absolute tensile strength and stiffness, regardless of the wood species, raw material, quality of veneers or pre-treatment and duration applied, as can be seen in Fig. 5 and 6. It can be seen that samples made from birch wood tend to have higher mean values than samples made from beech wood. This goes for strength as well as stiffness. A negative effect of the densification process is the increase in scattering, regardless of the wood species and quality. However, this can be explained by the sample preparation. On the one hand, individual veneers were tested. On the other hand, testing samples were cut along the veneer edges and not exactly along the grain direction. Since some veneers tended to wrap after chemical treatment and densification, the fiber deviation increased after densification. This in turn affects the mechanical properties. However, this fact should not detract from the impressive results, as low values were always measured for samples with high fiber deviation. Moreover, birch samples with almost no fiber deviation exhibit tensile strength and stiffness values of over 400 MPa and 40 GPa, respectively, which again shows the high potential of the applied densification process. An even more impressive and important result was that the average strength and stiffness of the densified veneers within a wood species were almost the same regardless of the quality of the raw material (with the exception of $NaOH_2$ SD beech). Thus, the high potential of the combination of the applied process with the use of veneers made from small diameter roundwood was already demonstrated at this stage of the $HolzF^3$ project. In addition, the improvement of the average absolute as well as the specific tensile properties extends the applicability of the presented raw material. Since the mobility sector is a space-limited sector (the dimensions of cars are usually predetermined), the improvement of the absolute values is advantageous. This increases the potential to use wood for structural components in cars. While at the same time increasing the specific mechanical properties, wood does not lose its lightweight character. Thus, an intelligent substitution of existing structural components in cars that are currently made of metals with wood can improve product sustainability and reduce the overall CO_2 footprint of cars.

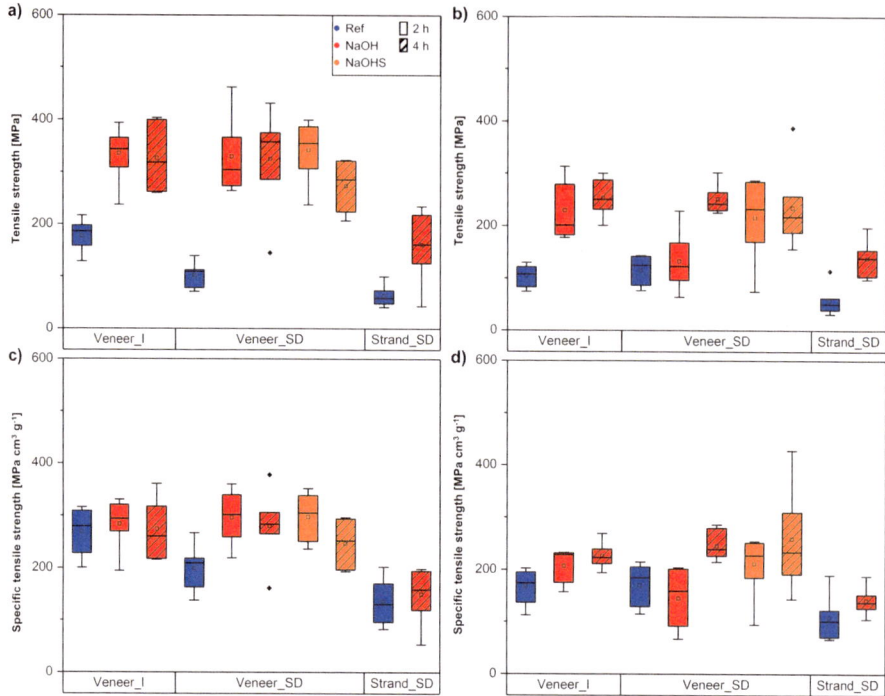

Fig. 5. Mechanical characterization of untreated references (Ref), sodium hydroxide and densified (NaOH), and sodium hydroxide + sodium sulfite and densified (NaOHS) industrial (I) veneers, small diameter roundwood (SD) veneers and small diameter roundwood (SD) strands prepared from birch (a & c) and beech (b & d) wood. Samples were conditioned at 20 °C and 65% relative humidity and tested by longitudinal tensile tests: Strength (a & b); Specific strength (c & d). (n = 7 for each variant)

In a next step, the HolzF³ project will focus on the use of small diameter round wood strands to contribute to potential complex 3D shaped wood components. Therefore, in a first step, the established densification process for veneers was applied to small diameter birch and beech roundwood strands. In order to evaluate the potential for improving the mechanical properties, the samples were tested in longitudinal tensile tests and compared with reference strands and veneers. Figure 5 and 6 show that untreated strands are clearly inferior to veneers, regardless of the wood species or veneer quality. These results are not surprising, as the mechanical properties of wood decrease with the degree of comminution. However, similar to veneers, strands also show a clear improvement in average tensile properties after chemical pretreatment and densification. It was thus possible to improve the average tensile properties of strands up to the properties of industrial veneers or even to exceed them in the case of tensile modulus.

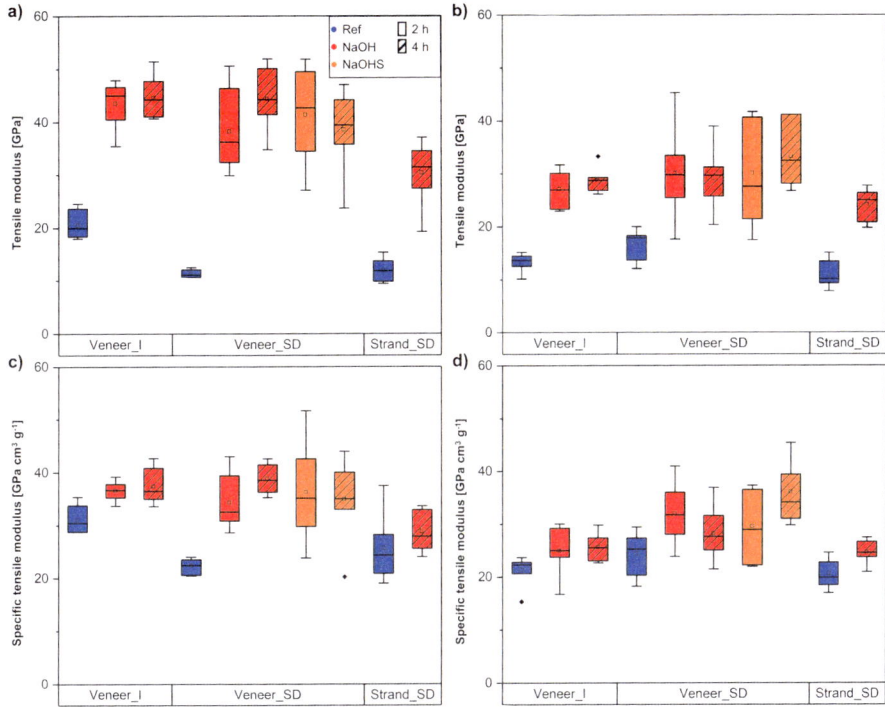

Fig. 6. Mechanical characterization of untreated references (Ref), sodium hydroxide and densified (NaOH), and sodium hydroxide + sodium sulfite and densified (NaOHS) industrial (I) veneers, small diameter roundwood (SD) veneers and small diameter roundwood (SD) strands prepared from birch (a & c) and beech (b & d) wood. Samples were conditioned at 20 °C and 65% relative humidity and tested by longitudinal tensile tests: Modulus (a & b); Specific modulus (c & d). (n = 7 for each variant)

The results in Fig. 5 and 6 show the high potential of the enhancing in mechanical properties of wood by means of the process used in HolzF[3]. Regardless of the raw material, the average tensile strength and stiffness could be improved along or even beyond the pure scaling of the densification, as can be seen from the specific tensile strength and modulus values in Fig. 5c and 5d, and 6c and 6d, respectively. In addition, compared to other processes in the literature [12], chemicals could be reduced or even eliminated and the energy requirement in terms of extraction temperature and duration could be reduced. This in turn can have a positive influence on the environmental costs of the potential product.

5   Demonstrators

As demonstrators for the investigation of the application properties of the delignified and densified wood semi-finished products in the context of the challenges addressed in Sect. 2, two components from the automotive sector from the product range of the

Volkswagen Group were initially selected. The first component is the lower hinge rein-forcement of a car driver's door (Fig. 7). It represents a structural body component, which must therefore be installed before the cathodic dip coating (KTL) treatment of the body in white. The main function of that part is to add static stiffness to the door for load cases like door sag and dynamic load cases like overpressing. In addition to the complex mechanical loads, there are therefore requirements for this component with regard to temperature resistance to up to 200 °C in the coating drying process and, due to the location in an area where water can also occur, also with regard to moisture, corrosion and aging resistance.

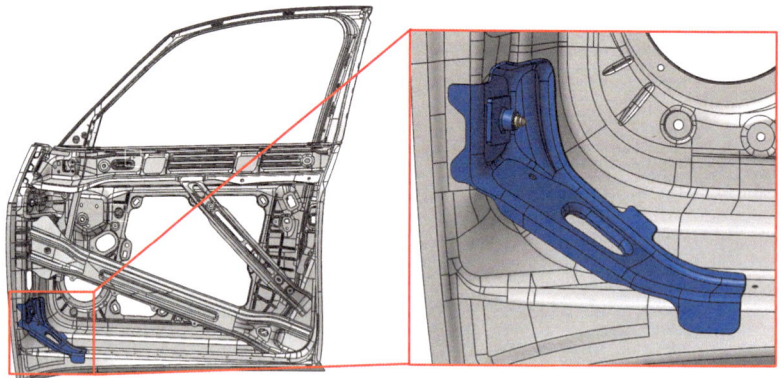

Fig. 7. Planned demonstrator part: lower hinge reinforcement of a car driver's door

As a second component from the automotive industry, a semi-structural interior com-ponent was selected with the cover of a center console that is currently made of aluminum casting (Fig. 8). In addition to the load-bearing capacity of mechanical loads (espe-cially in cases of misuse), this results in special requirements with regard to emissions, flammability, UV resistance and, if necessary, the adhesion of decorative coatings.

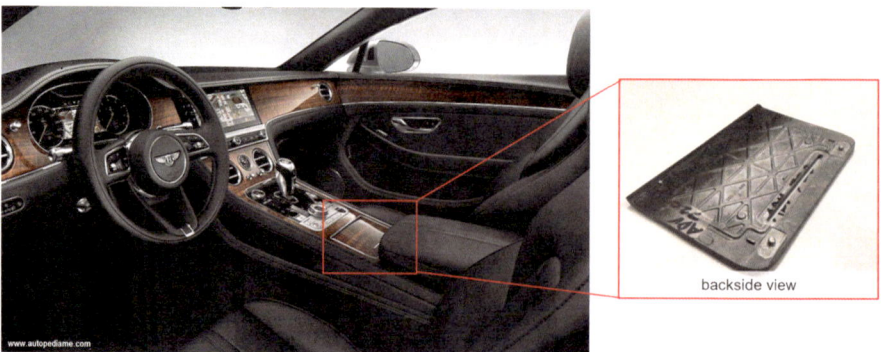

Fig. 8. Planned demonstrator part: center console cover

In addition to the components from the automotive industry, an interior component from the aerospace sector is also under discussion, which is to be designed and calculated by the project partners FACC and Luxner.

In the further course of the HolzF^3 project, it is now to be found out to what extent the delignified and compacted wood with an appropriate adhesive system is suitable for meeting the aforementioned requirements of the demonstrator parts.

6 Conclusions

In the previous projects, the high application potential of wood-based materials in structural vehicle applications has already been demonstrated. The production of complex three-dimensional geometries in combination with a high mechanical load-bearing proved to be a particular challenge. In the investigations carried out so far, partial delignification and subsequent densification have achieved a distinct increase in the absolute and density-specific tensile strength and modulus for industrial available veneers and veneers made of small diameter roundwood wood of birch and beech.

Acknowledgements. This project is funded by the Forest Fund, an initiative of the Federal Ministry of Agriculture, Forestry, Regions and Water Management, and is carried out as part of the Think.Wood program of the Austrian Wood Initiative (FFG 893356). The contribution by the company partners is gratefully acknowledged.

Waldfonds
Republik Österreich

Eine Initiative des Bundesministeriums
für Land- und Forstwirtschaft, Regionen
und Wasserwirtschaft

References

1. IPCC, 2023: Sections. In: Climate Change 2023: Synthesis Report. Contribution of Working Groups I, II and III to the Sixth, Assessment Report of the Intergovernmental Panel on Climate Change [Core Writing Team, H. Lee and J. Romero (eds.)]. IPCC, Geneva, Switzerland, pp. 35–115, https://doi.org/10.59327/IPCC/AR6-9789291691647
2. Reimer, L.; Kaluza, A.; Cerdas, F.; Meschke, J.; Vietor, T.; Herrmann, C.: "Design of Eco-Efficient Body Parts for Electric Vehicles Considering Life Cycle Environmental Information" Sustainability 2020, 12, 5838. https://doi.org/10.3390/su12145838
3. Koffler, C., Rohde-Brandenburger, K.: "On the calculation of fuel savings through lightweight design in automotive life cycle assessments", Int J Life Cycle Assess 15, 128–135 (2010). https://doi.org/10.1007/s11367-009-0127-z
4. Grosse, T.; Fischer, F., Berthold, D.; Schauerte, O. "Wood materials—suitable for structural automotive applications?"; European Wood-based Panel Symposium, Hamburg, 2018

5. Kohl, D.; Böhm, S.: "Integration von holzbasierten Multimaterialsystemen in Fahrzeugstruk-turen durch geeignete Fügetechnologien und in Rohbaufertigungen durch geeignete Prozess-ketten", Schlussbericht des Teilprojekts innerhalb des BMBF-Verbundprojekts "Holz-formteile als Multi-Materialsysteme für den Einsatz im Fahrzeug-Rohbau (HAMMER)" FKZ 13N12039 2016

6. Müller, U.; Jost, T.; Kurzbock, C.; Stadlmann, A.; Wagner, W.; Kirschbichler, S.; Baumann, G.; Feist, F, Pramreiter, M.: "Crash simulation of wood and composite wood for future automotive engineering."; WOOD MATER SCI ENG. 2020; 15(5): 312–324

7. Grosse, T. et al.: "Structural assemblies based on sustainable wood-based material systems to reduce mass and environmental impact in road and rail vehicle construction—synonym: (For(s)tschritt)"; final report, 536 pages, 2020, https://doi.org/10.2314/KXP:1755150091

8. Kopp, M.; Prix, A.; Moshammer, Dr.; Pomberger, S.; Luxner, M. H.: "Eco-Design für Schienenfahrzeuge"; 48. Tagung „Moderne Schienenfahrzeuge", 17.–19. September 2023, Graz, Austria

9. Piazza, G.: „Innovative Holzanwendungen für die Mobilität der Zukunft"; Cluster Innovativ Holzbau-Donnerstag, 23.03.2023, Ostfildern, Germany

10. Feser, T.; Piazza, G.: "Beispiele von holzbasierten Bauteilen", WerkstoffPlus Auto, 20.02.2024, Stuttgart, Germany

11. Jakob, M. et al.: "The strength and stiffness of oriented wood and cellulose-fibre materials: A review", Progress in Materials Science, Volume 125, 2022, https://doi.org/10.1016/j.pma tsci.2021.100916

12. Frey, M. et al.: "Delignified and Densified Cellulose Bulk Materials with Excellent Tensile Properties for Sustainable Engineering", ACS Appl. Mater. Interfaces 2018, 10, 5, 5030–5037, https://doi.org/10.1021/acsami.7b18646

13. Song, J. et al.: "Processing bulk natural wood into a high-performance structural material"; Nature. 2018 Feb 7;554(7691):224–228. https://doi.org/10.1038/nature25476

14. Jakob, M., Czabany, I., Veigel, S. et al.: "Comparing the suitability of domestic spruce, beech, and poplar wood for high-strength densified wood", Eur. J. Wood Prod. 80, 859–876 (2022). https://doi.org/10.1007/s00107-022-01828-0

Influence of Two Closed Recycling-Loops on the Mechanical and Material Properties of Expanded Polypropylene Beads

Sören Handtke[1(✉)], Jörg Hain[1], Fabian Fischer[1], Tim Ossowski[2], Sven Hartwig[3], and Klaus Dröder[2]

[1] Volkswagen AG, Berliner Ring 2, 38440 Wolfsburg, Germany
soeren.handtke1@volkswagen.de

[2] Institute of Machine Tools and Production Technology, Technische Universität Braunschweig, Langer Kamp 19B, 38106 Braunschweig, Germany
{t.ossowski,k.droeder}@tu-braunschweig.de

[3] Institute of Joining and Welding, Technische Universität Braunschweig, Langer Kamp 8, 38106 Braunschweig, Germany
s.hartwig@tu-braunschweig.de

Abstract. Car manufacturers are currently facing the challenge of making their products and production more sustainable in order to comply with sustainability requirements and legislation. For example, the latest draft legislation in the European Union requires a general use of 25% recycled material for plastic parts, of which 25% should be post-consumer recyclate (PCR). One way to make products more sustainable is to reduce the carbon footprint of fossil-based plastic parts. This can be achieved by using lightweight parts with minimum material input and therefore low density on the one hand, and recycled materials on the other. Particle foam parts have the potential to meet both requirements. Expanded polypropylene (EPP) parts are currently used in vehicles for their lightweight potential, insulating and energy absorbing properties.

To achieve sustainable particle foam parts made from EPP, the recycled content must be increased while meeting material and mechanical property requirements. For this purpose, three simulated life cycles with two recycling loops were carried out on virgin EPP. The virgin material was processed to specimens in a steamfree process and then artificially aged to replicate automotive life cycle stresses. Then, the mechanical properties were evaluated, mechanical recycling was carried out and an EPP particle foam with 100% recycled content was produced and processed. Meanwhile, the material properties, such as thermal and chemical properties, were determined and analysed comparatively.

After each recycling cycle the influences of different aging mechanisms and processing steps on the properties could be demonstrated. The molecular chain structure shows a clear dependence on the number of recycling steps, resulting in a more inhomogeneous distribution. Compression properties show a decrease of 13,7% after the first recycling step and 39,25% after the second recycling step.

Overall, the recyclability of EPP material has been demonstrated, potentially allowing the use of higher levels of recycled content.

© The Author(s) 2025
K. Dröder and T. Vietor (Eds.): CD 2024, Proceedings, pp. 179–193, 2025.
https://doi.org/10.1007/978-3-658-45889-8_14

Keywords: Circular Economy · Lightweight plastic parts · Material properties · Bead foams · Expanded polypropylene

1   Introduction

Global plastics production in 2022 reached 400,3 million tonnes (Mt), while European production in the same year was 58,7 Mt. Of this, 80,3% were fossil-based, 13,1% from post-consumer recyclates (PCR) and 1% bio-based. Polypropylene (PP) accounts for 15,4% of European plastics production [1]. The production and processing of PCR has been increasing continuously for years. Against the backdrop of increased public awareness of sustainability issues and the possibility of stricter legal requirements in the future, industrial companies are facing the challenge of using more PCR in their plastic products. For example, the European Union's draft legislation calls for a recycling rate of 25% for plastic products in cars. In addition 25% of this proportion must come from PCR.

The processing of plastics into automotive parts accounts for about 8% of all plastics processing in the European Union. Plastic parts make up around 12–15% of the average mass of a car, of which PP makes up the largest proportion at 26% [1]. Despite their small mass share, lightweight plastics are increasingly being used in cars due to their reduced density compared to compact materials. The reasons for this are, on the one hand, the goal of a lower vehicle mass and thus potentially lower fuel/energy consumption and, on the other hand, minimised material use and thus a lower CO_2 footprint.

One form of lightweight plastics are so-called particle foams. These are used in a wide range of applications in the automotive exterior and interior, e.g. in the area of energy absorption as bumper cores, as acoustic and thermal insulation applications as luggage coverings, but also as containers in production and logistics. Typical foam part densities in automobiles are in the range of 15–350 g/l, which means that a high lightweight construction potential can be realised [2, 3]. In terms of mechanical properties, particle foam parts are characterised above all by good compression properties [4].

The processing of expanded particle foam beads is currently carried out in a well-established process using a steam-based technology. The steam-based process is characterised by a hollow chamber tool design and the use of steam for temperature control of the mould and for welding the particle foam beads. The steam pressure curve exhibits an exponential behavior, according to the temperature-dependent gas pressure of water so that a steam pressure of approx. 3,6 bar is required for temperatures of approx. 140 °C. Using water vapour as an energy carrier medium, the temperature required for welding the particle foam beads is transported into the moulded part [5–8]. Current research is focussing on steam-reduced or steam-free processing of particle foam beads. In the field of steam-free processing technologies, radiation-based concepts using dielectric heating of plastics by applying electromagnetic waves, such as microwaves or radio frequency, are being investigated [9–12]. Another possibility is variothermal mould temperature control, in which a bead bond is generated by thermal conduction. Such a mould technology is also used in this work (see Sect. 3.2).

Current issues, particularly in the automotive industry, involve finding solutions for the sustainable use of polypropylene-based particle foam parts. Approaches to this include the reprocessing and recycling of EPP waste in order to realise a closed material cycle in the sense of a circular economy. Using the automotive industry as an example, this means fulfilling the potential legal requirements mentioned above and at the same time meeting the requirements for the mechanical properties of the components also with recycled materials. For the automotive industry, the challenge is to ensure the properties of recycled plastic components over this period of time, given the average life expectancy of an automobile of approx. 10 to 15 years. In the case of post-consumer recyclates, this means a period of 30 to 45 years with double or even triple recycling. However, the challenges that arise during the reprocessing of EPP to EPP from recyclates have not yet been sufficiently considered. Difficulties lie, among other things, in the identification of the reclaimed material, i.e. the material that is to be reprocessed. This may, for example, result from different manufacturers, contain various additives (flame retardants, colour pigments) but can also contain external substances or add-on parts made from injection-moulded PP.

2   Theory

Most studies on the influence of aging on material and specimen properties relate to the analysis of the influence of aging on injection-moulded polypropylene specimens [13, 14]. The effects of recycling foams or explicitly particle foams made of polypropylene and their moulded parts are increasingly coming into focus. Publications on the aging behaviour of thermoplastic foams are mainly only available for polyurethanes (PU) [15]. With regard to the aging behaviour of polypropylene foams, the patent by Delabroye et al. is an exception [16]. An extrusion foam was developed that resists aging for up to 48 h. A recent study was published by Weingart et. al. Comparing the thermo-oxidative aging of expanded polycarbonate (EPC) specimens with that of EPP specimens. The aging was carried out in a convection oven at various temperatures and time steps and mechanical parameters were then analysed. The pressure characteristics of EPP decrease after aging, while the bending properties increase [15].

3   Experimental Process

The aim of this study was to simulate the aging and load influences to which a particle foam component is exposed during its life cycle in the automotive sector. The aim of the investigations is to analyse the effects of the load on the mechanical properties of specimens on the one hand and the influence of recycling and reprocessing on material properties on the other. For this purpose, simulated life cycles containing two recycling stages were carried out, consisting of the individual steps described in the following chapters. This allows for the continuous comparison of specimen and material properties and precise analysis of the effects of aging and recycling process, as shown in **Fig. 1**. The aging section of the virgin material specimens is divided into four series (see Sect. 3.3).

These four series were processed separately over the two recycling cycles according to the aging experienced in the previous cycle. As a result, each series always experienced specific aging over two recycling cycles.

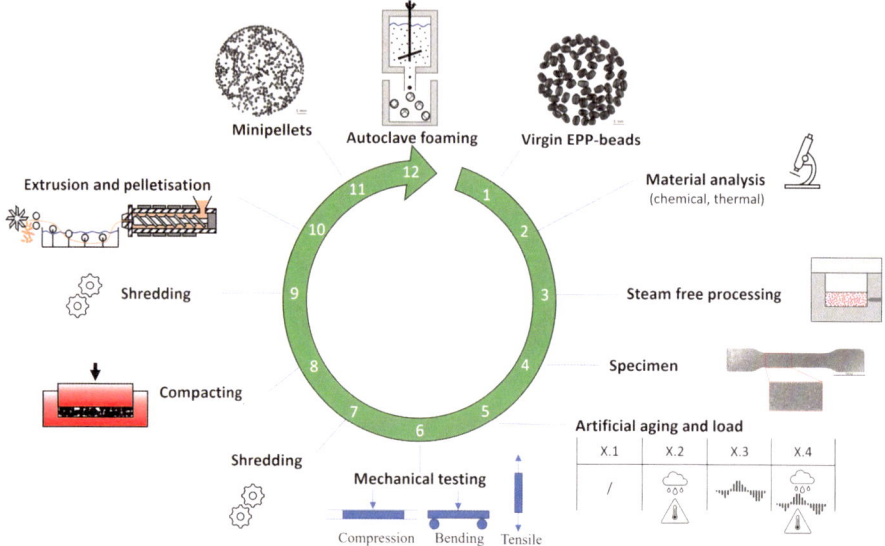

Fig. 1. Artificial life cycle with process steps and position of analyses

3.1   Material

For the processing of specimens, particle foam beads made of expanded polypropylene (EPP) with a bulk density of 33–37 g/l and dimensions of ca. 3–5 mm were used.

3.2   Steam-Free Variothermal Processing

In this study, the EPP material is processed into specimens using a steam-free variothermal tool. The tool consists of an upper and a lower mould. The design of the tool with an immersion edge allows for the processing of already expanded, non-propellant beads by the gap-filling method. Compressed air ejectors for residue-free demoulding of the part and temperature sensors are integrated in the two mould halves. The temperature control unit consists of two temperature circuits to maintain different temperatures. The channels are located close to the cavity to ensure that the heating and cooling is as close to the contours as possible and is defined and uniform. The processing temperature was in the range of 140–142,5 °C. The dimensions of the processed specimen in the form of a tension rod is shown in **Fig. 2**. The compression specimens were manually cut from the centre section.

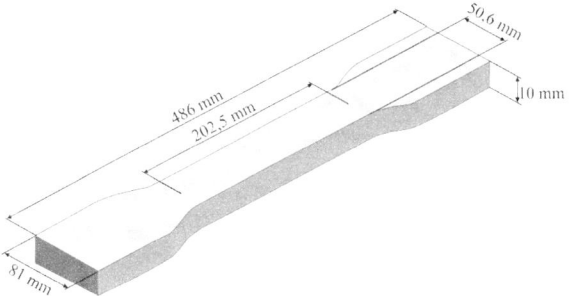

Fig. 2. Dimensions of processed EPP-specimen

3.3 Artificial Aging

The artificial aging includes changing ambient temperatures and humidity as well as vibrations induced by the vehicle's driving profile. The specimens produced are divided into four series in order to be able to differentiate between the influence of various aging effects, see **Table 1**. The first series is not subjected to any artificial aging. The second series undergoes artificial aging in the form of climatic cycling. The climatic cycling load consists of a two-part, twelve-hour cycle, which requires temperatures of both 80 °C and −40 °C. Humidity was controlled at 80% for the first six hours. It remained uncontrolled for the second six hours of the cycle because of negative temperatures. A high frequency shaker test was carried out for the third series to simulate the effect of vibration through the vehicle's driving profile. In this test, the specimens were placed on a shaker for 36 h at frequencies between 5 and 200 Hz. The fourth series undergoes a combined aging process consisting of a preceding climatic cycling and a subsequent high frequency load, including temperature cycling. The abbreviations 0R (virgin material and no recycling), 1R (particle foam beads after one recycling stage), 2R (particle foam beads after two recycling stages) stand for the recycling stage of the particle foam beads.

Table 1. Overview of processed series and different aging

		Series 1	Series 2	Series 3	Series 4
Load / Aging type	Climate changing	/	✔	/	✔
	High frequency	/	/	✔	✔
	Abbreviation	VS 1.1_0R	VS 1.2_0R	VS 1.3_0R	VS 1.4_0R
		VS 1.1_1R	VS 1.2_1R	VS 1.3_1R	VS 1.4_1R
		VS 1.1_2R	VS 1.2_2R	VS 1.3_2R	VS 1.4_2R

3.4   Mechanical Properties

The four series presented above (see **Table 1**) are tested using tensile and compression tests. The aim of the mechanical characterization is to determine the influence of the simulated aging by comparing the mechanical properties of the series. Since there is no defined standard for the testing of particle foam specimens, they are tested on the basis of existing standards (Tensile properties: DIN EN ISO 527–1; Compression properties: ISO 844). The mechanical characterization is carried out on eight specimens in a standard atmosphere (23 °C), shortly after aging (see Sect. 3.3) and without separate pre-treating of the specimens.

3.5   Mechanical Recycling

In the next step the specimens were recycled to obtain minipellets, which were required for the subsequent foaming process. The entire recycling process is based on a real, production-related plastics recycling process. For this reason the material was not dried beforehand. The specimens were first shredded with a mill using a 3 mm sieve, compacted and shredded again under the same conditions. They were then extruded using a twin-screw extruder at a die temperature of 190 °C. The melt strand had a diameter of 3 mm at the nozzle and was drawn through a 3 m long water bath for cooling. A pelletizer drew in the melt strand and reduced it to a diameter of about 1 mm before cooling. The rotating cutter granulated the melt strand into mini-pellets with a diameter and length of about 1 mm. The four differently aged series are processed separately, resulting in four minipellet batches each cycle.

3.6   Autoclave Foaming

After mechanical recycling, the particle foam beads were produced again from the recycled PP minipellets using an autoclave batch foaming process. The minipellets were fed into a stirring autoclave filled with water and surfactants and saturated with a propellant gas (CO_2) at 145 °C and 75 bar for 30 min. The expansion of the minipellets into recycled particle foam beads was then initiated through a pressure and temperature drop by opening the valve. The four batches were foamed separately, resulting in four batches of recycled beads.

3.7   Material Properties

The three material grades (virgin beads, single-recycled beads, double-recycled beads) were analytically characterised in order to investigate the influence of aging, recycling and foaming on the material. For this purpose, we conducted Size Exclusion Chromatography (SEC) measurements to analyze the chemical structure of the chain molecules. Additionally, we performed Differential Scanning Calorimetry (DSC) and Thermo-Mechanical Analysis (TMA) measurements to examine the thermal properties. Three measurements per series were carried out for each analysis method.

High temperature SEC measurements were performed on three beads. The material was dissolved in trichlorobenzene at a concentration of 1 g/L at 160 °C for 2 h. Polystyrene standards were then used for calibration and an infrared detector (IR6) was used to determine the molecular weight distribution, the number average (M_n) and the weight average (M_w). The polydispersity index (PDI) was determined according to formula (1).

$$PDI = M_w / M_n \tag{1}$$

DSC measurements were carried out to investigate the melting and crystallization behaviour of the samples using a DSC 204 F1 Phoenix from NETZSCH. The first and second heating curves and the intermediate cooling curve were recorded in the temperature range of $-20 -200$ °C. The heating/cooling rate was 10 K/min and nitrogen was used as the purge gas.

TMA measurements were carried out on individual beads to investigate the thermal stability of the virgin and recycled particle foam beads. A TMA 402 F1 Hyperion from NETZSCH was used for this purpose. A single bead was placed under the glass plunger and loaded with a measuring force of 0,01 N. First, the temperature was held isothermally at 10 °C for 15 min to obtain a stable initial length. Then the temperature was heated to 160 °C at a heating rate of 5 K/min.

After determining the material properties of the recycled particle foam beads, they were processed again into specimens using the method described in Sect. 3.2.

4   Results

First, the results of the mechanical characterizations of the EPP specimens are shown according to the mechanical test method. Then, the influences of the aging and recycling on the material properties are pointed out according to the chemical and thermal test methods.

4.1   Mechanical Properties of EPP Specimens

The Tensile properties of the processed EPP specimens are shown in **Fig. 3**. The maximum tensile stress is pictured in boxes. There are clear differences in the properties of specimens made from virgin EPP beads (unfilled boxes) and recycled beads (shaded boxes). In the case of the differently aged specimens made from virgin material beads, the completely aged specimens (VS1.4) exhibit characteristic values at a similar level to the unaged specimens (VS1.1). In contrast the climate-change-aged specimens (VS1.2) and high frequency-excited specimens (VS1.3) show a decrease in the mean values of approx. 18% and 16,5% respectively.

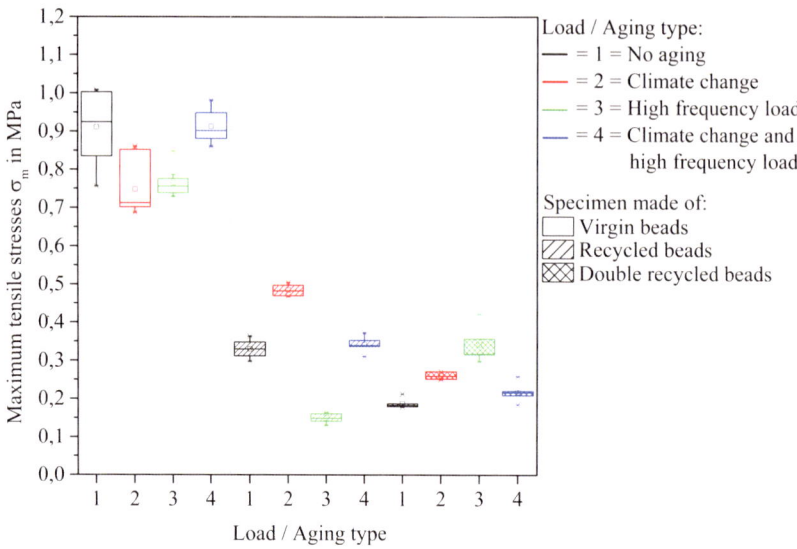

Fig. 3. Comparison of tensile stresses of virgin and recycled EPP specimens

The properties of the single and double recycled specimens are at a significant lower level. The properties of the completely aged specimens (VS 1.4_1R and VS1.4_2R) are also close to those of the unaged specimens (VS 1.1_1R and VS 1.1_2R). The properties of the climate-change-aged specimens (VS1.2_1R and VS1.2_2R) and high frequency-excited specimens (VS1.3_1R and VS1.3_2R) show ambiguous behaviour. In the case of the single-recycled specimens, the properties of specimens exposed to climate change are significantly higher than those of the high frequency-excited specimens (224%) and also higher than those of the unaged specimens (47%) and the completely aged specimens (42%). In the case of the double-aged specimens, the separately aged specimens have higher properties than both the unaged and the completely aged specimens. The lower tensile strengths of the single and double recycled test specimens are probably due to the reduced material properties. The TMA analyses in Sect. 4.2 have shown that the thermal stability of the individual particle foam beads at processing temperatures has been significantly reduced by recycling. Due to greater shrinkage than virgin beads, it may not be possible to ensure sufficient contact between the adjacent bead surfaces, which is necessary for creating a cohesive bead bond and force transmission.

The Compression properties of the processed EPP specimens are shown in **Fig. 4**. The compression stress at a deformation of 10% is pictured in boxes. There are also clear differences in the properties of specimens made from virgin material and recycled EPP particle foam beads. A drop in the properties of all aged series compared to the unaged specimen series can be seen for both the differently aged specimens made from virgin material beads and those made from recycled beads. It can also be seen in all cycles that the specimens, exposed to climate change, have the lowest properties, while the specimens, exposed to high frequencies, have higher properties than the completely aged series or are approximately at the same level (VS1.3<->VS1.4). Looking at the completely aged series in comparison to the unaged series, it is noticeable that there is only a decrease in the mean values of approx. 6% for the specimens made from virgin material beads, while the properties of the specimens from the first recycling cycle already show a decrease in the mean values of 16%. In the second recycling cycle, the completely aged specimens show approx. 13% lower properties.

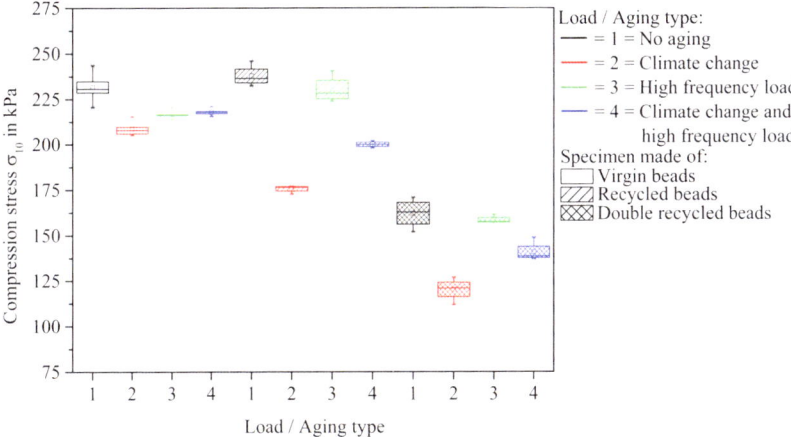

Fig. 4. Comparison of compression stresses of virgin and recycled EPP specimens

In summary a decrease in the compressive stresses of single and double recycled specimens can be determined. It is remarkable that a significant decrease in the properties of all aging series could only be determined for the specimens from the second recycling cycle. Possible reasons for this may be found in the material properties of the double-foamed particle foam beads, which could be confirmed by the material analyses carried out in the following chapter. The main reason for this could be the embrittlement of the material. It is possible that the material degradation may have reached a critical level as a result of the second recycling cycle, which has a negative effect on the mechanical properties of the specimens.

The reason why the properties of the completely aged series are often at a similar level to the unaged series could lie in the artificial ageing of the VS 1.4 series. In these series, high frequency aging was carried out after climatic cycling. A temperature change profile was used in the same way as for climatic cycling. The constant temperature of 80 °C for four hours may have led to partial evaporation of the moisture on the surface and in the spaces between the specimens caused by the preceding climatic cycling. This can reduce or even prevent a potentially damaging effect of moisture on the properties, as in case with the VS 1.2 series. Furthermore, the temperature influence of thermo-oxidative aging may have led to crystallite rearrangements and post-crystallisation at the bead interfaces, which can counteract a decrease in the properties.

One influence of the individual steps of the recycling cycle on the mechanical properties could be the production of mini-pellets of non-uniform size. All other steps were
carried out in exactly the same way as described above. A possible non-uniformity in the
size of the minipellets can also lead to a non-uniform size of the expanded beads. As a
result the particle foam beads made from recycled material may have different properties
during processing into specimens, which can result in reduced mechanical properties.

In summary, a decrease in properties can be observed for both of single and double recycled specimens. However no prediction can be made regarding the levels of
the properties for the test specimens that have undergone climate change or high frequency ageing. Possible reasons for this could be the individual processing steps that
take place in the recycling process. The minipellets from different batches can have
different dimensions, which may lead to different levels of sorption with the blowing
agent CO_2 in the high pressure autoclave. The subsequent expansion process can also
result in different bead properties due to the mentioned deviations. These can affect, for
example, the density, the surface quality of the bead, the internal cell structure (cell size,
type of distribution, cell wall thickness, cell web thickness). In addition, the thermal
properties of the individual particle foam beads in particular can have a strong influence
on the bead bond during processing.

4.2   Material Properties of EPP Material

The Chemical properties of the three material grades (virgin beads, beads recycled once,
beads recycled twice) were characterised using SEC analysis and thermal properties
were determined using DSC and TMA analysis.

SEC curves are presented in **Fig. 5a**) and enlarged in **Fig. 5b**). For the purpose of
clarity, only representative curves of the unaged and completely aged beads of the first
and second recycling cycle are shown in comparison to the virgin beads. The mean
values of the number averages (M_n) and weight averages (M_w) are also shown. It can
be seen that as the aging and recycling cycles increase, the number averages decrease.
The weight of the beads remains relatively constant during the first recycling cycle, but
decreases during the second cycle compared to the virgin material. **Table 2** presents the
polydispersity index values for the virgin material, as well as the aged and recycled EPP
particle foam beads. It can be clearly seen that the average value of the index is higher
for the completely aged recycled beads than for the unaged recycled beads. A higher
PDI index indicates a broader molecular weight distribution and higher non-uniformity
of the chain molecules, which can be caused by aging-related degradation phenomena.
Another possible explanation for the altered molecular weight distribution of the recycled
beads may be the high mechanical stress on the chain segments during expansion in the
autoclave process.

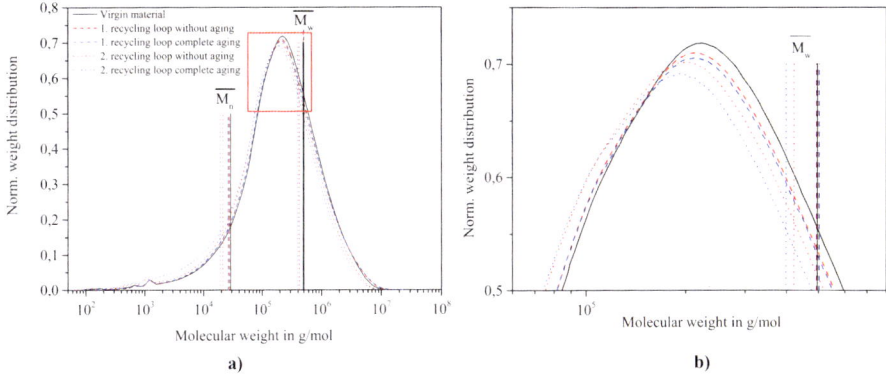

a)   b)

Fig. 5. SEC curves of virgin and recycled EPP Beads

Table 2. Figures of polydispersity index (PDI) of aged and recycled EPP Beads

Aging / Load type	Virgin material /	1. Recycling loop		2. Recycling loop	
		Without aging	Complete aging	Without aging	Complete aging
#1	17,71	18,09	19,87	22,59	21,13
#2	17,76	18,95	20,46	17,94	21,47
#3	16,61	17,79	17,83	20,04	20,48
Average	17,36	18,28	19,39	20,19	21,03

DSC curves of the first heating are shown in **Fig. 6** with the middled resulting peak temperatures. Only the curves of the unaged and completely aged beads of the first and second recycling cycle are shown in comparison to the virgin beads. Particle foam beads produced using the autoclave process often exhibit a double melt peak with two corresponding peak temperatures $T_{m,low}$ and $T_{m,high}$. It can be seen that it was possible to produce recycled beads with a double melting peak. However, the characteristics of the two melting ranges differ from those of the virgin material. For example, the higher melting peak $T_{m,high}$ in the recycled beads is shifted to lower temperatures. It is also significantly steeper, which indicates the presence of crystallite structures of the alpha-2 fraction in a different form than in the virgin material. The appearance of the double melt peaks is generally highly dependent on the manufacturing process in the high pressure autoclave.

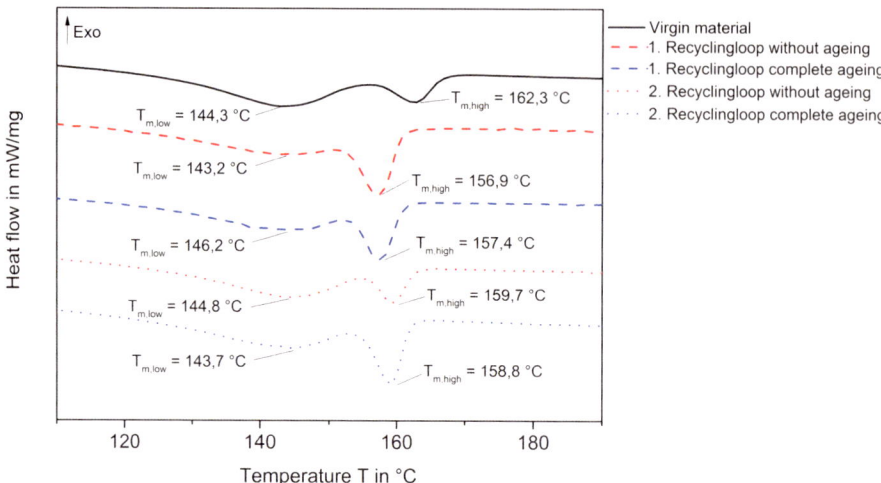

Fig. 6. Comparison of DSC curves of virgin, single-recycled and double-recycled EPP beads

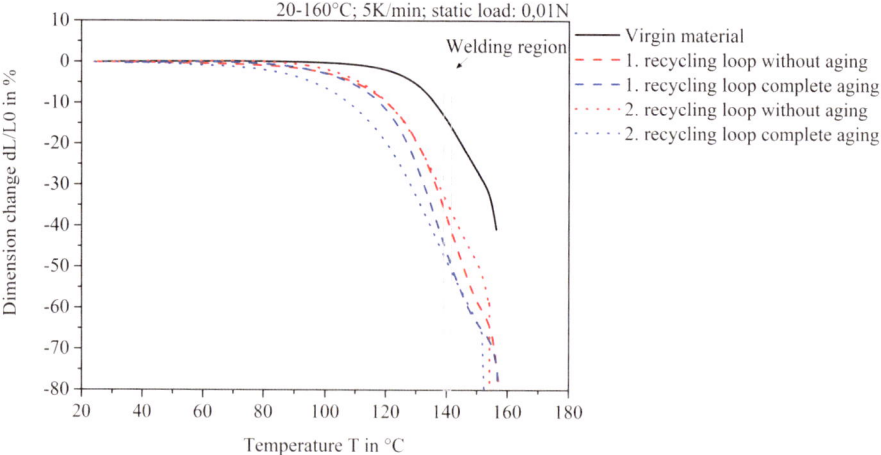

Fig. 7. TMA curves of virgin and recycled EPP beads

TMA curves are shown in **Fig. 7**. The curves of virgin and recycled beads exhibit qualitative similarities, however, their temperature-dependent shrinkage properties differ significantly. The shrinkage process of the virgin beads initiates at temperatures of approximately 120 °C, whereas the recycled beads exhibit a negative change in length even at lower temperatures. **Table 3** demonstrates that the single recycled beads without aging experience a shrinkage of 35,6%, while the beads with complete aging show a shrinkage of 45,1%, within the range of processing temperatures. In contrast, the virgin material beads only show a shrinkage of 13,1%. The double-recycled beads without aging exhibit a similar level of shrinkage to the single-recycled beads. The double-recycled and completely aged beads demonstrated an earlier onset of shrinkage than the single-recycled beads. Nevertheless, they ultimately achieved a similar change in length as the single-recycled beads.

The greater shrinkage of recycled beads compared to virgin beads in the processing temperature range has a significant impact on the production of EPP parts. Due to the shrinkage of the beads, the outer skins of the beads have a smaller contact area with each other, which can make it difficult or prevent the formation of a cohesive bead bond. This phenomenon could also be an indication of the significantly lower mechanical properties of the recycled specimens, see Sect. 4.1. Possible reasons for the lower temperature stability of the recycled beads may be due to the degradation of the polymer material itself or to the structure of the (inner) cell structure (thickness of the outer skin, thickness of the cell webs, distribution of the cells).

Table 3. Dimension change of virgin, single- and double-recycled beads

Series	dl/L0 at beginning of welding region in %	dL/L0 at end of welding region in %
Virgin material	−13,1	−17,2
1. Recycling loop without aging	−35,6	−43,2
1. Recycling loop complete aging	−45,1	−52,7
2. Recycling loop without aging	−33,6	−39,0
2. Recycling loop complete aging	−48,2	−54,2

5 Conclusions

In this study three simulated life cycles with two closed recycling-loops were carried out on polypropylene particle foam specimens. In each cycle, these test specimens were subjected to four load and aging profiles: no aging, climate change aging only, high frequency load only, climate change aging and high frequency load. In this way, the influences of automotive loads from the driving profile and environment on the mechanical properties of the particle foam specimens and, on the other hand, the effects of aging, loads and material recycling on the material properties were continuously examined. The tensile characteristics decreased through the first and second recycling. The compression properties only decreased after the second recycling. The material properties were strongly influenced by material recycling. The higher melting peak of $T_{m,high} = 162,3\,°C$ for virgin beads has reduced to $T_{m,high} = 158,8\,°C$ for double-recycled beads with complete aging. The recycled beads also showed a significantly earlier shrinkage than virgin beads. At a processing temperature of approx. 140 °C, virgin beads show a shrinkage of approx. 13%. Single-recycled beads from previously completely aged specimens show 45,1% and double-recycled beads with the same aging and load show a shrinkage of 48,2%. Aging and recycling cycles also have a strong impact on the molecular chemical properties. The recycled beads have significantly more inhomogeneous chain distributions, which is confirmed by an increase in the PDIs from 17,4 for virgin beads to 21 for double-recycled beads, the specimens of which were previously completely aged.

The aging and recycling cycles carried out have shown that the material properties are reduced by degradation mechanisms. In addition to the artificial aging and loading of the specimens, these influence the material properties, for example through reduced dimensional stability at processing temperature. As the degradation process of polypropylene and its products is well understood, the application of additives can prevent or reduce the degradation of material properties. On the one hand, further research should investigate the effects of these additives on the dimensional stability of the particle foam beads after a recycling process. In addition, adjusting the autoclave processing parameters can significantly affect the internal cell structure. It is therefore essential to investigate their effect on the thermal stability of a particle foam bead.

References

1. Plastics Europe, Automotive—The world moves with plastics.
2. M. Avalle, G. Belingardi, R. Montanini, Characterization of polymeric structural foams under compressive impact loading by means of energy-absorption diagram, International journal of impact engineering (25) (2001) 455–472.
3. P. Rumianek, T. Dobosz, Nowak R., P. Dziewit, A. Aromiński, Static Mechanical Properties of Expanded Polypropylene Crushable Foam.
4. N.J. Mills, Polymer foams handbook: Engineering and biomechanics applications and design guide, 1st ed., Elsevier; Butterworth-Heinemann, Amsterdam, Heidelberg, Oxford, 2007.
5. D. Raps, N. Hossieny, C.B. Park, V. Altstädt, Past and present developments in polymer bead foams and bead foaming technology, Polymer 56 (5) (2015) 5–19. https://doi.org/10.1016/j.polymer.2014.10.078.
6. V. Srivastava, Srivastava R., A review on manufacturing, properties and application of expanded polypropylen, MIT International Journal of Mechanical Engineering 4 (1) (2014) 22–28.
7. W. Zhai, Y.-W. Kim, C.B. Park, Steam-Chest Molding of Expanded Polypropylene Foams. 1. DSC Simulation of Bead Foam Processing, Ind. Eng. Chem. Res. 49 (20) (2010) 9822–9829. https://doi.org/10.1021/ie101085s.
8. D. Dörr, J. Kuhnigk, V. Altstädt, Bead Foams, in: S.T. Lee (Ed.), Polymeric Foams: Innovations in technologies and environmentally friendly materials, CRC Press, Boca Raton, Florida, 2022, pp. 51–90.
9. F. Prissok, M. Harms, M. Schuette (BASF SE) WO2016146537A1, 2016.
10. M. Zubair, R. Ferrari, O. Alagha, N.D. Mu'azu, N.I. Blaisi, I.S. Ateeq, M.S. Manzar, Microwave Foaming of Materials: An Emerging Field, Polymers (Basel) 12 (11) (2020). https://doi.org/10.3390/polym12112477.
11. K. Schneider, B. Gothe, M. Drexler, J. Siltamaeki, H. Weiger, A. Seefried, D. Drummer, The effect of dielectric and thermal properties of plastic mold materials on the high frequency welding of three-dimensional foam components, Polymer Engineering & Sci 62 (10) (2022) 3400–3411. https://doi.org/10.1002/pen.26112.
12. V. Romanov (Kurtz GmbH) WO2017125412A1, 2017.
13. I. Schwarz, M. Stranz, M. Bonnet, J. Petermann, Changes of mechanical properties in cold-crystallized syndiotactic polypropylene during aging, Colloid and Polymer Science 279 (5) (2001) 506–512. https://doi.org/10.1007/s003960100488.
14. M. Rjeb, A. Labzour, A. Rjeb, S. Sayouri, Y. Claire, A. Périchaud, TG and DSC studies of natural and artificial aging of polypropylene, Physica A: Statistical Mechanics and its Applications 358 (1) (2005) 212–217. https://doi.org/10.1016/j.physa.2005.06.023.

15. N. Weingart, D. Raps, M. Lamka, M. Demleitner, V. Altstädt, H. Ruckdäschel, Influence of thermo-oxidative aging on the mechanical properties of the bead foams made of polycarbonate and polypropylene, Journal of Polymer Science 61 (21) (2023) 2742–2757. https://doi.org/10.1002/pol.20230267.
16. C. Delabroye, L.T. Nguyen, J.-F. Koenig, M. Heekmann, C.P. Park, A.M. Chatterjee US 2005/0004285 A1.

Circularity and Integration of Functions for Metal Surfaces by Plating on Plastics in Automotive Series Applications

Felix A. Heinzler[✉]

BIA Group, Lotharstr. 6, 42655 Solingen, Germany
felix.heinzler@bia-group.com
https://www.bia-group.com/

Abstract. High-quality and sustainable surfaces are an important step in design-ing the automotive interior and exterior of the future. Next generation car designs have to consider the aspects of sustainable production and long-life application in the field in combination with recycling concept after the use case. Clever material combinations and new processing technologies for circular products have to be established. Electroplated polymers have several advantages for every step in this discussion to enable sustainable decor elements with integrated functions.

Using a process developed by the BIA Group and industrial partners, metal and plastic in electroplated decorative components can be separated with a very high degree of purity in the polymer fraction. Without any chemical treatment a degree of 99,8% purity in the polymer fraction is achieved and can directly be used for new products. By focusing plated polymer parts, the substrate material is limited to ABS plating grades and ensures a very low probability of contamination by other polymers. The metallic fraction is recycled by the well know processes for metal and steel recovery.

First products with a recycled content of 30% are now in series production. Because of the high purity of the material, a down-cycling is not necessary and these products are high quality surface parts in the automotive interior. The quality matches the known standard of parts with virgin material. The recycling content has no influence on the adhesion of the coating or the resistance to environmental influences. Specification requirements like DBL 1665, TL 528 or GS standards are passed.

With the integration of ambient lighting and touch controls on real metal surfaces, these components are equipped for functional integration into the environment in the interior.

Keywords: Circularity · Electroplating · Recycled plating grades · Touch with real chrome surfaces

1 Introduction

The core function of surface technology is to improve product properties and extend the service life of components. This applies both to purely functional parts, which have to be resistant to corrosive media or transfer loads in a wear-resistant manner, as well

© The Author(s) 2025
K. Dröder and T. Vietor (Eds.): CD 2024, Proceedings, pp. 194–205, 2025.
https://doi.org/10.1007/978-3-658-45889-8_15

as to decorative elements that should maintain a high surface quality in the handle and functional areas. Efficiency in the processes has also played a crucial role, as circular materials and specialty chemicals determine the costs of the processes. Accordingly, the currently focused topics of a sustainable economy can be addressed very well in surface technology.

The future of production lies in the focus on efficient processes and circular economy. The efficient use of raw materials is always an important factor, but is now supplemented by reuse and processing for direct availability and reduced carbon footprint. It is important to close local cycles in processes and products. The BIA Group has been very active in these areas in recent years and shows ways in which electroplated plastics offer decisive advantages and closed cycles can be implemented in the automotive industry.

2   Advantages of Electroplated Polymer Parts

The advantages of plated plastic components in the automotive sector for decorative applications are diverse. Thanks to the free shaping in injection molding and the finishing with real metals, an ideal combination of lightweight construction, functionality and value is achieved. The processes developed at BIA for ambient lighting or the integration of touch controls on the real metal components offer further advantages for future applications. In addition to these well-known topics, plated plastics also offer advantages when it comes to efficiency in processes, GWP (Global Warming Potential) and recycling. This is illustrated by a comparison of alternative processes like painting and PVD, with data collected by the BMUB as state of the art in [1, 2].

The core index for comparing coating systems is the energy used per coated square meter in [kWh/m^2] in addition to the square meters per hour as capacity. In [1] the current state of the painting industry is presented and is subsequently used as conventional painting for comparisons. Table 1 summarizes the comparison of the energy-based processes. The basis for evaluation of both technologies is a coating system for automotive application and chrome optics.

Table 1. Comparison of different coating technologies regarding energy consumption and scrape rate.

		conv. Painting	conv. Plating
Electric energy	[kWh/m^2]	34–36,5	ca. 19
Thermal energy	[kWh/m^2]	18–20,2	ca. 14
Scrape rate	%	ca. 25–35% (BIA intern 15–25%)	< 10%

Next to the energy consumption, scrap in production is a factor that has a significant impact on all necessary resources. Every percentage of reduced scrap can be shown directly on the balance sheet as an improvement. Statistics according to [1] show that an average scrape rate of 25% can be expected when painting high-quality surfaces. This is significantly higher compared to average plating and would lead to a further reduction in resources of around 10%. In recent years, the focus has naturally also been on efficiency and energy in the area of painting. A state-of-the-art painting system installed exemplary at BIA in Solingen can reduce thermal and electrical energy use by around 35% each. New painting technologies with single-layer systems and good appearance and properties further reduce the impact on production. So, both technologies for coating plastics parts can be produced efficiently using current technology.

New projects in electroplating also show further optimization potential for reducing the necessary energy. Using new anode technology, optimizations in the electrolyte formulation and optimizations in the line periphery can save up to 35% of the coating energy in a specific bath. A corresponding BMU flagship project (NKa3–003543) is currently being implemented by the BIA Group with a new electroplating line in Solingen. In addition to chrome-free pre-treatment and chrome(VI)-free coating, various topics on energy efficiency are also being implemented in large series within the project [3].

In a direct comparison, plated plastics offer corresponding advantages. The higher efficiency in production has been demonstrated. In addition, there is the high-quality surface created by real metal coating. Compared to painted surfaces, this is an emotional surface that is considered to be of higher quality in terms of its appearance and higher in its durability in the field. Long-lasting surfaces are important when considering the product life cycle and lead to a longer service life and higher value of the vehicle.

3   Commodity Recycling Grades for Plating Application

Plating companies as well as material distributers have looked intensively into the use of recycled materials for plated plastics. A first step is the use of materials in injection molding that already contain a proportion of recycled content from the manufacturer. However, during subsequent coating, it should be noted that standard acrylonitrile butadiene styrene (ABS) recycled materials lead to problems in the coating process and the adhesion of the metal layer to the substrate due to the uncontrolled butadiene distribution [4, 5]. A correspondingly targeted development and evaluation is necessary. Together with the company BARLOG Plastics GmbH as a material developer and distributer, a PC/ABS was tested that contains a post-consumer recycled content in the polycarbonate (PC) component. From a plating perspective, this is not relevant for the coating and leads to very good results in the component appearance and the final adhesion tests. An alternative is a PC/ABS from Covestro AG, in which the polycarbonate portion is made from bio-based phenols. The bio-based component also reduces the CO_2 input of the material, which is shown in a corresponding balance sheet at over 30% [6]. Through these substrate materials, a positive influence on the product carbon footprint (PCF) of the component can be introduced and quantified directly during product development.

If multi-component components are considered in the next step, polycarbonate from bio-based phenol sources or materials with recycled content can also be used here. The selective component has no influence on the plating ability or adhesion of the coating. Thanks to the selective properties in the plating, light guides and mechanical holding structures can be integrated directly during injection molding. They do not have to be masked separately in the coating step. This was tested positively in a near-production test on a housing with electroplating components (Fig. 1) and a high proportion of selective surfaces. Qualification tests such as warm storage or climate change tests also showed no abnormalities and have been passed. Using these two approaches, reductions in the product carbon footprint of electroplated parts can be considered in the component design.

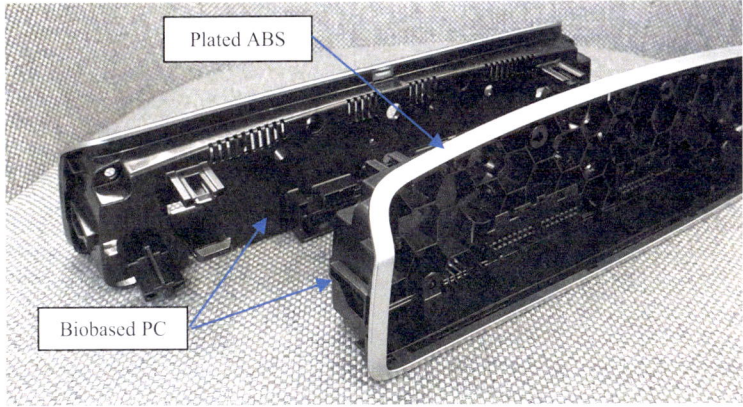

Fig. 1. Multicomponent part with ABS for plating and polycarbonate from bio-based sources

4   Recycling of Plated Polymer Parts

The strength of electroplated plastic components lies in their excellent recyclability. Due to the metal content on the components, they have a higher material value than other coatings (paint, PVD). With other coatings, the components cannot be separated from the substrate with sufficient purity. The remaining lacquer particles cannot be separated from the polymer. For PVD coatings the metallic content is to small ($< 5\mu$m for PVD, $> 25\ \mu$m for plating) and in addition the coating system of PVD it combined with protective paint layers, which leads to insufficient purities of the recycled polymer. This usually leads to down-cycling of the recycled materials, if they can be reused at all. Examples of the use of painted plastic components were presented at the K trade fair 2022 in Düsseldorf with processing into pallets or laundry baskets.

Since metals can be reprocessed almost infinitely, the proportion of recycled copper in industry in Europe is already 70–80%. The recycling process requires up to 85% less energy than primary production. Accordingly, the metal components of electroplated plastics are already being recovered and fed back into the various processes. The plastic has so far been lost. [7].

Together with partners the BIA Group has developed a process that enables the processing of both recyclable materials. Plastic and metal can be separated in advance using a specialized, mechanical process and separated due to the magnetic nickel components in the electroplating layer. The metal is made available to the industry again and the plastic is cleaned and granulated as recycled plastic (Fig. 2).

Fig. 2. Left: Plated applications in the interior; right: separated polymer and metal fraction after the recycling process

The degree of purity of the separated fraction at the end of the process chain is over 99.5%. The recycled content can be mixed with virgin material without any problem, as both fractions are of so-called "plating grade" polymers. These are ABS recipes specially tailored to electroplating, which ensure secure adhesion of the metal and very good surface quality. Since the recycled content also uses these substrates from the source "electroplated components", a high degree of purity and quality of the raw materials is automatically guaranteed. Downcycling, i.e. the use of recycled plastics for lower-quality products or even thermal recycling, is therefore not necessary for plated plastics. The surface quality was validated on a test specimen tool. The test specimen tool is design for injection molding as a cross geometrie and is used for polymer substrate material qualification for different coating technologies for Mercedes Benz AG (Fig. 3).

Fig. 3. Sample specimen for substrate material and coating technologies

Figure 4 shows the raw part with 50% recycling content and remaining impurities after the separation process on the left side. Optimizations to reduce the impurities are ongoing and first promising results are seen on the right part of Fig. 4.

In the first step, this can be implemented with post-industrial recycled material for the waste generated during the coating process. However, current analyzes also show that recycling of post-consumer components can be implemented if the parts are returned after the lifetime of the vehicle.

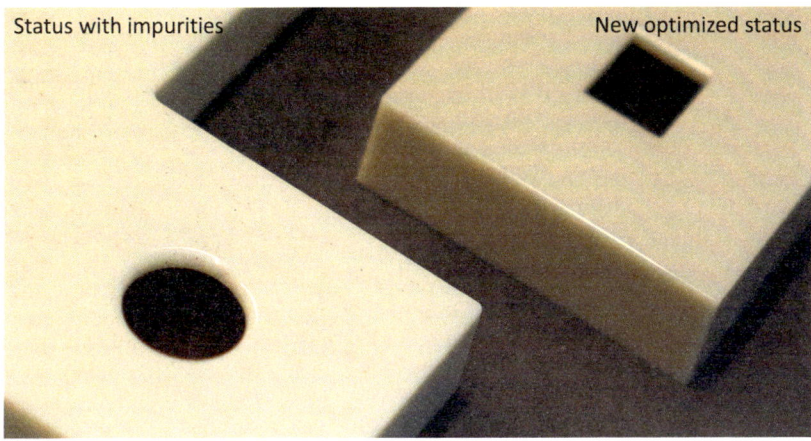

Fig. 4. Test parts with 50% recycled material content and remaining impurities. (Left: Status with impurities; Right: Reduced impurities as actual status)

In addition to the optical quality, the specifications in the area of adhesion between the substrate material and electroplating must of course also be evaluated. A detailed view of the surface of the raw part of the test sample is shown in Fig. 5 and provides further information. Figure 5 shows that the impurities are residues of the metal fraction that remain in the recyclate and then reach the component surface during injection molding. A SEM picture of the etched polymer surface as shown on the right side of Fig. 5 clarifies two influences. On the one hand, the etched structure is absolutely comparable to virgin material and therefore enables a very good adhesion. On the other hand, the remaining impurities have been analyzed by an EDX mapping and which confirms the metallic layer system as remaining particles. Theses impurities can lead to surface defects in the metallic coating and optical defects that lead to scrap.

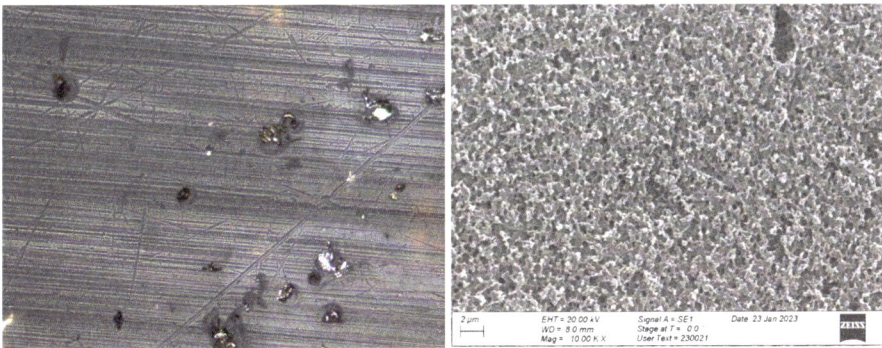

Fig. 5. Surface of a raw part with impurities and etch surface for good adhesion of the coating (SEM)

5 Series Parts with Recycling Content

In addition to the development of recycling processes, the qualification of the components and assessment of the quality are also crucial. The general material properties and the high degree of purity must prove itself in the production of series components. In an exterior strip, a door panel and a control element in the interior, various recycled materials were evaluated for their behaviour in injection molding and electroplating. Before the discussion of the general optical quality after the coating and of course the production scrape rate, the specification requirements regarding adhesion and clime change stability are shown.

Different components have been tested according to various manufacturer specifications. The results for the components based on electroplated types are consistently positive and the tests were passed for each component. However, problems also arise here when ABS recyclates are used without taking the electroplating requirements into account. The materials with plating grades as recycling content (exPlating)provide good adhesion and test results, standard commercial re-ABS leads to cracks and bubbles on the parts. This behavior was verified again on the cross test specimen for remap re-ABS with 50% recycling content and commercial standard re-ABS. Table 2 summarizes the results of different climate change tests and temperature shock tests.

Table 2. Test results of parts with different recycling ABS types according to automotive standards

Test	exPlating re-ABS REC50	Commercial re-ABS REC50
DBL 1665 Shocktest 110 °C/3 Cycles	10/10 ok	0/10 ok (cracks)
TL528 PV1200 – 8 Cycles	10/10 ok	0/10 ok (bubbles)
DBL 1665 AKLV 10 Cycles Exterior	10/10 ok	0/10 ok (bubbles)
DBL 1665 AKLV 42 Cycles Interior	10/10 ok	–

The cracks and bubbles mainly occur in geometrically demanding areas that are heavily stressed by the temperature changes. On the one hand, the adhesive base for commercial re-ABS types is not sufficient, and on the other hand, the strong movement during temperature changes leads to tension in the coating with corresponding cracks when overloaded. An example is shown in Fig. 6.

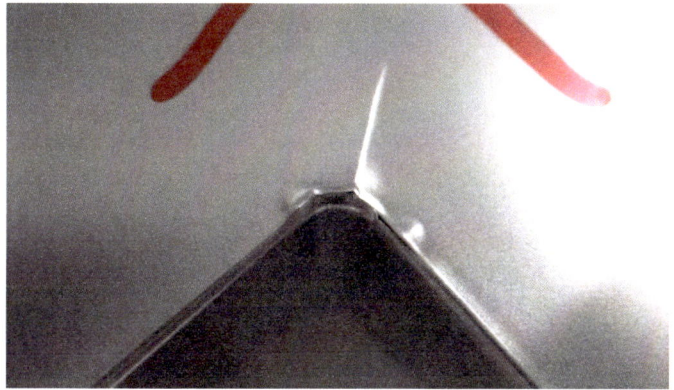

Fig. 6. Cracks in the coating after climate change test (DBL1665 AKLV 42 Cycles)

The mechanical properties have been validated on test specimens and correspond to the requirements at new product level [8]. A high level of varietal purity is always a prerequisite for a successful recycling approach. The already mentioned use of plating grade in ABS as a substrate plays a crucial role, which generates an advantage in the circular economy. The property of the specified grade purity affects not only the plating component itself, but the entire assembly. The simpler an assembly is constructed, the easier it is to dismantle it into pure materials. Since electroplated components often have a function in the vehicle in addition to their pure value, the assemblies also have a correspondingly complex composition.

Last important step is the optical quality and analysis of the scrape rate in series production. The recyclate was mixed accordingly with the standard materials like Novordur P2MC, Xantar C CP200 and Bayblend T45 PG. The following Fig. 7 shows an excerpt from the test series and the corresponding evaluation for optical errors after electroplating. Materials from the Remap GmbH are created with recyclate from electroplating scrap and represent the material types specified in the BIA Group for electroplating. The re-ABS types are other commercial ABS types (not plating grades). As a reference part, a door frame with satin chrome finish based on a trivalent chrome process has been chosen. The part is shown in Fig. 8.

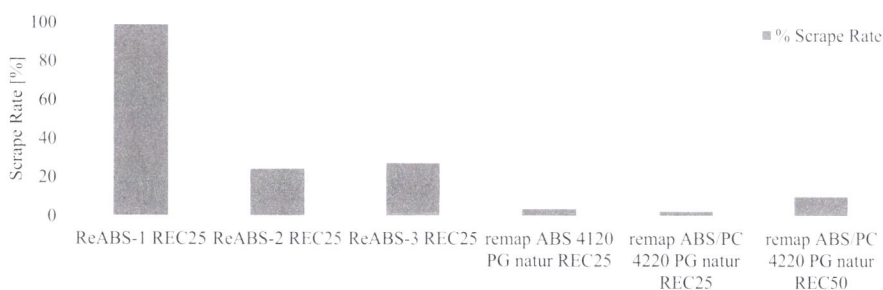

Fig. 7. Scrape rate for different Re-ABS types and content in plating

The commercial ABS recyclates, which are not designed for electroplating applications, immediately lead to an increased scrape rate. The proportion of pores on the components and streaks increases. These materials cannot be used for high-quality applications. However, the controlled materials from the Remap GmbH are very easy to integrate back into the process. With a recycled content of 25% on a mixture basis of both ABS and ABS/PC, the reject rate is comparable to that of a series material and is also less than 5%. A slight inaccuracy arises within the pattern due to rinsing processes and material changes in the injection molding, which resulted in understandable pickles and streaks. At this point, prototype productions with 30% recycling content have proven the low scrape rate under series conditions. Next to the purity of the material, the surface finish in the plating is mandatory. On a mirror like bright chrome finish, small pickles and variations more likely to be visible. The state of the art and design satin chrome finished have physical structures on the surface on their own and can cover small imperfections. Figure 8 shows the door panel with the material remap ABS/PC 4220 REC50, with 50% recycling ABS content from former electroplated parts and no visible imperfections on the surface.

Fig. 8. Door panel with 50% recycling ABS content and BIA TouchChrome controls

6   Sustainable Real Metal Surfaces with High-Quality Touch Controls

A typical example of complex assemblies are vehicle controls. A classic switch has a single stored function. The advantage of a switch is the driving safety aspect. Because a single switch only controls one function, it can be felt directly. The typical mechanics provide immediate feedback when the switch is operated. Backlit switches, which also have a tactile symbol, can be operated safely without having to pay too much attention to them. This functionality is associated with mechanical effort, which in turn creates corresponding costs in the assembly but also a corresponding value in the feedback. New approaches tend to substitute these values by touch screens and software solutions to integrate all applications within one command window. This is reducing the costs for components but has a very negative impact on the safety, intuitive controlling and general value appearance of the interior. [9].

An alternative are capacitive switching areas, which are not arranged in software on screens, but rather as sensors in a defined position behind a control bar. The goal of reducing the number of components remains the same, but thanks to the fixed position and, if necessary, sensing aids, operation can be carried out intuitively while driving. The sensor technology is integrated into a film, back-injected or attached behind the decorative panel. However, this sensor principle is shielded on real metal surfaces and is not yet suitable for use on high-quality decorative components.

BIA solved this problem without losing focus on the goal of sustainable construction and innovative functional integration. A high-quality, touch-sensitive surface has been developed that is also easy to recycle. This includes reducing the number of necessary components and the component complexity by substituting the mechanics and, finally, considering the material mix in the assembly.

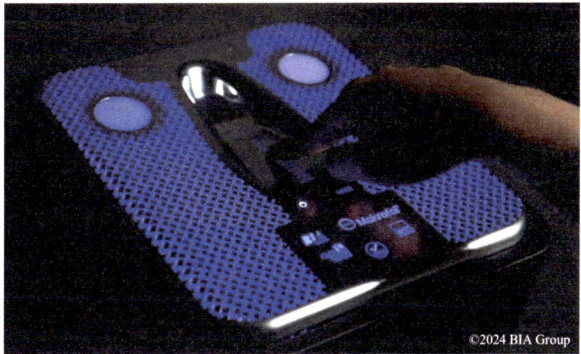

Fig. 9. BIA TouchChrome Prototype with ambient light, seamless design and touch functions

Although the continuous electrical conductivity of the metal layers of a chrome-plated component prevents the use of typical capacitive technology, a touch-sensitive surface can still be achieved using DFA sensors that are attached to the underside of the component [10]. The patented deep field analysis (DFA) technology enables a direct detection of finger movements through the plating layer systems with a very clear accuracy. A separation of signals for "selecting" and "activating" the function is possible as well as slider functions.

The sensor system is mounted on the back after the surface has been plated or is directly integrated on the controller PCB [11]. This decouples the manufacturing process for the electroplating part from that of the sensor system. Since no individual, mechanically decoupled components occupy the functions, an efficient assembly design with few elements but many functions can be achieved. The proportion of injection molds required is significantly reduced. At the same time, there are advantages through the integration of several functional areas on one surface in the area of seamless design. The BIA TouchChrome technology can be combined very well with other technologies such as BIA AmbientLight and BIA TextureChrome, which means that the button can be supplemented with a translucent, structured symbol.

Figures 8 and 9 show respective applications of the technologies presented. Figure 9 combines the integration of a touch control with a plated real metal surface and lighting switched using the example of a headliner unit. Figure 8 shows an application in the door area in which the component is made from 50% recycled content (remap 4220 REC50) for the electroplating component. Touch functions are also included here for exemplary control of the ambient lighting. To demonstrate the flexibility of the technology, the touch fields were only indicated by laser marking on the chrome surface. This allows various functions to be flexibly displayed using symbols and integrated into the interface.

7 Conclusions

Chrome-plated components combine the advantages of real metal surfaces with the properties of plastics. With an average mass proportion of 20% metal and 80% plastic, a chrome-plated plastic component is rather lightweight with a low energy requirement while driving. The corrosion and media resistance as well as the easy-to-clean properties underline the longevity of chrome surfaces in various areas of application. Even if a long period of use is crucial for a sustainability consideration, the production and "end-of-life" phases should not be missing from the complete assessment of a component.

The manufacturing process using injection molding and electroplating is already superior to a painting process in terms of resource conservation and energy efficiency. The internal material cycle closure in the manufacturing process and the recovery of raw materials from production waste also make a decisive contribution to saving raw materials. Using this technology can make an important contribution to circularity in automotive components with high-quality surfaces. The use of real, valuable metal surfaces plays an important role in the application in terms of value and service life, but also in the circular economy through the equivalent processing and preservation of the valuable materials for production. The recycled material can be used for standard electroplating processes as well as new chrome(VI)-free pretreatments and trivalent surface finished.

The goal of creating controls with an innovative, high-quality and at the same time sustainable design can now be achieved using BIA TouchChrome technology. The use of capacitive pressure sensors makes it possible to equip a high-quality chrome-plated surface with a touch-sensitive switching function. Thanks to tactile and translucent symbols, the buttons can be easily recognized or felt, which offers enormous advantages in terms of driving safety. Recyclability is improved by component simplification because the touch component group consists exclusively of the chrome-plated component and the touch electronics. As a high-quality, durable and fully recyclable technology, the galvanization of plastics in vehicle construction is an important contribution to future vehicle generations.

References

1. Heinzler, F. A.; Hochwertige Oberflächen im Fokus der Klimabilanz, WOMAG 3, 2022
2. Abschlussbericht BMUB-Umweltinnovationsprogramm—Innovatives und umweltfreundliches Lacksystem für Automobil-Chrom-Optik; 2016
3. BMU Umweltinnovationsprogramm; Effiziente, chrom(VI)-freie Galvanisierung von Kunststoffen für die Automobilindustrie; https://www.bmuv.de/pressemitteilung/bundesumweltministerium-foerdert-umweltfreundliche-galvanisierung-von-kunststoffteilen, 2021
4. Heinzler, F. A.; Eine Studie zu Beschichtungstechnologien in Bezug auf Qualität, Designoptionen und Ressourceneffizienz, ZVO Oberflächentage 2021, Berlin

5. Häp, M.; Welitschko, W.; Heinzler, F. A.; Touchsensitive verchromte Kunststoffelemente; JOT Special Galvanotechnik 2022, 2022

6. Bruckel, Frank; Carbon footprint data for COV specific grades, LCA data for different versions of Bayblend®T45 PG, Covestro AG, 2023

7. Auskunfts- und Beratungsstelle für die Verwendung von Kupfer und Kupferlegierungen; Ressourcenschonung dank Recycling, www.kupfer.de; 2024

8. Zapf, D.; Recycling von metallisierten Kunststoffen durch hochenergetische Impulsbehandlung; ZVO Oberflächentage 2022, Leipzig

9. Martin Seeger: Die Berührungsdauer eines Button-Klicks bei stationären Touchscreens, Mensch und Computer, Stuttgart: Oldenbourg Wissenschaftsverlag, S. 243–252, 2015

10. Ling Zhu, Texas Instruments: "Capacitive Touch Through Metal Using MSP430™ MCUs With CapTIvate™ Technology", 2017

11. Keith Curtis, Microchip Technology Inc.: "AN1325, mTouch™ Metal Over Cap Technology", 2010

Impacts of Circular Product Life Cycles on the Paintability of Plastics

Katharina Tonn[1]([envelope]), Max Juraschek[2], and Christoph Herrmann[2,3]

[1] Volkswagen AG, Berliner Ring 2, 38440 Wolfsburg, Germany
katharina.tonn@volkswagen.de

[2] Institute of Machine Tools and Production Technology, Chair of Sustainable Manufacturing and Life Cycle Engineering, Technische Universität Braunschweig, Langer Kamp 19B, 38106 Braunschweig, Germany

[3] Fraunhofer Institute of Surface Engineering and Thin Films, Riedenkamp 2, 38108 Braunschweig, Germany

Abstract. Due to their flexible and advantageous set of properties, plastics are used widely and in large quantities in the automotive industry. The use of recycled plastics is a promising approach to conserve fossil resources, reduce negative environmental impacts and achieve sustainability goals.

However, the quality of recycled material can vary depending on the individual and often unknown conditions it was exposed to in previous life cycle(s) (stages) and prevent the direct substitution of virgin material in many applications. Each life cycle stage of a plastic component or product, as for instance manufacturing, use or disposal, influences up to impairs the structure and properties of the potential recyclates by causing degradation and introducing impurities. In addition to quality fluctuations performance losses can arise and complicate the application of recycled plastics, for example by deteriorating the paintability of components with recycled content.

To reduce defects or reworking efforts, additional process adjustments can be required for using plastic recyclates in order to react to process deviations and varying material conditions.

Against this background, an overview of the various impacts and interdependencies of the life cycle stages and processing steps on painted injection-molded components and their material properties is presented and resulting challenges are derived. Subsequently, an approach is introduced, presenting a method to address these challenges. The approach takes the conditions and requirements of the automotive industry, as well as the implications of mechanical recycling into consideration, to establish circular material flows.

Keywords: Plastics · Recycling · Product properties · Paintability · Product life cycle · Automotive

1 Introduction

Plastics are characterized by a flexible portfolio of properties. The properties are on the one side influenced by the structure and composition of the material. On the other hand, process parameters and external influences along the product life cycle can lead

K. Dröder and T. Vietor (Eds.): CD 2024, Proceedings, pp. 206–214, 2025.
https://doi.org/10.1007/978-3-658-45889-8_16

to changes in properties. [1] The versatility of plastics enables broad applications across diverse industries, as evidenced by the continuous increase in worldwide plastic production to approximately 400 million tons over the past 70 years [2]. Thermoplastic plastics constitute the majority, owing to features such as reversible malleability that affords design, processing, and reusability freedoms [3, 4]. However, analogous to the demand for plastics, the quantity of plastic waste also exhibits a comparable trend, with a substantial portion being subjected to thermal processing or ending up in the environment e.g. in the world's oceans [5–7].

One promising approach to mitigate waste and negative environmental impacts, conserve fossil resources, and simultaneously meet demands is to close material cycles and establish a circular economy. Material cycles can be closed through the recycling of plastic waste, transforming linear product life cycles (cradle to gate) into circular product life cycles (cradle to cradle) [8, 9]. Mechanical recycling is a suitable method for thermoplastic plastics, as recycled granulates can allow to produce more with higher efficiency with respect to energy, cost and resource than through chemical recycling, leveraging the reversible malleability of thermoplastics [4]. However, the quality of mechanically recycled materials is largely influenced by the history of input materials and conditions of previous life cycle stages, whereas chemically recycled plastics show no differences from virgin material [10–12]. Consequently, when closing material cycles using mechanical recycling, impacts on material quality increase, with stresses in the usage stage and processing conditions during recycling influencing the material beyond the typical stages of linear product life cycles [1, 13].

The multitude of influencing factors affects the quality of resulting recyclates, leading to variations and losses in properties as well as component defects, resulting in waste and rework [11, 14]. For instance, in the use of recyclates in painted applications, increased occurrences of delamination, yellowing, and blistering are observable. To proactively implement measures, information on the quality of recyclates is imperative.

Against this background, a methodology to characterize recyclates early in the production process regarding their paintability is presented, thereby reducing efforts related to waste and rework.

2   Theoretical Background

2.1   Plastic Parts Manufacturing

High-quality plastic components often have to meet not only functional requirements, but also fulfill optical and haptic demands, while being resistant to light and weather conditions. In order to align with these requirements, plastics can be coated, for example by painting. [4, 15].

The production of painted and injection-molded thermoplastic plastic components can be simplified into the four steps of synthesis, compounding, shaping, and coating of plastics before the utilization of the components. Synthesis refers to the generation of plastics through various polymerization methods that describe the formation mechanisms of plastics. All polymerization methods share the common feature that monomers are linked to form polymers by breaking double bonds. [4, 16].

Polymers are, in most cases, not processable and applicable after polymerization due to damage caused by heat, oxygen, media exposure, and radiation during processing and use, as well as not meeting the optical, mechanical, or processing-related requirements of applications [1, 4, 16]. To protect against damage and meet application-specific requirements, additives are added to the polymer during compounding. Examples include plasticizers, reinforcing agents, lubricants, colorants, antioxidants, or stabilizers. [4, 15, 16].

The result of the compounding is a plastic compound, which, during subsequent shaping in the injection molding process, exists as granules. Shaping in injection molding requires a flowable state, so the granules are melted in the plasticizing unit of the injection molding machine. The resulting melted plastic is then injected into the mold cavity under defined pressure. The plasticized material solidifies as it cools in the mold, with temperature-induced shrinkage compensated by injecting molten material. After complete solidification, the component is ejected from the open mold and, if necessary, undergoes post-processing. To protect the component from UV exposure and meet optical requirements, coating through painting follows the shaping process. [4, 15].

To ensure adequate adhesion between plastic and the coating system, the coating process begins with surface pretreatment, including cleaning and activating the component surface. Activation is particularly relevant for non-polar plastics, such as polypropylene, where functional groups are built up. These functional groups influence the wetting of the surface and the adhesion between paint and substrate by enabling interactions and bonds between paint and plastic substrate. Following pretreatment, the coating system is applied, with multiple layers of paint possible. [16, 17].

The production of painted and injection-molded plastic components concludes with the curing of the paint, opening the use stage [4, 16].

2.2 Mechanical Recycling

To close material cycles and establish a circular economy, the product life cycle must be expanded to include the recycling of waste. According to DIN EN ISO 14024, plastic waste is classified into two categories, namely 'Pre-Consumer Waste' and 'Post-Consumer Waste,' depending on the source. 'Pre-Consumer Waste' arises before use and is separated from the waste stream during the manufacturing process. In contrast, 'Post-Consumer Waste' is generated after use and can no longer be used for its intended purpose. Unlike Pre-Consumer Waste, the life cycle of Post-Consumer Waste includes the usage stage, resulting in differences in history and impact. [11, 18].

Despite the different conditions, the basic recycling of these waste types is similar. Mechanically recycled material is the outcome of a multi-step process, including pre-processing, shredding, sorting, cleaning and drying, compounding, and regranulation. Pre-processing involves collection and sorting steps before the actual recycling processes. The collection system varies significantly across countries and regions, challenging the establishment of homogeneous waste stream conditions. During sorting, various tools such as filters, airstreams, magnetic or sensor-based units separate the waste stream and remove foreign materials by detecting properties such as density or electrical conductivity. Density-based sorting is widely used but limited by overlaps in the density ranges of polymers due to additivation. Following sorting, the proper recycling process begins with shredding, followed by precise sorting steps. Shredding involves cutting the materials in various ways using cutting mills or a rotary cutter. The method employed affects the material differently, subjecting it to various stresses and inducing polymer chain changes. Subsequent washing and drying steps remove adhesives and contaminants using specific solvents. Limited knowledge about the nature of impurities may lead to inappropriate solvent choices, resulting in reduced washing process effectiveness. During compounding and regranulation, the melt is typically filtered and degassed to remove further impurities. Depending on the quality of the recyclate and the product's requirements, additives or co-polymers are added, the mixture is homogenized, and finally extruded to produce recycled granulate. The efficiency and resulting quality are influenced by every step of the recycling process. Different methods, inhomogeneous material streams, undefined impurities, and limited knowledge contribute to property reduction and fluctuation. [11, 14] To mitigate these uncertainties, transparent processes and states are necessary [11, 19].

3   Challenges in Establishing Circular Life Cycles

Circular material cycles, in addition to the mentioned advantages, pose technical challenges that hinder comprehensive establishment, especially in demanding applications. These challenges stem from the influence of the conditions of each life cycle stage on plastic properties. [1, 11] Based on existing literature, the influenceability throughout the life cycle is presented in the following and summarized in Fig. 1 focusing on significant characteristics and features that determine the paintability of plastics.

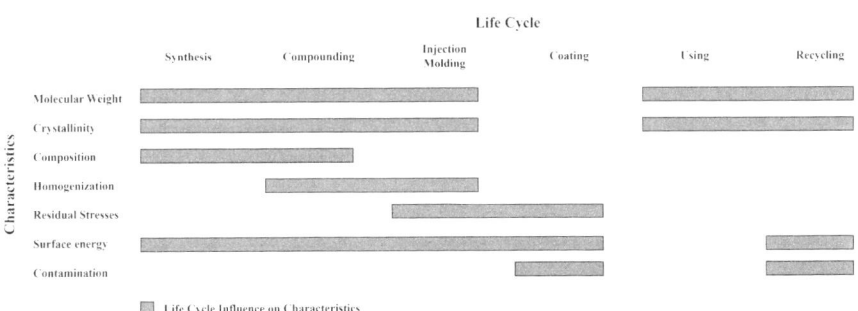

Fig. 1. Influence of the life cycle stages on significant paintability-determinig characteristics based on [1, 11–16, 19–25]

Within the synthesis step, the foundation for the molecular structure is laid, with polymer chains forming and arranging based on processing conditions. In addition to polymer architecture, synthesis determines the molecular weight or molecular weight distribution and the crystallinity of the plastic, influencing mechanical and rheological properties reflected in processability, strength, and stiffness as well as the surface energy as an adhesion-determining parameter. For instance, high-molecular-weight plastics exhibit high wear resistance and dimensional stability, while low-molecular-weight plastics show significant creep tendency and act as plasticizers. [16].

During compounding, additives such as stabilizers and flame retardants, and if necessary, fillers, are introduced to influence the resistance and various properties of plastics and determine the final composition. To homogenize the materials, the raw materials are melted, affecting polymer architecture and molecular structure again. Besides reorientation of polymer chains, thermal degradation reactions and damage from heat and shear forces may occur. Inadequate homogenization can result in the formation of agglomerates, causing surface irregularities and heterogeneous or weak properties. [1, 16].

In the subsequent shaping step in the injection molding process, these characteristics are also influenced, as the material undergoes melting and solidification. Furthermore, the stress state, such as residual stresses, is formed in the component through tempering. [1, 15, 20] The cooling rate further determines crystallinity, which not only affects mechanical properties but also influences surface adhesion [15, 21]. Surface quality is influenced by the tool's surface condition as well as injection molding parameters such as pressure and filling volume, affecting the occurrence of defects and resulting unevenness [15].

In the subsequent coating process, activation measures influence surface energies and, thus, wetting and adhesion properties. By applying inappropriate energy levels the activations can initiate surface degradation or promote migration of additives, resulting in surface deposition with weakening adhesion effects. Intermediate drying processes can cause similar migrations, with permeability determined by crystallinity. Heat input can also influence the stress state and relieve residual stresses. [15, 22, 23].

In the subsequent usage stage, mechanical stresses, temperature, UV radiation, moisture, and chemicals promote material aging and the degradation of polymer chains [1].

The degradation is intensified by subsequent mechanical recycling to close material cycles. Shredding, melting, and regranulating have similar effects on the structure as preceding melt-processing steps. Additionally, contaminations, such as paint particles, can be introduced, negatively impacting the properties of the plastic. [1, 11, 13, 24, 25].

Contaminations and aging effects can lead to performance losses and defects in circular materials. Defects such as yellowing, delamination, or blistering occur more frequently when using recyclates in painted applications. [12, 19] Furthermore, the different histories, processing conditions, and compositions of materials contribute to property variations, so constant qualities cannot be assumed [11–14]. To reduce efforts for rework and defects, early information on the qualities and conditions is necessary, and analysis methods to predict the performance and processing properties of recyclates need to be developed [12, 19].

4   Approach for Evaluating the Paintability of Plastics

The challenges in establishing circular material cycles and the use of plastic recyclates are e.g. performance losses and property variations of the recyclates, making their application in demanding contexts, such as painted component areas, difficult or even prevented. To reduce defects and rework and to be able to take early actions, information about the conditions and qualities of the recyclates is necessary. Figure 2 outlines an approach for the characterization and assessment of recyclates with regard to their paintability.

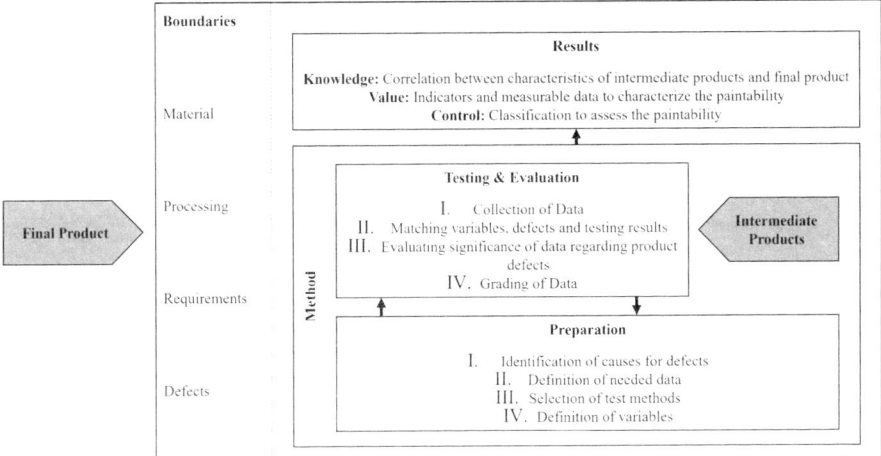

Fig. 2. Approach for evaluating the paintability of plastics

The boundaries for the evaluation arise from the requirements for the painted end product and the defects like delamination or yellowing that should be avoided through early characterization. Additional conditions emerge from the processing and material composition. Based on this, the characterization and data collection are carried out on the intermediate products like granulate and unpainted body to identify the influences of the intermediate steps and derive correlations with the final product properties.

Initially, the causations for the defects, such as surface energy deviations in case of delamination, need to be identified. From the identified causations, the data and metrics are derived that provide quantitative information about the condition. Subsequently, methods are identified and defined that allow the corresponding data collection on the intermediate products. To identify the impact of different intermediate product conditions, variables such as recyclate content and constants such as injection molding parameters need to be defined before testing the conditions and collecting the data. In the subsequent evaluation, the test results are compared to the defects and variables, classified, and relationships are identified to ultimately derive the suitability of the collected data for assessing paintability.

The arising results can be categorized as knowledge, value, and control. The correlations between the characteristics of the intermediate products and the properties of the end products create a quantitative understanding, for example, of the impacts of process steps and life cycle stages on the paintability of the final product. Deriving indicators and measurable data provides values for characterizing paintability and allows control by being assessable through classification, enabling efficient measures, balancing variations, and reducing failures and rework.

5   Conclusion and Outlook

The establishment of circular material cycles reduces waste, negative environmental impacts, and conserves fossil resources, but it also poses challenges that hinder its implementation and the use of recyclates. The reasons lie in the multitude of influencing factors that affect plastics across their life cycle stages, shaping their properties. Property variations and losses result, necessitating early information about material qualities to efficiently initiate measures and reduce rework and returns.

Building on the influences of life cycles on material characteristics, this paper presents an approach for evaluating the paintability of an injection-molded end product based on the characteristics and features of intermediate products.

The characterization of paintability on the intermediate products enables an early assessment of paintability within an early stage of the production process reducing rejects or reworking efforts caused by defects, like delamination and blistering. In addition to production, material and component development also benefit from the results, as new materials can be evaluated at an early stage regarding their suitability for painted applications, thereby reducing development cycles.

To validate the approach, data for selected plastics must be collected according to the described methodology, and the feasibility in mass production must be assessed. Further work should investigate the transferability to other end product properties, such as adhesiveness, to cover as many use cases as possible with a methodology and enable efficient and broad quality monitoring.

6   Disclaimer

The results, opinions and conclusions expressed in this publication are not necessarily those of Volkswagen Aktiengesellschaft.

References

1. Ehrenstein, G. W., Pongratz, S.: Resistance and Stability of Polymers. Hanser Publishers, Munich (2013).
2. Statista, https://de.statista.com/statistik/daten/studie/167099/umfrage/weltproduktion-von-kunststoff-seit-1950/, last accessed 2023/11/15.
3. OECD.Stat, https://stats.oecd.org/Index.aspx?DataSetCode=PLASTIC_USE_7, last accessed 2023/11/10.
4. Baur, E., Drummer, D., Osswald, T.A., Rudolph, N.: Saechtling Kunststoff-Handbuch, 32nd edn. Carl Hanser Verlag, Munich (2022).
5. Our World in Data, https://ourworldindata.org/explorers/plastic-pollution?tab=table&country=~OWID_WRL&Metric=Plastic+waste+generation&Sub-metric=++&Per+capita=false&Share+of+world+total=false&Source=OECD, last accessed 2023/11/10
6. OECD.Stat, https://stats.oecd.org/viewhtml.aspx?datasetcode=PLASTIC_WASTE_5&lang=en, last accessed 2023/11/10.
7. Statista, https://www.statista.com/statistics/1339124/global-plastic-waste-generation-by-application/, last accessed 2023/11/10.

8. Ellen Macarthur Foundation: TOWARDS THE CIRCULAR ECONOMY: Economic and business rationale for an accelerated transition (2013).
9. Kara, S., Hauschild, M., Sutherland, J., McAloone, T.: Closed-loop systems to circular economy: A pathway to environmental sustainability?. In: CIRP Annals,vol. 71, pp. 505–528 (2022).
10. Jepsen, D., Reihlen, A., Wirth, O., Sander, K.: Reach und Kunststoffrecycling: Handreichung für eine sachgerechte Umsetzung der Reachanforderungen für Betreiber von Recyclinganlagen, Umweltbundesamt (2011).
11. Niessner, N.: Recycling of plastics. Hanser Publishers, Munich (2022).
12. Plastverarbeiter.: https://www.plastverarbeiter.de/qualitaetssicherung/analytik-fuer-ein-bes seres-recycling.html, last accessed 2024/03/10.
13. Bunjes, A.: Systematic investigation into the molecular and material properties of impact-modified PP recyclates for automotive applications. In: VDI Berichte Nr. 2418e, p. 601, VDI Verlag GmbH, Düsseldorf (2023).
14. Eyerer, P., Schüle, H.: Polymer Engineering 3. 2nd edn. Springer Vieweg Berlin, Heidelberg (2020)
15. Eyerer, P., Schüle, H.: Polymer Engineering 2. 2nd edn. Springer Vieweg Berlin, Heidelberg (2020)
16. Eyerer, P., Schüle, H.: Polymer Engineering 1. 2nd edn. Springer Vieweg Berlin, Heidelberg (2020)
17. Ehrenstein, G. W.: Strukturverhalten. Carl Hanser Verlage, München (2021).
18. DIN EN ISO 14021: Umweltkennzeichnungen und -deklarationen - Umweltbezogene Anbietererklärungen. Beuth Verlag, Berlin (2021)
19. Berg, E., Dahlmann, R., Schön, M.: Verunreinigungen auf der Spur. Kunststoffe 11, 60–63 (2022).
20. Al Menen, B., Ekinci, A., Oksuz, M., Ates, M., Aydin, I.: Effect of processing parameters on the properties of two-component injection molded recycled polypropylene/ethylene propylene diene monomer automotive parts. The International Journal of Advanced Manufacturing Technology 1–2 (2023)
21. Rasche, M.: Handbuch Klebtechnik. Carl Hanser Verlag, München (2012).
22. Ernst, E., Reußner, J., Poelt, P., Ingolic, E.: Surface stability of polypropylene compounds and paint adhesion. Journal of Applied Polymer Science 97(3), 797–805 (2005)
23. Jacobasch, H.-J., Grundke, K., Schneider, St., Simon, F.: The influence of additives on the adhesion behaviour of thermoplastic materials used in the automotive industry. Progress in Organic Coatings 26(2–4), 131–143 (1995).
24. Yao, S., Taominaga, A. Fujikawa, Y., Sekiguchi, H., Takatori, E.: Inner Structure and Mechanical Properties of Recycled Polypropylene. Niheon Reoroji Gakkaishi 41(3), 173–178 (2013).
25. Gonzales-Gonzales, V.A., Neira-Velazquez, G., Angulo-Sanchez, J.L.: Polypropylene chain scissions and molecular weight changes in multiple extrusion. Polymer Degradation and Stability 60(1), 33–42 (1998)

Life-Cycle Engineering for Circular Economy

Sustainable Production with Circular Equipment - Opportunities and Challenges in Practice

Philipp Blanke[✉], Aileen Blondrath, Oliver Petrovic, Michael Riesener, and Werner Herfs

Laboratory for Machine Tools and Production Engineering (WZL), RWTH Aachen University, Campus-Boulevard 30, 52074 Aachen, Germany
{p.blanke,a.blondrath,o.petrovic,m.riesener,
w.herfs}@wzl.rwth-aachen.de
https://www.wzl.rwth-aachen.de

Abstract. The importance of sustainability and reducing CO_2 emissions to tackle climate change has long been recognized in the automotive industry. A key area of consideration is production itself. Until now, existing systems and equipment have often been put into storage or scrapped when products are changed, even though they could still be put to good use in other production lines. The potential lies in counteracting the waste of resources through circular resources, so that production itself becomes more sustainable. At the same time, production will become even more flexible, as it should be possible to book equipment as required and only pay for it during the utilization phase. This can become a competitive advantage for companies as it enables them to react to rapid product changes and fluctuating demand. The paper looks at the opportunities for companies to use circular equipment using the example of clamping fixtures but also highlights the current challenges that prevent their use in practice. In addition to identifying the potential of digitalization and the linking of individual life cycles, the complexity of a life cycle assessment of operating resources is shown on the one hand and what is possible with the result on the other. The new "Equipment as a Service" business model for renting equipment will also be highlighted and the associated issues of responsibility and billing will be discussed.

Keywords: Production equipment · Sustainable automotive production · Equipment as a service · Reconfiguration

1 Introduction

Managing climate change and, in this context, the importance of the circular economy is a major topic for the production industry[1]. The relevance of extending the scope of consideration to the entire value process has already been recognized. This means that production itself must also be included. One focus here can be the equipment used in production to manufacture products. Due to the constantly increasing customer requirements for innovations and the associated rapid introduction of new product variants,

© The Author(s) 2025
K. Dröder and T. Vietor (Eds.): CD 2024, Proceedings, pp. 217–228, 2025.
https://doi.org/10.1007/978-3-658-45889-8_17

these also have a major impact on the associated production facilities. These are currently disposed of after use, which offers great potential for optimization in terms of saving resources [2]. This also includes the operating materials used, some of which are individually adapted to the respective production plant. However, great potential is seen here for further use, but also for the transfer of this equipment and the corresponding use by another user through appropriate adaptation. In order to achieve a circular economy, this can be used as an approach at the equipment level.

The potential, but also the challenges, extend across the entire product life cycle [3]. In the end, it must be a benefit for the customer and must not lead to higher costs, downtimes, or quality problems. It is therefore important to solve the current problems along the value chain and create further added value. The challenges and potentials in the respective phases are discussed using the example of clamping fixtures. Based on the evaluation and disassembly of a component, it becomes clear which materials have which influence. Engineers can be given direct feedback to design more sustainably, as there is still a lack of knowledge and understanding of how a particular choice of material affects $CO2$, for example. Another important factor is the new business model (renting instead of buying) and the associated questions: Who is responsible for the operating resources, or how is the billing done?

2   Fundamentals and Methods

For a common understanding, this section contains the definition of circular production and the equipment used in production with a focus on the use case of the clamping fixture.

2.1   Circular Economy

Resource scarcity, stricter environmental regulations and changing consumer expectations are pushing companies to develop alternative concepts to traditional linear production [4]. Many research projects see circular economy concepts as promising solutions [5]. Activities on the road to a circular economy can be structured using the 9R framework, whereby reduce, reuse, recycle and remanufacture are seen as particularly important for sustainable product development [6, 7]. In addition, remanufacturing is a promising approach for recycling used products and extending the lifespan of a product. The resources already used, and the energy value added during the original production of the products are retained. In addition to resource efficiency, this also increases productivity and profitability [8].

According to Fontana et. al., the intensity of research into the circular economy concerning production resources is low [4]. The existing approaches focus on product design[9], condition monitoring[10] and the automated dismantling of production equipment [11].

2.2 Production Equipment

According to the DIN 6300 standard, production equipment includes all material resources required to produce a product. Production equipment is clearly distinguished from production materials, which do not influence the end product [12]. The categorisation of operating equipment according to DIN 6300 within production equipment is shown in Fig. 1. This categorisation provides a systematic overview of the various elements that act as production equipment in the production process.

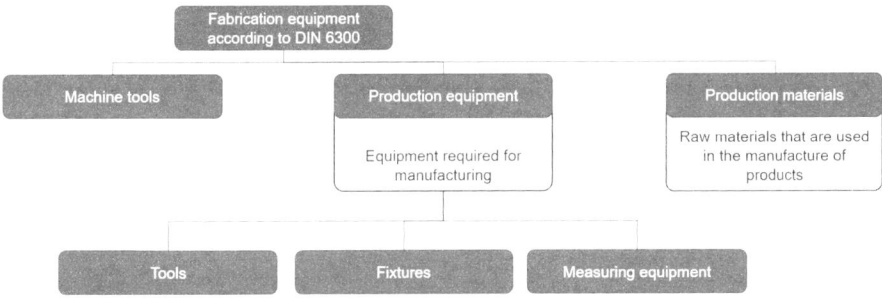

Fig. 1. Classification of equipment in the category of production equipment according to [13]

2.3 Fixtures

Alongside tools and measuring equipment, fixtures are an essential category of equipment in industrial manufacturing processes. These technical aids are used to hold a workpiece in a stable position, fixed alignment, or specific location. Their main function is not the direct execution of manufacturing processes, but rather to ensure precise positioning between different workpieces. Typically, fixtures are used primarily in multi-dimensional machining processes [14].

Various clamping principles can be used to realise fixtures. In the automotive industry, fixtures are primarily used in serial welding and assembly processes. They fulfil a crucial role in automated industrial production processes by ensuring that consistently functioning components can be manufactured. The use of fixtures in these processes helps to ensure the quality and reliability of the manufactured parts. Their use therefore not only enables efficient production but also standardised and repeatable production of assemblies and components.

3   Application

The opportunities and challenges in practice for sustainable production equipment are divided into five levels of consideration and are shown in Fig. 2. These five levels are further explained in the following section.

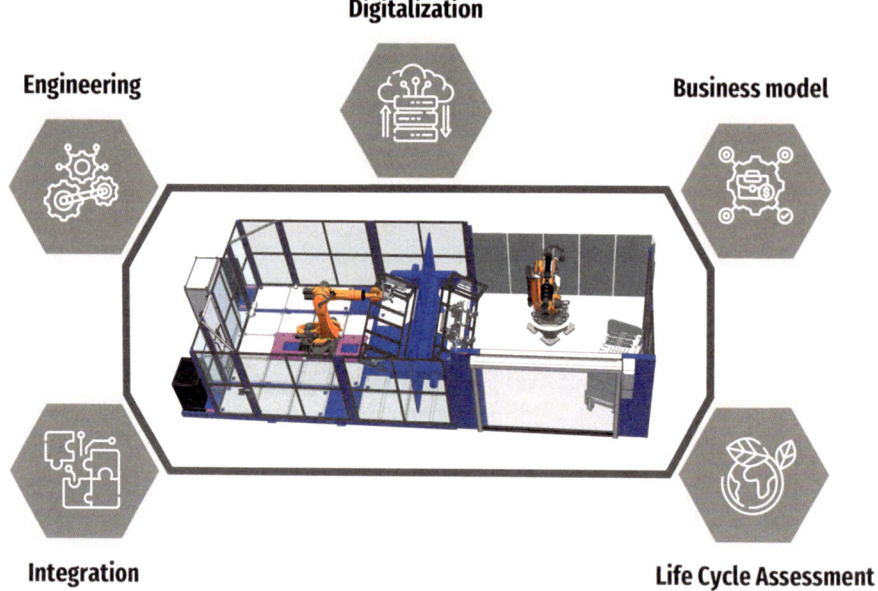

Fig. 2. Opportunities and challenges for circular equipment for the given use case

Based on a use case, a manufacturing cell, the levels of consideration relate to various aspects of this system. The production cell itself can be viewed as a closed system that has been specially developed for carrying out a joining process in the context of assembly production. Within this scenario, four different operating resources work together to successfully realise the manufacturing process.

The main players in this manufacturing cell are the three production equipment. A specially developed **gripper** on a handling robot and a **welding fixture** are two of these resources. The process begins with the provision of individual components, which are placed in a **load carrier**. These individual components are then precisely picked up by the handling robot and placed in the welding fixture. The specially designed gripper ensures precise handling and positioning of the components.

The actual joining process is carried out by a high-precision welding robot. This robot uses its welding tools to join the individual components together to create the complete assembly. The highest levels of precision and efficiency are considered to ensure high-quality production.

These levels of consideration make it possible to understand in detail the interaction of the operating resources in the specific contextualization of a use case production cell. The integration of handling and welding robots as well as the use of a load carrier and the provision of components by a shooter-like load carrier not only increases the efficiency of the process but also ensures precise and reproducible production. This structuring and automation of the production process helps to optimize the quality of the assemblies produced and increase the overall performance of the production cell.

The use case follows the primary approach of sustainable resources and a flexible production cell. The following chapters will therefore address the engineering of sustainable resources and a corresponding evaluation on the one hand and the business model behind a flexible production cell through a corresponding data-driven production network on the other.

3.1 Engineering

While the focus in the engineering of production equipment has so far been on implementing customer requirements as cost-efficiently as possible, other factors must be also taken into account for the development of circular equipment. Here, equipment manufacturers must weigh up how far the solution should be tailored to a specific application or whether a generally valid solution would be better in the context of a circular economy. The universal variants are often not aligned with company-specific guidelines, especially from the automotive industry. This also means that the use of operating resources across customers is made more difficult, although it is more and more requested by the industry.

Three key questions need to be clarified as part of the design process:

– Is a viable, modular structure of the production equipment possible and what does this look like?
– How can modifications be made as easily as possible?
– Which materials are used?

The first question focuses on how the equipment, which itself consists of individual components, can best be structurally designed so that it can be repaired and upgraded in line with the overarching goal of the circular economy, thus maximising the lifespan of the product. The second question addresses the point of repairability and upgradeability again and requires that the equipment can be easily assembled and disassembled. This also has a direct positive effect on user-friendliness and maintainability for the end customer. Easy accessibility and interchangeability of parts without specialised tools or specialist knowledge are of special interest here. The final question relates to the material. The material used must not only fulfil the technical requirements but must also be selected responsibly with regard to its environmental impact.

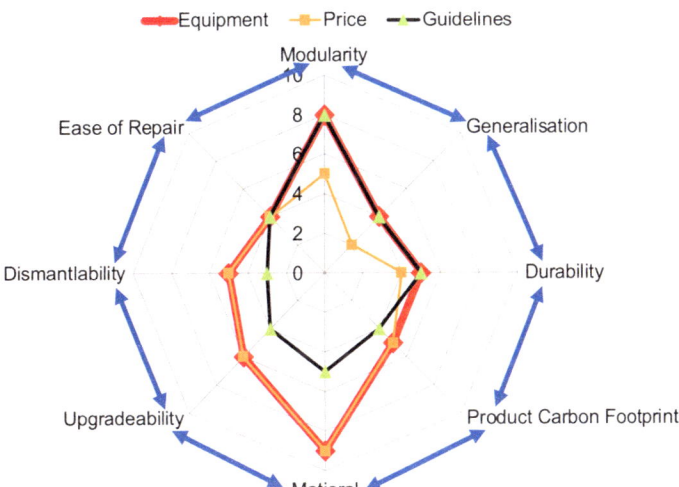

Fig. 3. Optimisation problem during the design of circular production equipment

Figure 3 illustrates the trade-off between eight factors during the design of circular production equipment. Due to dependencies, changes to one factor can have both positive and negative impacts on the others. In addition, each of these factors can be implemented to varying degrees, hence they are each assigned a value between zero and ten for quantification reasons. A ten means that this factor has been fully implemented. Further constraints, such as policy and price, often exist on top of this, resulting in an optimisation problem that needs to be solved. This leads to situations where the factors are constrained by either the price or the policy.

3.2   Integration

How (circular) production equipment can be integrated into existing and new production facilities is the subject of numerous research projects. To ensure that companies make use of the new approach, the efforts for the customer must be minimised. This raises the question of how intelligent an operating resource must be. Figure 4 Illustrates the two extremes. The left-hand side shows what is called simple equipment, which has no logic or control of its own. Every signal and every pneumatic connection must be integrated and parameterised in the existing system, in this case, the controller and valve terminal, which leads to increased integration costs. On the other hand, the costs for the operating resources are minimal. The opposite is shown on the right a smart equipment. It has its own controller, which controls the valves and subsequently the clamping device independently. The equipment only needs to be supplied with the required media (in this case: compressed air, electricity, and network) and then operates autonomously. This offers the advantage that the equipment manufacturer can guarantee the desired behaviour and is not dependent on any third party that it cannot control. However, it presupposes that the subordinate platform has the appropriate functionality to enable dynamic integration into a system and the cost are much higher.

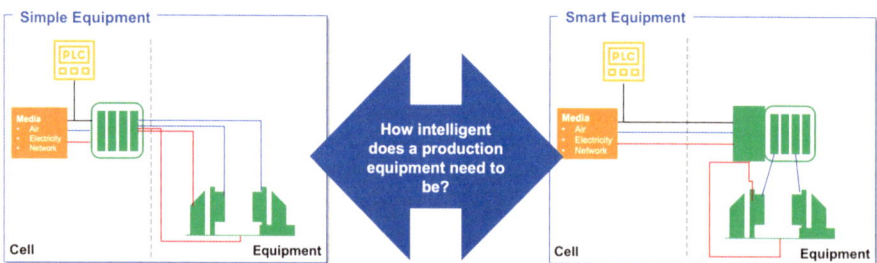

Fig. 4. How intelligent does a production equipment need to be?

One possible approach could be the development of a standardised platform with appropriately standardised connections. This platform is currently being developed in the ongoing research project.

3.3 Digitalization

Within this approach, digitalisation is an important enabler, as data must be exchanged between equipment manufacturers and end users in all life cycle phases. In the engineering phase, a link with the user's operation backbones is necessary so that the equipment fits into the existing system and can be integrated with minimal effort. During the utilisation phase, data from the equipment must be transmitted to the manufacturer for billing and maintenance. It is therefore essential to have a higher-level platform that enables data exchange and provides the corresponding services. Since every company has different IT systems and data structures, it is necessary to create a common data basis and understanding. Open standards, such as the Asset Administration Shell, are ideal for this purpose. The existing sub-models, which represent standardisation within the AAS, provide a good basis for mapping individual elements and activities but are still insufficiently defined in many cases. These gaps need to be closed so that a corresponding interaction, as outlined in Fig. 5, is possible. This shows, on the one hand, which services are required in the various entities and, on the other hand, the different ways and levels in which they must interact with each other.

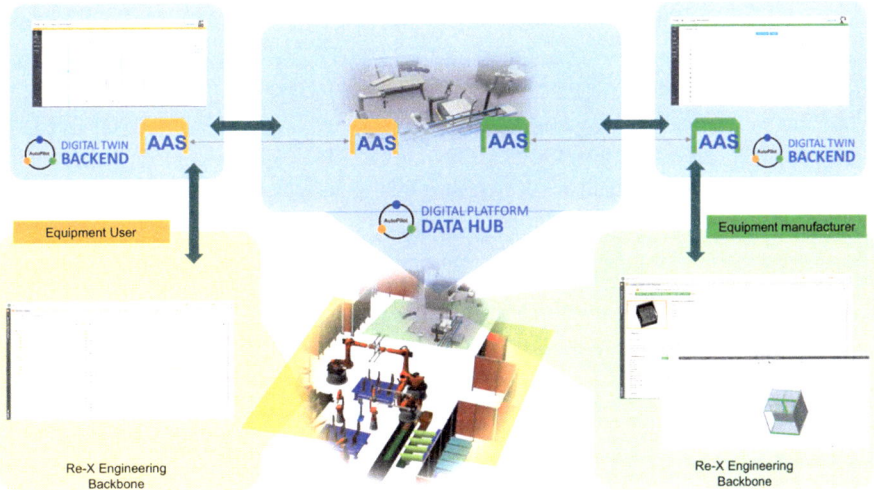

Fig. 5. System architecture of a digital platform for circular equipment

3.4 Life Cycle Assessment

Life Cycle Assessment (LCA) is a methodical approach to analysing the environmental impact of products or processes along their entire life cycle. The analysis ranges from the extraction of raw materials through production to utilisation and disposal (see Fig. 6). This holistic approach enables the LCA not only to identify environmental impacts but also to derive targeted measures to minimise the ecological footprint and resource consumption, as well as to identify and compare granular ecological and economic hotspots along the life cycle of a product.

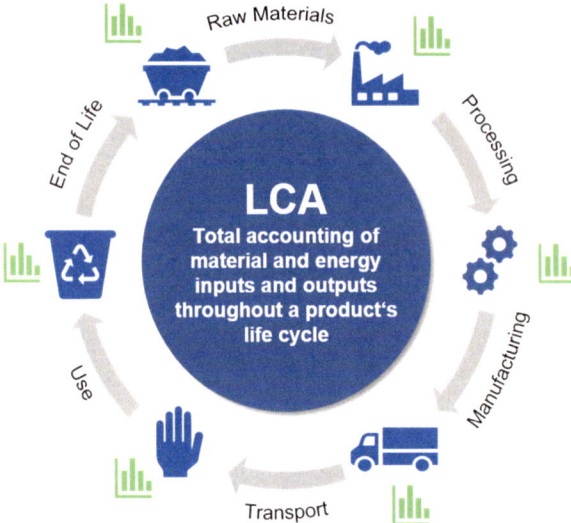

Fig. 6. System architecture of a digital platform for circular equipment. (Source: Capgemini Engineering)

A key challenge in the LCA analysis approach is the sometimes-extensive data gaps, which make it difficult to take a holistic view and increase the potential for distortions in the analysis. Data must be collected over the entire life cycle, as shown in Fig. 7. Another critical aspect is the heterogeneity of the available data, which can come in different forms and formats. This leads to considerable additional work in data preparation and integration, as the same information from different sources has to be standardised. The data processing challenges are further compounded by the need to convert the available data into the specific formats of the LCA models. This conversion requires a deep understanding of the model structure and a precise adjustment of the data to ensure a coherent basis for the analysis. Among other things, this is necessary to achieve comparable results.

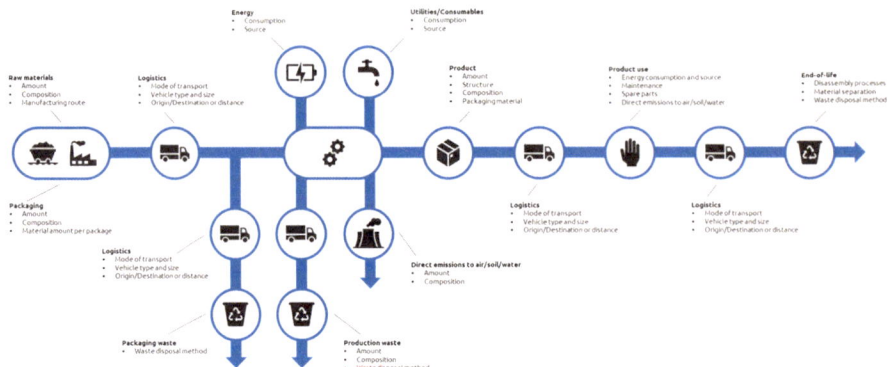

Fig. 7. The data required for the LCA comes from a variety of sources along the product life cycle. (Source: Capgemini Engineering)

However, some opportunities can improve both data quality and the analysis process. The elimination of data gaps can be achieved through an extended integration of LCA data requirements in different company departments. Comprehensive data exchange between production, procurement, logistics and other relevant areas enables holistic data collection across the entire life cycle of a product. This not only helps to eliminate data gaps but also promotes a more comprehensive and accurate analysis of environmental impacts. The introduction of digital twins and the automation of data collection and processing offer another promising opportunity. By creating digital replications of physical products and processes, data can be captured and continuously updated in real-time. Automating data collection and processing not only reduces manual effort but also enables a faster response to changes in the production process. This leads to a more efficient and cost-effective implementation of LCA.

According to the use case described in Sect. 3, an LCA was carried out for the load carrier by the project partner Capgemini Engineering. It should be particularly emphasized here that a load carrier made of plywood was developed in line with the focus on sustainable resources. This load carrier is compared directly with a conventional steel load carrier. In Fig. 8, the steel load carrier is declared as the status quo, while the plywood load carrier represents the innovation. Figure 8 shows the corresponding calculations for CO_2 emissions in kilograms per unit based on three levels of consideration. Firstly, the individual life cycle phases that the load carriers go through, from raw material extraction, logistics and production through to the end of their life, are shown in direct comparison. This shows that the status quo load carrier made of steel has significantly higher CO_2 emissions per unit during raw material extraction. The middle figure shows a comparison of the individual components that make up the load carriers. Both load carriers consist of steel components and steel connecting elements, as well as wheels and packaging, but to varying degrees. The proportion of steel components in the status quo load carrier is significantly higher due to its complete steel construction than that of the innovation load carrier, due to the fact that the latter is largely made of wood.

The last diagram is intended to show a direct comparison of the CO_2 consumption of the individual materials in the two load carriers. Here too, due to the different materials of the two load carriers, the status quo load carrier with a significantly higher proportion of steel elements also has a correspondingly higher CO_2 consumption. In contrast, the Innovation load carrier consists mainly of wooden elements and therefore has a significantly higher consumption for the material "wood". Overall, however, it is clear from all three figures that the CO_2 consumption of the innovation load carrier has significantly lower CO_2 emissions per unit, both along the life cycle phases and in a direct comparison of the components. This is mainly due to the high proportion of wooden components and the associated lower CO_2 emissions than steel.

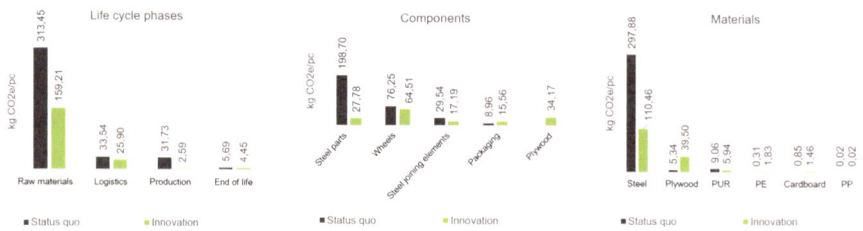

Fig. 8. Influence of the material on the life cycle phases and components. (Source: Capgemini Engineering)

3.5 Business Model

A new, customised business model is required to implement the presented approaches for the circular operating resources. A change in industrial processes must be brought about by aiming to create a closed-loop system. The business model therefore aims to maximise the use of operating resources in production by implementing innovative solutions for the highest possible degree of reuse.

The existing business models for operating resources face several challenges that affect both environmental and economic aspects. One of the central problems is the low utilisation rate of existing operating resources. Inadequate utilisation not only impairs economic profitability but also contributes to inefficient use of resources. A sub-optimal product life cycle exacerbates this problem, as complete production lines, including the associated equipment, are often completely disposed of. These short life cycles increase the need for new purchases and therefore the consumption of resources. Another significant obstacle is the lack of end-of-life concepts for operating equipment. Insufficient planning for the end of the useful life makes recycling or disposal difficult, which leads to considerable environmental impacts. At the same time, the high capital commitment in conventional business models places a financial burden on the user, as cost-intensive operating resources must be procured. Subscription-based business models for recyclable operating resources can offer added value here and promote sustainability.

This is increasingly focussing attention on the need for sustainable business models in the context of circularity for operating resources. Subscription-based approaches offer a wide range of potentials that promote the circular economy and reuse of equipment and thus save resources. Improving the utilisation of operating resources is a central aspect of subscription-based business models. By creating access instead of ownership, the efficiency of resource utilisation is maximised. Users have the opportunity to utilise resources only when they are needed. The self-interest of users in long product life cycles is strengthened by subscription-based models. Subscription-based models offer users the advantage of gaining access to high-quality products without tying up large amounts of capital. This leads to greater planning security, as users can calculate exactly what costs will be incurred in a certain time. At the same time, the equipment manufacturers are responsible for the end-of-life concepts or reuse.

Such an approach, where the equipment remains the property of the manufacturer and can be rented directly from the manufacturer as required and only the actual use is paid for, is called "Equipment as a Service".

The most important components of the new business model approach are determined with the help of the business model canvas. In addition to the value proposition, the required infrastructure and the potential customer base are also identified. Based on this description, the further development towards an "Equipment-as-a-Service" business model is pursued. Figure 9 shows the results of the business model canvas with the corresponding inputs.

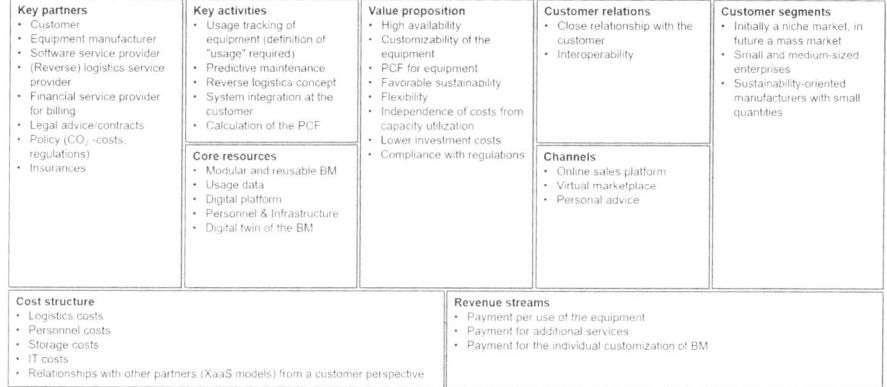

Fig. 9. Business Model Canvas for the "Euqipment as a Service" approach

4 Conclusion and Outlook

The use of circular production equipment is a promising approach to making the production of the future sustainable. It can also be integrated into existing production facilities and also provides an opportunity to take into account fluctuating demand and the ever-shorter product life cycles.

Using the example of a manufacturing cell, five aspects (engineering, integration, digitalisation, business model and LCA) were discussed that must be considered when developing and introducing circular production equipment. It became clear that some fundamental questions still need to be answered in the individual aspects to create verifiable added value for the customer. Despite the challenges that still exist, there are also many opportunities and possibilities that are currently not possible. For this, it is necessary that this innovative business model is implemented in its full complexity with a real use case so that it can serve as the basis for further products.

Acknowledgements. This project received funding from the Federal Ministry of Education and Research under funding code 02J21E000. The responsibility for the content of this publication lies with the authors.

References

1. E. Lindahl, J.-E. Dahlin, and M. Bellgran, "A framework on circular production principles and a way to operationalise circularity in production industry," Cleaner Production Letters, vol. 4, p. 100038, 2023, https://doi.org/10.1016/j.clpl.2023.100038.

2. W. Walla and J. Kiefer, "Life Cycle Engineering – Integration of New Products on Existing Production Systems in Automotive Industry," in pp. 207–212.
3. C. Brecher, O. Petrovic, Y. Dassen, P. Blanke, M. Trinh, and S. Wurm, "Sustainable Production-as-a-Service," 2023.
4. A. Fontana et al., "Circular Economy Strategies for Equipment Lifetime Extension: A Systematic Review," Sustainability, vol. 13, no. 3, p. 1117, 2021, https://doi.org/10.3390/su13031117.
5. E. Suzanne, N. Absi, and V. Borodin, "Towards circular economy in production planning: Challenges and opportunities," European Journal of Operational Research, vol. 287, no. 1, pp. 168–190, 2020, https://doi.org/10.1016/j.ejor.2020.04.043.
6. J. Potting, M. P. Hekkert, E. Worrell, and A. Hanemaaijer, Circular Economy: Measuring innovation in the product chain, 2017.
7. B. Xu, S. Liu, and H. Wang, "Developing remanufacturing engineering, constructing cycle economy and building saving-oriented society," J Cent. South Univ. Technol., vol. 12, no. 2, pp. 1–6, 2005, https://doi.org/10.1007/s11771-005-0002-4.
8. A. P. Barquet, H. Rozenfeld, and F. A. Forcellini, "An integrated approach to remanufacturing: model of a remanufacturing system," Jnl Remanufactur, vol. 3, no. 1, 2013, https://doi.org/10.1186/2210-4690-3-1.
9. PengjiaWang, Y. Liu, S. K. Ong, and A. Nee, "Modular Design of Machine Tools to Facilitate Design for Disassembly and Remanufacturing," Procedia CIRP, vol. 15, pp. 443–448, 2014, https://doi.org/10.1016/j.procir.2014.06.085.
10. A. Zacharaki et al., "RECLAIM: Toward a New Era of Refurbishment and Remanufacturing of Industrial Equipment," (in eng), Frontiers in artificial intelligence, vol. 3, p. 570562, 2020, https://doi.org/10.3389/frai.2020.570562.
11. J. Liu, Z. Zhou, D. T. Pham, W. Xu, J. Cui, and C. Yang, "Service Platform for Robotic Disassembly Planning in Remanufacturing," Journal of Manufacturing Systems, vol. 57, pp. 338–356, 2020, https://doi.org/10.1016/j.jmsy.2020.10.005.
12. DIN 6300:2009-04, Vorrichtungen für die Fixierung der Lage von Werkstücken während formändernder Fertigungsverfahren_- Benennungen und deren Abkürzungen, Berlin.
13. S. Hesse, Betriebsmittel Vorrichtung: Grundlagen und kommentierte Beispiele, 2nd ed. München: Hanser Verl., 2012. [Online]. Available: http://ebooks.ciando.com/book/index.cfm/bok_id/291650
14. F. Fiedler et al., "Jigs and fixtures in production: A systematic literature review," Journal of Manufacturing Systems, vol. 72, pp. 373–405, 2024, https://doi.org/10.1016/j.jmsy.2023.10.006.

On the Systematic Selection of CE Strategies for End-of-Life-Products: A Guide for Practitioners

Julia Dvorak[1(✉)], Leonie Stanzl[1], Tobias Lachnit[1], Martin Benfer[1], Frank Balzereit[2], and Gisela Lanza[1]

[1] Karlsruhe Institute of Technology, Wbk Institute of Production Science, Kaiserstraße 12, 76131 Karlsruhe, Germany
julia.dvorak@kit.edu
[2] Ford Werke GmbH, Henry-Ford-Straße 1, 50735 Köln, Germany
https://www.wbk.kit.edu, https://www.ford.de

Abstract. Motivated by the goal of reducing resource consumption, industries around the globe are rethinking their linear production systems to move towards a circular economy (CE). CE strategies must be selected individually for each product and each instance. The scope of the decision logic developed in this work was constrained to reuse, repair, remanufacture, recycle, and recover. Following this, factors that influence the selection of a strategy are identified based on literature and expert interviews and mapped to the selected CE strategies. A criteria clustering occurs according to PESTEL categories (political, economic, sociological, technical considering product and process, environmental and legal). A pairwise comparison of criteria is made based on which a hierarchy is established. In addition, strategic factors were assigned a higher order in the hierarchy. This order provides the basis for the decision trees to be created. In a first decision tree, suitable CE strategies are identified at a high level based on the product properties for which the applicability is validated in a second tree. This procedure may be followed in an iterating manner for the whole product and its components. In the logic, criteria can be either strategic (for example, process setup) or operational (decision at the product instance level). The termination criterion is the selection of the CE strategy 'recover' or product disposal. The selection of CE strategies was validated using the example of batteries used in electric vehicles. For this purpose, two scenarios with differing product states were developed, and the decision logic was applied to both.

Keywords: Circular production · CE strategy selection · Batteries · End-of-life-products

1 Introduction

In alignment with sustainability, the circular economy (CE) concept is gaining importance for industry. From product development, circularity extends to the end customer, and various recycling stages can be realised from different life cycle phases to other

K. Dröder and T. Vietor (Eds.): CD 2024, Proceedings, pp. 229–242, 2025.
https://doi.org/10.1007/978-3-658-45889-8_18

life cycle phases. Therefore, several CE strategies are defined. These strategies specify what should be done with end-of-life (EOL) products. [1] analyse various strategies and examine which scope is covered in each case. A hierarchical arrangement is made for value retention. It is of greater interest for companies to select a CE strategy early along the product lifecycle. Even a cost-driven approach can be favoured here, as it is essential to differentiate between the depth of disassembly and the process selection for a specific CE strategy.

Therefore, this work develops an easy-to-use decision logic that identifies suitable CE strategies based on impact factors determined beforehand from literature and expert interviews. On the one hand side, a structured decision approach is provided. On the other hand, it aims to provide an overview of criteria that serve as a starting point to make such a decision. In the following, a brief insight into CE and decision trees is given, and the current state of research is analysed. The developed methodology is presented, and the applicability is underpinned with a validation using the example of an electric vehicle battery. The content is summarised and critically assessed, and further steps are outlined in a final section.

2   Foundations

The CE encourages a rethinking of linear economic patterns. There are numerous definitions in the literature, but these have not yet been standardised [2]. An attempt at a meta-definition was made by [3], in which, based on the definition of the Ellen MacArthur Foundation, the following definition is given: "The circular economy (CE) is a regenerative economic system which necessitates a paradigm shift to replace the ‚end of life' concept with reducing, alternatively reusing, recycling, and recovering materials […] to promote value maintenance and sustainable development, creating environmental quality, economic development, and social equality, to the benefit of current and future generations. It is enabled by an alliance of stakeholders (industry, consumers, policymakers, academia) and their technological innovations and capabilities." [3] The CE affects all areas of society and ranges from the macrosystem level and, thus, the national level to the microsystem level, influencing individual entities [2].

2.1   A Brief Insight into CE Strategies

In industrial production, two aspects are outstanding in terms of circularity. On the one hand, a resource-conserving value chain should be realised, and, at the same time, sustainable products should be created. Resource savings can be realised in processes and product design, for example, using renewable materials for energy generation or in products themselves [4]. The necessary process steps in a reverse value chain are reclaiming the products, inspecting and classifying the products to determine the type of reuse, restoring the products using a selected CE strategy and reintegrating the products into the value chain [5]. Depending on the CE strategy applied, the return occurs at different points in the production process, as shown in Fig. 1.

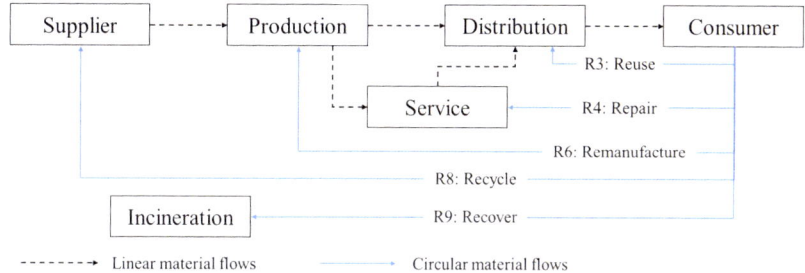

Fig. 1. Material flows in the product life cycle (based on [6] and [7])

Various sources define many strategies, beginning with product development and, thus, the redesign of products in their early stages until recycling or energy recovery of an end-of-life product [1]. A classification was made according to the speed of the circular loops, and these were assigned to different strategies, indicating in which timeframe a particular strategy could be executed [1]. The scope of strategies considered varies across different approaches [1]. Since the strategies reuse, repair, remanufacture, recycle and recover are of greater interest to the methodology developed, they will be explained in further detail and differentiated.

The reuse strategy refers to the direct utilisation in a second lifecycle of the entire product or individual components [6]. The decisive factor here is that no or very few energy inputs or work processes need to be carried out on the product and that it is fully functional [8]. By contrast, to reuse, repair means that specific work steps are carried out to restore the product to a functional condition [8]. Building upon this, remanufacturing implies that a product is used for the same function [6]. Nevertheless, complete remanufacturing of the product requires that it is returned to its original condition, and thus, a warranty claim also exists [7]. Recycling implies that the raw materials of a product are kept within the loop [8]. If only the energy is recovered by combustion from combustible materials, this is called recovery [6].

2.2 Decision Trees as Decision Support

Decision trees consist of a root, several inner nodes and several leaf nodes [9]. A decision, which can vary in complexity, is made at each node. The answers to these questions provide direct links to the child nodes. Values from a small set and linear or logical combinations of several characteristics can serve as answers. The simplest response form is a yes or no answer [10]. The sequential traversal of the decision tree ends when a leaf node is reached. Decision trees can be used to visualise chained decisions and transparently present the decision options in a complex, multi-stage decision-making process [11]. This clarifies the relationships between the individual aspects and makes the results easy to understand and comprehend [11]. In addition, decision trees are flexible and an easy-to-use method [12].

3 State of the Art

Following a systematic literature search using the PRISMA methodology [13], 15 relevant publications were identified in which approaches for the systematic selection of CE strategies are presented. "circular economy" AND "recovery strategies" AND "decision" AND ("end-of-life products" OR "product returns") was used as a final search string in Google Scholar. The period considered was limited to the timespan between 2000 and 2023, and it has been limited to English publications. In Fig. 2, an overview of the procedure is given. After identifying relevant literature, publications were screened based on their title, and 36 relevant publications were identified. After scanning keywords and abstracts, all 36 publications were considered relevant. Using a forward-backwards search, six additional publications were identified. Using this set of publications, the content was screened with a focus on the used decision logic and a total of 9 publications were selected. This was further enhanced by a forward-backwards search, and a total of 16 publications were selected.

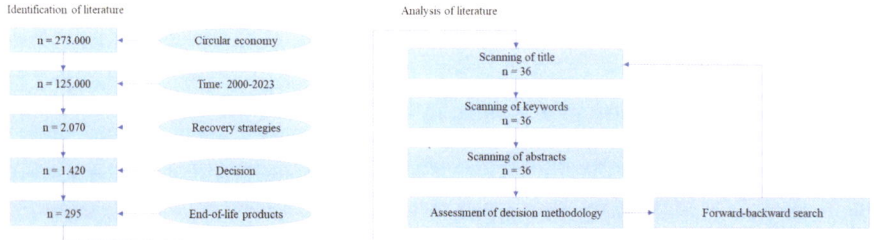

Fig. 2. Approach to systematic literature review

Multiple criteria were employed to assess these approaches. First, the methodology was evaluated from an application perspective, and the criteria of required prior knowledge, comprehensibility and simplicity of implementation were used. Two levels were included in the analysis to determine the scope of the respective logic. On the one hand, which framework can be considered at the content level concerning the PESTEL (Political - Economical - Sociological - Technical - Environmental - Legal) criteria were examined. Market-related and business-related factors were added as further dimensions. On the other hand, it examined which CE strategies were considered in each case to provide a comprehensive analysis of the opportunities along the value chain. In addition, it is evaluated if a validation was made considering a real use case. An overview of the state of the art is provided in Table 1.

Several of the contributions provide a good basis in terms of the applicability of the approaches (e.g. [8, 14–18] and [19]). Other approaches are more complex in application (e.g. [20–25] and [26]), but still cover a broad range of CE strategies and PESTEL criteria. In other cases, a different focus is chosen. For example, in [27], an approach focusing more on steps before the actual product processing, such as sorting, is described. [27] enlarge the scope and consider the whole supply chain, whereas [28] remain at a use-case-specific level. To conclude, an easy-to-use approach must be formulated to select the most suitable CE strategies for a used product and its components. Within this approach, it is essential to include impact factors from all eight defined categories. This enables a high-quality, suitable and appropriate decision to be made quickly and easily by the decision-makers. To do justice to the concept of the CE, it is also necessary to consider CE strategies arranged in a high hierarchical order as decision alternatives. In this way, the lowest possible consumption of natural resources and environmental impact and simultaneously the longest possible product life extension can be realised.

Table 1. State of research

Legend:
- ● Criterion fulfilled
- ◐ Criterion partially fulfilled
- ○ Criterion not fulfilled
- ✓ Factor or strategy considered

			[8]	[14]	[15]	[16]	[27]	[17]	[20]	[21]	[22]	[18]	[28]	[23]	[25]	[24]	[26]	[19]
Application	Previous knowledge		●	●	◐	◐	◐	●	○	○	○	●	◐	○	○	○	○	◐
	Traceability		●	◐	●	●	◐	●	○	○	○	●	◐	○	◐	○	◐	●
	Simplicity		●	●	◐	◐	○	◐	○	○	○	●	○	○	○	○	◐	◐
PESTEL	Political								✓									✓
	Economic		✓	✓	✓	✓	✓	✓	✓	✓	✓		✓	✓	✓	✓	✓	✓
	Social				✓	✓			✓		✓			✓				✓
	Technical		✓	✓	✓	✓		✓	✓	✓	✓		✓	✓	✓	✓	✓	✓
	Environmental		✓	✓	✓	✓	✓		✓		✓		✓	✓		✓	✓	✓
	Legal					✓											✓	✓
	Business			✓	✓	✓					✓			✓		✓		
	Market						✓											
CE strategy	Reuse		✓		*Depending on use case*	*Depending on use case*		✓		✓	✓	✓	*Depending on users*	*Depending on hypothesis*		✓	*Depending on experts*	✓
	Repair		✓	✓					✓		✓							✓
	Remanufacture		✓	✓			✓	✓	✓		✓	✓			✓	✓		✓
	Recycle		✓	✓			✓	✓	✓	✓	✓	✓			✓	✓		✓
	Recover																	✓
	Validation		✓		✓	✓	✓	✓	✓						✓	✓	✓	✓

4 Decision Logic for Ce Strategy Selection

A central aspect of CE is to close resource loops to maximise the value and benefits of a product and its components [29]. [17] have developed a universally applicable decision-making method for selecting the most suitable CE strategy, which can be applied to both product and component levels. This work aims to create an easily implemented approach to guarantee traceability, transparency and simplicity. Following the approach of [17], the method applies to both product and component levels to achieve the most appropriate selection for either. To arrive at the most suitable strategy for a product or its components, the objective is to consider as few as possible but as many as necessary impact factors as possible during the decision-making process. In conclusion, this work can be seen as an extension of the work of [17]. As relevant impact factors vary by product and component, a universally applicable decision logic has been developed that can be adapted to specific scenarios of different products or companies. To develop the decision logic to select the most appropriate CE strategies for a used product and its components, the approach depicted in Fig. 3 is chosen.

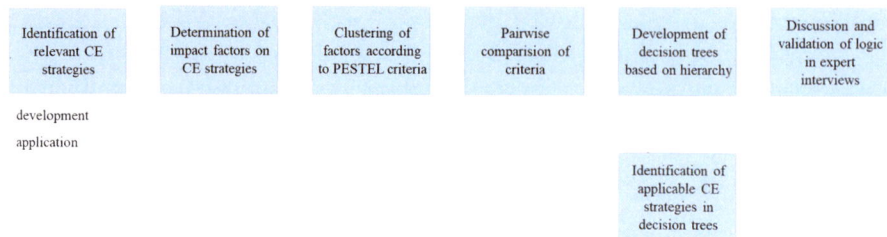

development

application

Fig. 3. Approach for decision logic development

4.1 Description of Approach

First, the relevant CE strategies are identified. In this work, the CE strategies reuse, repair, remanufacture, recycle, and recover are considered, and the hierarchical order is based on the 9R model of [6]. Second, impact factors on selecting CE strategies for a used product and its components are determined since the extant literature does not offer a list of universal impact factors on CE strategies. Impact factors were derived from the literature in the first step. Therefore, different combinations of the keywords (i.e. circular economy, critical success factors, product design, recovery strategies,…) were used in different search engines. Using the snowball principle, further literature was screened. The extracted factors were then clustered and presented to experts from the industry. In the first step, their relevance was questioned; in the second step, experts were asked for further factors. Due to the multitude of impact factors and their dependencies, a structural approach is employed in the third step. This approach comprises the six categories of the PESTEL analysis and expands them with two categories. Thus, the identified impact factors are classified into eight categories: political, economic, sociological, technical, legal, environmental, business- and market-related impact factors. To comprehend the dependencies of the impact factors, a pairwise comparison is made, and thereby, a hierarchy is created. However, the hierarchy is biased by the use case and must be adjusted when a new use case is considered. A universally applicable hierarchy cannot be created due to the subjective assessment of the importance of the factors, which depends on the group of decision-makers, the product, the company and external circumstances. For instance, decision-makers in production rate the materials employed as highly significant for the decision-making process, whereas those in the planning department view them as less relevant (according to expert discussions within the research project ZirkulEA (17th July 2023)). However, the analysis of the impact factors indicates starting with broad, externally determined factors before moving on to more specific ones and finally to the product's properties and the production process. The final step is the development of the decision logic to select the most appropriate CE strategies for a used product and its components. This work adopts decision trees as a decision method to provide a user-friendly, transparent and traceable decision method. The sequence of the decision trees follows the hierarchy of the CE strategies and thus depends on their circularity level. An initial decision tree is created to determine the CE strategy that achieves the highest level of circularity and appears possible for a specific product. Additionally, decision trees for each CE strategy, reuse, repair, remanufacture, recycle, and recover, are formulated to assess their feasibility. The decision trees include many identified impact factors from different categories, focusing on technical, business- and market-related impact factors. The logic can be applied iteratively by initially decid- ing on the complete product and assigning one CE strategy to the whole product or using the decision logic to determine a strategy for each component or sub-component. The basis for strategy selection is the 9R-model of [6], which is modified for the context of this work. The first three CE strategies, refuse, rethink and reduce, are irrelevant to the present context because they cover product manufacture and utilisation. Developing a user-friendly and easy-to-use decision logic requires a reduction of complexity and, therefore, a targeted selection of relevant CE strategies based on previous research [17, 19]. Thus, the scope of the decision logic is constrained to the CE strategies of reuse, repair, remanufacture, recycle and recover.

4.2   Clustering of Factors

In analogy with prior research, this work employs the six categories of the PESTEL analysis to classify the impact factors on CE strategies (e.g. [19]). Additional categories are introduced as this work identifies factors that influence selecting the most appropriate CE strategies for a used product and its components. These categories comprise influ- ences from the company, the sales market and demand. Therefore, the identified impact factors are classified into eight categories: political, economic, sociological, technical, legal, environmental, business-, and market-related impact factors. Most of the identi- fied impact factors belong to the category of technical factors. Furthermore, the specific application heavily influences a product's technical factors and properties. Its compo- nents and production processes significantly affect the selection of suitable CE strategies for a given product. Therefore, this work distinguishes between technical impact factors related to the product and the process, as previously identified in research [19, 21]. The allocation of impact factors into the eight defined categories is based on existing liter- ature and expert consultation. Collaboration with experts from industry and research is particularly relevant, as factors cannot always be categorised due to a lack of literature or assignment to varying categories in the existing literature. For instance, the impact factor 'pressure from environmental activists or parties' is assigned to the environmen- tal category in [30] and to the sociological category in [19]. However, expert interviews indicate that this factor is perceived in practice as a political impact factor, which is why, in this work, the categorisation based on the expert opinions is chosen.

The comparison of the identified factors and the creation of a hierarchy is a nec- essary step to develop the decision logic and better comprehend the impact factors' dependencies. Developing a separate assessment and sequence of impact factors for each application is necessary. The analysis of the impact factors and their dependencies indicates that it is generally advisable to start by examining broad, externally deter- mined factors before moving on to more specific ones. In the end, the properties of the product and the production process should be considered. Figure 4 depicts the proposed hierarchy of impact factors. Broad, externally determined factors are sociological and market-related factors such as the reason for returning the product, customer needs, the reason for buying the product or demand. Next, certain technical product-related fac- tors must be fulfilled for specific CE strategies to be feasible. Additionally, technical process-related factors must be verified to implement specific CE strategies. It is essential to consider operational factors, such as product quality or functionality, and strategic impact factors, such as production facilities or demand.

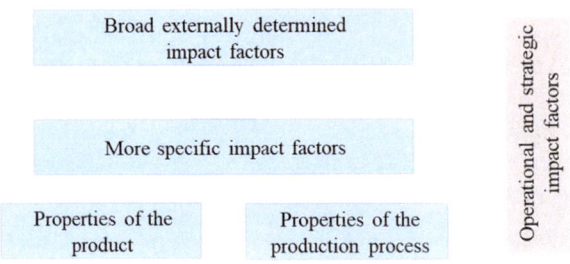

Fig. 4. Proposed hierarchy of impact factors

4.3 Logic Development and Decision Trees

The final step is the development of the decision logic to select the most appropriate CE strategies for a used product and its components. In this work, decision trees are used to encounter the multitude of impact factors and the complexity within the decision-making process.

An initial decision tree, as well as individual decision trees for the five CE strategies, reuse, repair, remanufacture, recycle and recover, have been created in this work. Figure 5 shows the arrangement and sequence of the decision trees. To determine the level of the decision-making process, the decision-makers need to answer a question outlined in the initial decision tree. Returning to this question to apply the decision logic iteratively for whole products and components is possible.

The approach depicted in Fig. 5 can be understood as follows: The initial decision tree (1) is used to determine the CE strategy that achieves the highest level of circularity, based on [6], and appears possible. The initial decision tree's questions focus on general product features and external factors such as, for example, demand. Product properties like structure [19], modularity, dismantling ability [17] and functionality [14], available product information, demand and legal regulations (expert discussions within the research project ZirkulEA) are considered in the initial decision tree. As an output, the initial decision tree provides potential CE strategies. Once the CE strategy with the highest circularity, which appears possible, has been identified, its feasibility must be assessed further using the respective decision tree of the CE strategy (2, 3, 4, 5). The questions within these decision trees focus on technical process-related impact factors such as essential production facilities [14] or expertise and provided instructions [7]. Moreover, the suitability of the product's condition for the CE strategy is examined [7, 28]. After assessing the feasibility of the identified CE strategy, decision-makers are ultimately directed towards the decision tree of the next relevant CE strategy examined. These connections are shown by the arrows in the graphic. It should be noted that the next selected CE strategy may not necessarily be the next in the hierarchy. For instance, the definition of reuse (2) already excludes the possibility of CE strategies repair (3) and remanufacture (4) applying to a reuse product and vice versa. The assessment of the feasibility of the CE strategies concludes with identifying the CE strategy with the lowest rank, 'recover' (6), as feasible or disposal as an option. At certain nodes in the decision trees, decision-makers need to assess, for example, a particular characteristic of a product, demand or necessary investments. These assessments depend on the group of decision-makers, the product, the company and external circumstances. Therefore, this work can be seen as a framework for further applicability. Criteria are listed and formulated on an abstract level, which may then be further specified depending on the use case. In addition, a methodological approach is given for ranking and decision tree building.

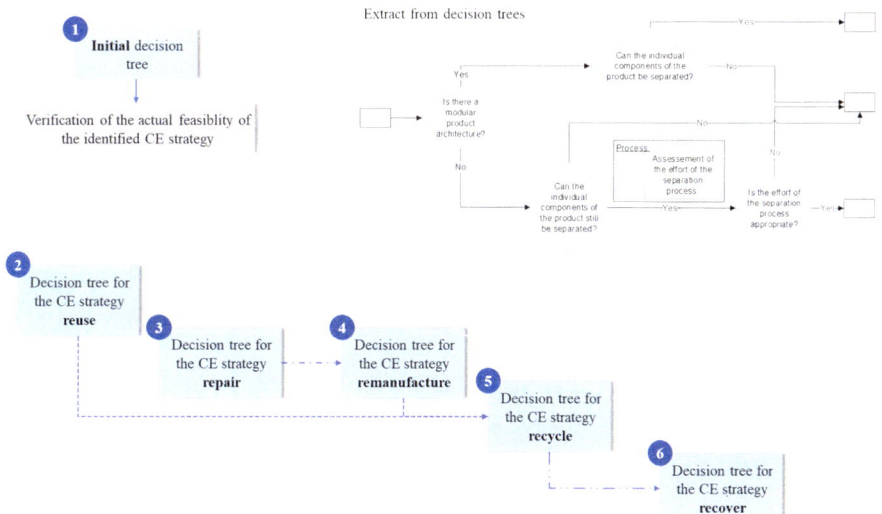

Fig. 5. Arrangement and sequence of the decision trees

Within each decision tree, depending on the way chosen, elements can occur repetitively on different branches of the tree. These repeating and content-related elements are combined into modules for complexity reduction and better visualisation. In this work, 19 modules are defined. For example, a module structure can be adapted to meet specific application requirements. One example is the module for product and component separation, which includes questions about the product architecture, the applicability and the expense of separation processes. Another example is the module for the analysis of the demand. This module analyses the demand for the novel product and the product to which a CE strategy was applied.

To apply the decision logic, the following steps must be followed. The first step is the determination of an objective function. This step has not yet been conducted, and the definition of such a target function is subject to future work. The second step implies the determination of required information and data for decision logic application. In the third step, the desired quality of a product or component must be specified. This must then, in the last step, be further detailed for each strategy and component.

5   Validation

The decision logic was applied with experts from the research project ZirkulEA using a real use case of a lithium-ion battery for electric vehicles. To validate the decision logic, the experts determined a scenario representing an expected standard case in the future. In this case, the battery is in good condition, meaning that the battery passed a few charging cycles and, at the same time, did not advance much in time-related ageing processes. To validate the decision logic in this work, it is assumed that the company possesses the required skills to execute the CE strategies, such as dismantling, refurbishment and assembly capabilities. However, the processes for battery remanufacturing are not fully defined yet. The decision trees passed for the full product are shown in Fig. 6.

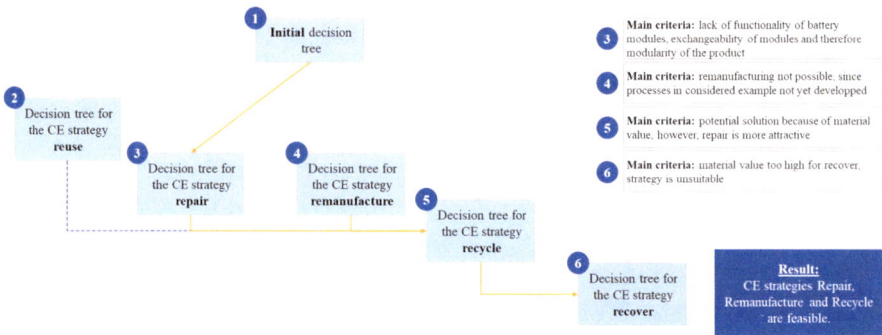

Fig. 6. Sequence of decision trees in the context of validation

First, the initial decision tree determines the CE strategies that are relevant to the use case. Therefore, an analysis of the product architecture is conducted. The battery has a modular product architecture, where individual components (i.e. battery modules, battery housing) can be separated. Moreover, the modules are assumed to be interchangeable in this battery type. Additionally, the stability of the components and materials used is given, facilitating non-destructive disassembly. An important factor to consider is the battery's functionality, which is found to be insufficient for direct reuse in the considered use case. The functionality of the battery is measured using its state of health as a reference. The state of health is defined as the ratio between the current and original performance of the battery and, therefore, expresses a changeable attribute of usage and time [31]. First, the product is considered as a whole.

The reuse strategy is possible in alternative applications such as external storage. A reuse is suggested as a state of health of 70–80% is reached [32]. However, based on experts' opinions, in the current scenario, the application of the strategy repair is more attractive than a **reuse**. To further prove this assumption, it is of greater interest to incorporate a target function into the model in the future.

Due to the battery's dismantlability and lack of functionality in this case, the CE strategy **repair** is possible. Its feasibility is further examined using the specific decision tree for the CE strategy repair. Second, the battery's quality, such as contamination, damages, and age, are evaluated per the CE strategy repair. These factors are considered appropriate, as is the ability to restore function and appearance. During the validation process, the experts highlight the standardised components and connecting elements, such as typical screw connections and seals, which require no extra effort when disassembling the battery. The authorisation requirements for batteries remain intact as a part of the CE strategy repair, which is why it is considered feasible. Repair, in this case, means that certain modules in which the deterioration occurs on a chemical level are exchanged. After the battery is repaired, its state of health is measured again. The validation process proceeds by assessing the feasibility of the CE strategy **remanufacture** for the battery. In the considered scenario, this is not possible since corresponding processes have not yet been developed. Next, the feasibility of the CE strategy **recycle** for the battery is evaluated. Due to the considerable material value and the quality of the materials assessed as appropriate, the CE strategy recycle presents a potential reprocessing solution for the battery. By contrast, the CE strategy **recover** is unsuitable and undesirable due to the high material value bound in elements like cobalt and nickel. Therefore, according to experts, batteries are more likely to be **recycled**, in the course of which extraction of those materials is enabled.

In the case that processing at the full product level is not possible, one could now continue to restart the logic for the battery modules and the battery housing. Overall, the experts evaluated the identified CE strategies repair and recycle as comprehensible and found repair to be the most favourable option. However, the validation process revealed that using the generally formulated decision trees requires considerable effort to map to a use case. The level of effort required is considered high due to the numerous steps, questions and assessments involved in applying the decision logic. Additionally, answering strategic questions within the decision logic is currently challenging due to the unknown process design and further product development. Obtaining the necessary information can be difficult, as some of it is unavailable, hindering the application of the decision logic. Furthermore, the profitability of the decision has not yet been considered in the decision trees. Thus, it is intended to extend the decision logic to include a cost-utility analysis.

6   Summary and Conclusion

Deciding what to do with used products is a challenge for companies. There are a variety of possible strategies that can be used. This work first collected relevant impact factors from the literature and was further enriched by expert interviews. These factors were then compared pairwise, and a sequence of decision trees was generated to systematically test the strategies, trying to identify the strategy with the highest circularity based on the ranking of the respective CE strategies. Based on some general questions, the first step is to select which of the strategies should be further investigated based on a specific use case.

The decision tree approach has been tested with experts using a battery as an example to determine which strategies can be selected based on the information provided. A further effort for validation is required. It was possible to evaluate that repair is the most desirable strategy if the modules can be separated. In addition, the battery can be repaired in use and with comparably few efforts. Provided that the functional fulfilment is not sufficient for reuse in its original case, alternative usage, such as, for example, using the battery as external storage, can be foreseen. To evaluate further options, the incorporation of a target function and a quantitative evaluation is essential and the focus of future work. The economic impact of a decision for a second use or a material recycling highly depends on the surrounding conditions within a system. This work creates a frame that can be filled further with an application case-specific view. In the further course of the research project ZirkulEA, a simplified version of the logic will implemented in a software demonstrator.

Acknowledgements. This research and development project was funded by the German Federal Ministry of Education and Research (BMBF) within the "The Future of Value Creation – Research on Production, Services and Work" program (funding number 02J21E130) and managed by the Project Management Agency Karlsruhe (PTKA). The author is responsible for the content of this publication.

References

1. J. Kurilova-Palisaitiene, E. Sundin, and T. Sakao, "Orienting around circular strategies (Rs): How to reach the longest and highest ride on the Retained Value Hill?," *Journal of Cleaner Production*, vol. 424, p. 138724, 2023, https://doi.org/10.1016/j.jclepro.2023.138724.
2. J. Kirchherr, D. Reike, and M. Hekkert, "Conceptualizing the circular economy: An analysis of 114 definitions," *Resources, Conservation and Recycling*, vol. 127, pp. 221–232, 2017, https://doi.org/10.1016/j.resconrec.2017.09.005.
3. J. Kirchherr, N.-H. N. Yang, F. Schulze-Spüntrup, M. J. Heerink, and K. Hartley, "Conceptualizing the Circular Economy (Revisited): An Analysis of 221 Definitions," *Resources, Conservation and Recycling*, vol. 194, p. 107001, 2023, https://doi.org/10.1016/j.resconrec.2023.107001.
4. T. Bergs and J. Brimmers, „Fertigung für eine Kreislaufwirtschaft," 2023.
5. F. Klenk, M. Potarca, B. Häfner, and G. Lanza, „Kreislaufwirtschaft in Produktionsnetzwerken," *Zeitschrift für wirtschaftlichen Fabrikbetrieb*, vol. 115, no. 10, pp. 668–672, 2020, https://doi.org/10.1515/zwf-2020-1151005.
6. J. Potting, M. Hekkert, E. Worrell, and A. Hanemaaijer, *Circular Economy: Measuring innovation in product chains*. The Hague, 2016.
7. S. Shahbazi, K. Johansen, and E. Sundin, "Product Design for Automated Remanufacturing— A Case Study of Electric and Electronic Equipment in Sweden," *Sustainability*, vol. 13, no. 16, p. 9039, 2021, https://doi.org/10.3390/su13169039.
8. M. Abdessalem, A. B. H. Alouane, and D. Riopel, "Decision modelling of reverse logistics systems: selection of recovery operations for end-of-life products," *IJLSM*, vol. 13, no. 2, p. 139, 2012, https://doi.org/10.1504/IJLSM.2012.048933.
9. U. Bankhofer and J. Vogel, „Datenanalyse und Statistik," 2008, https://doi.org/10.1007/978-3-8349-9654-1.
10. C. Kingsford and S. L. Salzberg, "What are decision trees?," *Nature biotechnology*, vol. 26, no. 9, pp. 1011–1013, 2008, https://doi.org/10.1038/nbt0908-1011.
11. C. Schawel and F. Billing, „Entscheidungsbaum," in *Top 100 Management Tools*, C. Schawel and F. Billing, Eds., Wiesbaden: Springer Fachmedien Wiesbaden, 2018, pp. 121–124.
12. B. de Ville, "Decision trees," *WIREs Computational Stats*, vol. 5, no. 6, pp. 448–455, 2013, https://doi.org/10.1002/wics.1278.

13. C. Köhler, Ed., *Basiswerkzeuge zur Erstellung wissenschaftlicher Arbeiten*. Wiesbaden: Springer Fachmedien Wiesbaden, 2020.

14. M. Abdessalem, A. Hadj-Alouane, and D. Riopel, "A decision support system for selecting recovery operations of in-use products," *The International Conference on Information Systems, Logistics, and Supply Chain (ILS 2008)*, 2008.

15. Y. A. Alamerew and D. Brissaud, "Circular economy assessment tool for end of life product recovery strategies," *Jnl Remanufactur*, vol. 9, no. 3, pp. 169–185, 2019, https://doi.org/10.1007/s13243-018-0064-8.

16. Y. A. Alamerew, M. L. Kambanou, T. Sakao, and D. Brissaud, "A Multi-Criteria Evaluation Method of Product-Level Circularity Strategies," *Sustainability*, vol. 12, no. 12, p. 5129, 2020, https://doi.org/10.3390/su12125129.

17. M. Benfer *et al.*, "A Circular Economy Strategy Selection Approach: Component-based Strategy Assignment using the Example of Electric Motors," 2022, https://doi.org/10.15488/12133.

18. D. A. Paterson, W. L. Ijomah, and J. F. Windmill, "End-of-life decision tool with emphasis on remanufacturing," *Journal of Cleaner Production*, vol. 148, pp. 653–664, 2017, https://doi.org/10.1016/j.jclepro.2017.02.011.

19. A. Ziout, A. Azab, and M. Atwan, "A holistic approach for decision on selection of end-of-life products recovery options," *Journal of Cleaner Production*, vol. 65, pp. 497–516, 2014, https://doi.org/10.1016/j.jclepro.2013.10.001.

20. N. Bognar, J. Rickert, M. Mennenga, F. Cerdas, and C. Herrmann, "Evaluation of the Recyclability of Traction Batteries Using the Concept of Information Theory Entropy," in *Cascade Use in Technologies 2018*, A. Pehlken, M. Kalverkamp, and R. Wittstock, Eds., Berlin, Heidelberg: Springer Berlin Heidelberg, 2019, pp. 93–103.

21. M. Dehghanbaghi, H. Hosseininasab, and A. Sadeghieh, "A hybrid approach to support recovery strategies (A case study)," *Journal of Cleaner Production*, vol. 113, pp. 717–729, 2016, https://doi.org/10.1016/j.jclepro.2015.11.064.

22. K. Meng, Y. Cao, X. Peng, V. Prybutok, and K. Youcef-Toumi, "Smart recovery decision-making for end-of-life products in the context of ubiquitous information and computational intelligence," *Journal of Cleaner Production*, vol. 272, p. 122804, 2020, https://doi.org/10.1016/j.jclepro.2020.122804.

23. K. K. Pochampally, S. Vadde, S. V. Kamarthi, and S. M. Gupta, "Beyond sensor-assisted diagnosis of used products," in *Environmentally Conscious Manufacturing IV*, Philadelphia, PA, 2004, pp. 138–146.

24. G. Vanson, P. Marangé, and E. Levrat, "End-of-Life Decision making in circular economy using generalized colored stochastic Petri nets," *Auton. Intell. Syst.*, vol. 2, no. 1, 2022, https://doi.org/10.1007/s43684-022-00022-6.

25. A. Raihanian Mashhadi and S. Behdad, "Optimal sorting policies in remanufacturing systems: Application of product life-cycle data in quality grading and end-of-use recovery," *Journal of Manufacturing Systems*, vol. 43, pp. 15–24, 2017, https://doi.org/10.1016/j.jmsy.2017.02.006.

26. S. Wadhwa, J. Madaan, and F. Chan, "Flexible decision modeling of reverse logistics system: A value adding MCDM approach for alternative selection," *Robotics and Computer-Integrated Manufacturing*, vol. 25, no. 2, pp. 460–469, 2009, https://doi.org/10.1016/j.rcim.2008.01.006.

27. S. M. Ayati, J. N. Uhrenholt, B. Waehrens, and J. Kristensen, "A decision model for re-engaging End-of-Life products into the forward supply chain," *NOFOMA: The Nordic Logistics Research Network*, 2020.

28. H. S. Phuluwa, I. Daniyan, and K. Mpofu, "Development of a sustainable decision framework for the implementation of end-of-life (EoL) options for the railcar industry," *Environ Dev Sustain*, vol. 23, no. 6, pp. 9433–9453, 2021, https://doi.org/10.1007/s10668-020-01035-y.

29. N. M. P. Bocken, E. A. Olivetti, J. M. Cullen, J. Potting, and R. Lifset, "Taking the Circularity to the Next Level: A Special Issue on the Circular Economy," *J of Industrial Ecology*, vol. 21, no. 3, pp. 476–482, 2017, https://doi.org/10.1111/jiec.12606.

30. T. Kaufmann, „PESTEL-Analyse," in *Strategiewerkzeuge aus der Praxis*, T. Kaufmann, Ed., Berlin, Heidelberg: Springer Berlin Heidelberg, 2021, pp. 19–28.

31. X. Kong, A. Bonakdarpour, B. T. Wetton, D. P. Wilkinson, and B. Gopaluni, "State of Health Estimation for Lithium-Ion Batteries," *IFAC-PapersOnLine*, vol. 51, no. 18, pp. 667–671, 2018, https://doi.org/10.1016/j.ifacol.2018.09.347.

32. A. Podias *et al.*, "Sustainability Assessment of Second Use Applications of Automotive Batteries: Ageing of Li-Ion Battery Cells in Automotive and Grid-Scale Applications," vol. 9, no. 2, p. 24, 2018, https://doi.org/10.3390/wevj9020024.

From Lab to Loop - Designing Sustainable Pathways Towards a Circular Economy in the Automotive Industry - A Digital Hub

Lisa Remke[1]([✉]), Marlon Philipp[2], Christoph Hoppe[3], and Benjamin Koch[4]

[1] Chemnitz University of Technology Chair of Business Information Systems, Chemnitz, Germany
lisa.remke@wirtschaft.tu-chemnitz.de

[2] Dortmund Social Research Centre, University of Technology, Dortmund, Germany
marlon.philipp@tu-dortmund.de

[3] Fraunhofer Institute for Software and Systems Engineering, University of Technology Braunschweig Chair of Business Information Systems, Braunschweig, Germany
christoph.hoppe@isst.fraunhofer.de

[4] Chemnitz University of Technology Department of Advanced Powertrains, Chemnitz, Germany
benjamin.koch@mb.tu-chemnitz.de

http://www.tu-chemnitz.de/wirtschaft/wi1/index.html,
https://www.sfs.sowi.tu-dortmund.de/,
http://www.tu-braunschweig.de/wi-dde, http://www.tu-chemnitz.de/projekt/alf/content.php?view=start&lang=en

Abstract. Climate change and resource scarcity pose existential long-term threats to humanity. To tackle these issues the adoption of circular economy (CE) measures presents a viable option to reduce the emission of greenhouse gases and preserve resources by reusing products and materials. Building on these insights, the DIONA project (Digital Ecosystem for a Sustainable Circular Economy in the Automotive Industry) aims to support the German automotive industries' circular efforts through the development and operation of a Digital Hub, which facilitates cross-company collaboration to advance and share sustainable CE practices. Consequently, we conducted a multi-pronged approach, involving a rigorous process of examining previous hub initiatives, reviewing case studies, conducting interviews, and initiating focus group discussions with industry leading experts from the automotive industry. Through this process, we articulate a strategic roadmap, where we define goals, identify, and schedule milestones and assort the activities necessary for a successful realization of the Digital Hub. The primary objectives of the hub encompass (1) fostering an open and collaborative culture, (2) encouraging companies to share best practices, pool resources, and (3) strengthening collective investments in sustainable initiatives. In addition, the roadmap considers stakeholders beyond the business community. It also addresses government agencies, research institutions, and Non-Governmental Organizations in the digital ecosystem and builds a holistic approach to a sustainable CE. The first results of the conducted studies suggest that a combination of virtual and physical actions is necessary to trigger collaboration. In this roadmap, we align a variety of stakeholders and provide valuable insights, into the cross-collaboration of twelve scientific

K. Dröder and T. Vietor (Eds.): CD 2024, Proceedings, pp. 243–253, 2025.
https://doi.org/10.1007/978-3-658-45889-8_19

research projects working together to shape a CE. Hence, the roadmap represents a pioneering effort to address pressing economic and environmental issues and offers a vision for a more sustainable and prosperous automotive industry.

Keywords: Circular economy · Digital hub · Collaboration · Sustainability · Transformation · Automotive industry

1   Introduction

In addition to ongoing short-term crises like the Russian invasion of Ukraine or the escalating conflict in the Middle East, human induced climate change and resource scarcity are pressing issues that require decisive action today [1]. Consequently, the UN ratified the Paris Agreement as well as the Sustainable Development Goals [2, 3]. Germany, as a country with one of the biggest economies, holds special responsibility to shift to sustainable means of production. According to the German sustainable development strategy, the shift towards CE aims to decouple consumption from growth and is a major area of future transformation [4]. This is especially true for the German automotive industry as it is one of Germany's biggest industries [5] with worldwide impact already undergoing transformation from combustion to electric powered cars. Nevertheless, this disruption holds the opportunity to alter the established linear economic system with its historically grown supply chains in the ongoing transformation process to support a more sustainable future. To tackle this issue the German Federal Ministry of Education and Research initiated the *MobilKreis* initiative to foster circular production in the automotive industry and respective emerging innovative business models. Consequently, twelve mostly engineering-driven research projects were founded thematically addressing a wide field of topics (e.g., battery logistics or effective remanufacturing) [6]. The interdisciplinary Project DIONA [7] is centered right in the middle of the *MobilKreis* Initiative as a 13th project. It aims to bundle the scientific results as well as foster cooperation and communication within *MobilKreis*, but also to the broader public. In addition, DIONA conducts its own research in the field of social science and information systems (IS) to sharpen the understanding of sustainability and CE within the community and discover relevant information flows in a CE. The overarching goal of DIONA is to bundle all its activities in one single place, a so-called Digital Hub.

Consequently, this paper outlines a strategic roadmap to create relevant scientific results, in designing, implementing, and operating a Digital Hub as well as creating relevant to foster CE in the automotive industry. Therefore, it is building on the theoretical foundation of sustainability, CE, and Digital Hubs (Sect. 2). Moreover, we explain our research design, the methods used (Sect. 3) as well as results (Sect. 4) and a discussion (Sect. 5). Finally, we conclude our paper by explaining its contribution, limitations and give an outlook on further research opportunities (Sect. 6).

2   Foundations

This section first addresses the vagueness of the sustainability definition and the resulting consequence of sustainability action. Second, the concept of CE as one transformation pathway for sustainable action is presented. Last, the concept of a Digital Hub is presented to raise awareness and foster collaboration on certain topics.

2.1   Sustainability

Sustainability is a crucial topic in current societal discussions, with sustainable development receiving significant attention since the release of the Brundtland Report in 1987 [8]. The report defines sustainable development as meeting the needs of present generations without compromising the ability of future generations to meet their own needs [8]. This definition suggests that sustainability has a dual foundation, with a focus on justice between present and future generations (intergenerational justice) and justice within present generations (intragenerational justice). The concept of sustainability is usually described in terms of these three overlapping dimensions. To be fully sustainable, an action needs to improve all three dimensions. Nonetheless, there are alternative depictions of sustainability where the dimensions interact and are prioritized differently. For instance, the three-pillar model, supports the sustainability roof, if one pillar (one of the dimensions) is removed, the roof collapses (see Fig. 1.) [9].

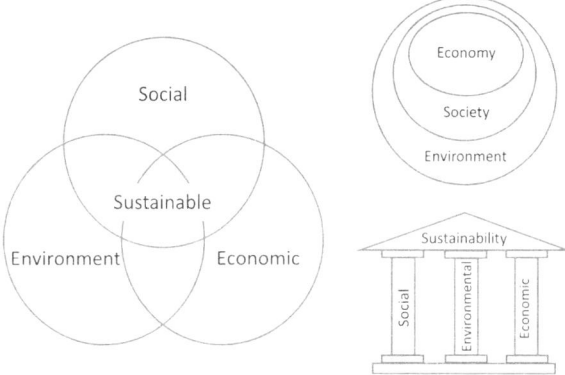

Fig. 1. The three dimensions of sustainability depicted as overlapping circle, as Three-Pillars Model and as dimensions that build on each other [8].

These examples illustrate that sustainability is a vague concept and that stakeholders can ultimately choose their preferred understanding of sustainability, even if this may circumvent conflicting objectives or the need to prioritize. This ultimately renders sustainability an essentially contested concept [10], which is regarded as significant by various stakeholders but differs in interpretation [11].

Consequently, it is essential to consider various stakeholder viewpoints on sustainability and ideally create a common understanding of sustainability to enable decisive sustainable action, building on a broad consensus.

2.2   Circular Economy

Methods of CE pose a possible viable strategy for addressing issues related to global sustainability by striking a balance between the environment and the economy [12, 13]. The primary goal of the CE is to maintain natural resources in the system for as long as feasible to protect their ecological and economic value. This can be achieved by giving products longer lives or repurposing resources for analogous or novel uses within the systems [14, 15]. By lowering resource input and ending material loops, the CE economic model aims to benefit business, society, and the environment [16]. Thus, unlike the linear 'take-make-waste' paradigm, CE tries to decouple growth from the consumption of limited resources through regenerative design. To achieve this, materials that have outlived their usefulness are recovered and recycled; products and their components are given a longer lifespan through remanufacturing, refurbishing, repurposing, and reusing. Moreover, products are used and manufactured more intelligently through reduction, reconsideration, and rejection [17]. We utilize Kirchherr et al.'s [16] definition, which represents the most recent conceptualization of CE: "*A circular economy describes an economic system that is based on business models which replace the 'end-of-life' concept with reducing, alternatively reusing, recycling and recovering materials in production/distribution and consumption processes, [...] with the aim to accomplish sustainable development [...] to the benefit of current and future generations*". As encouraged by the United Nations Sustainable Development Goals (SDGs), the advantages of CE can thereby possibly solve global environmental, social, and economic concerns [2, 18]. In fact, one of the main tenets of the EU Green Deal is the implementation of CE [19]. Even though CE implementation is still in its infancy [20], several significant elements hold the potential to convert the current linear model into a CE model [21]. Thus, it is important to identify already working best CE practices as well as supporting businesses in the development of circular innovations and cooperation.

2.3   Digital Hubs

This function can be fulfilled by Digital Hubs, which are also known as Digital Innovation Hubs in English scientific literature. Digital Hubs are platforms that generally serve the exchange and discussion of social problems by a large number of relevant economic, public, and civil actors [22]. Furthermore, Digital Hubs can help to create a collaborative working environment through online and offline offerings, such as working groups and events. In this form, Digital Hubs fulfil the role of a knowledge broker, as they connect heterogeneous actors [23]. According to a case study in Spain [24], Digital Hubs often emerge as the result of initiatives by large companies as they have the necessary monetary and technical competencies thus allowing smaller companies to collaborate within. In addition, as described before Digital Hubs can be actively founded by public bodies like the German Federal Ministry of Education and Research. Digital Hubs can also serve as test and experimentation spaces in which new, innovative products or technologies can be tested [22]. This brief overview shows the broadness of the Digital Hub term offering a wide range of design and service options. Due to its broadness, it seems necessary, to create an adapted and fitted version to the specific use. In the following chapters, the development of the DIONA Digital Hub is presented based on the corresponding roadmap and summarized results of studies that have already been carried out.

3 Methods

As shown in the theoretical foundation section of this paper, even though the concepts of sustainability and CE as tools to achieve sustainability remain vaguely defined, Digital Hubs pose a suitable solution to find a common notion as well as an agreed-on transformation path through stakeholder participation. Figure 2 shows the strategic roadmap for the design and implementation of the DIONA Digital Hub as a comprehensive knowledge platform that creates a shared vision of a feasible transformation towards a future CE in the automotive industry. As mentioned in the introduction, the foundations of the *MobilKreis-Initiative* as well as the DIONA Digital Hub are set by the internationally agreed on sustainability obligations and the resulting German Sustainability Strategy for the specific context of Germany (Fig. 2., 0). Building on this foundation, a variety of research was already and is still being conducted to create a successful Digital Hub.

As a first step (Fig. 2., 1), we identified and interviewed internal and external stakeholders about their individual sustainability understanding, relevant circularity indicators in the automotive industry, data and technologies, as well as necessary features to be included in the future DIONA Digital Hub by using semi-structured interviews. In addition, we conducted a systematic analysis of existing Digital Hubs, evaluated relevant characteristics, and created a Digital Hub taxonomy. Combining the Taxonomy as well as the insights from the interviews, we defined the requirements of a successful Digital

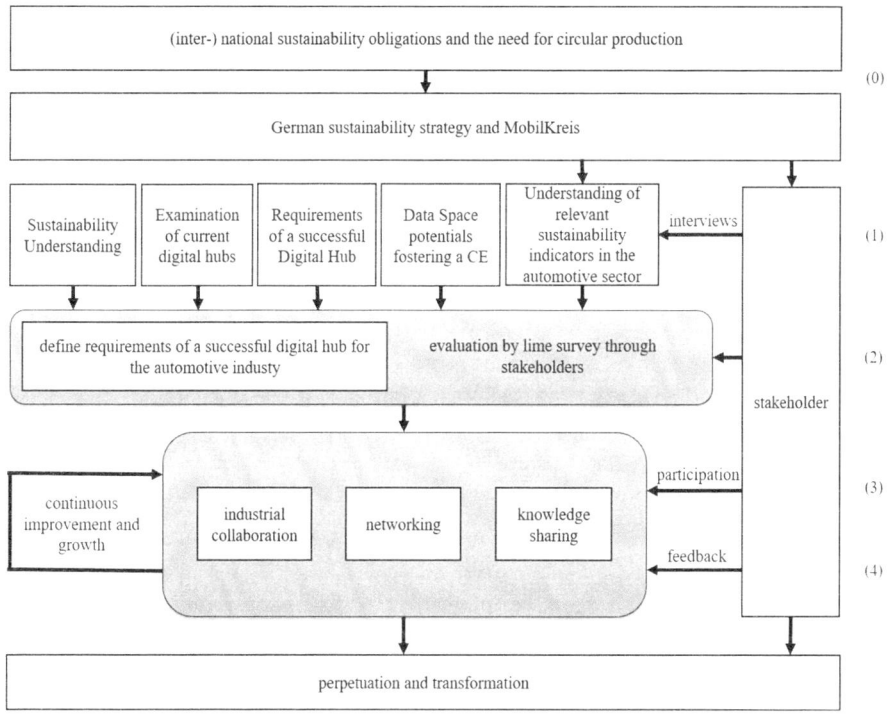

Fig. 2. Roadmap for designing a CE focused Digital Hub.

Hub for the automotive industry. To develop an application-oriented and user-friendly Digital Hub, the defined requirements were also evaluated in a second step through a stakeholder survey (Fig. 2., 2). As a next step in the foreseeable future, the DIONA Digital Hub will be established, and identified stakeholders will be invited to participate (Fig. 2., 3). The three main functions of the Digital Hub are: First, to foster circular collaboration, Second, networking and, Third, the sharing of knowledge between all participants. The fourth step of the strategic roadmap includes the operation and continuous improvement of the DIONA Digital Hub through stakeholder feedback after every activity (Fig. 2., 4). Through continuous effort, we hope the serve as a meaningful addition to transformation of the automotive industry by perpetuating the relevant knowledge and pairing the relevant stakeholders.

4   Results

In the following section summaries of already conducted studies to kick-start the DIONA Digital Hub along the strategic roadmap are reported. Detailed results and scientific discussions can be found in the corresponding circularity reports and other scientific papers.

Sustainability Understanding: From a comprehensive literature review, we were able to define the concepts of sustainability and sustainable transformation. Furthermore, semi-structured interviews were conducted with partners from the automotive sector, which revealed areas of agreement and disagreement between the individual and the corporate perceptions of sustainability. The results show that companies in the automotive industry focus primarily on economic aspects of sustainability in their understanding and measures, while individuals emphasize primarily ecological aspects of sustainability. Furthermore, this interview study highlights cultural barriers to a sustainable transformation within the German automotive industry that can be related to the prioritization of economic profitability. For example, an organizational culture that favors managers who provide short term economic gains instead of working towards long-term goals but also customer's (companies and consumers) buying decisions are primarily characterized by price and only secondarily by the sustainability of products.

Examination of current Digital Hubs: As part of the DIONA research, we additionally analyzed 17 innovation hubs and five platform ecosystems in Germany to gain empirical insights into existing practices. The research led to a taxonomy with meta-dimensions of service, communication, and website design. Under service, the dimensions of the target group, service focus, and price model were distinguished. Moreover, semi-structured interviews were conducted to compile a requirements catalog on the DIONA Digital Hub with both *MobilKreis* partners and external interviewees. The interviews revealed challenges and requirements for the operation of the DIONA hub, as well as a catalog of desired features and special considerations for SMEs that may wish to participate in the Digital Hub. The challenges mentioned encompass sparking interest among prospective participants and the provision of personnel for the Digital Hub. In addition, respondents highlighted the need for data protection measures and for operators to actively network to recruit potential participants. The provision of information such as projects, best practices, and success stories as well as informal and formal networking opportunities were also identified as desired features. Additionally, the interviewees highlighted the usefulness of regional structures for physical meetings in engaging SMEs when establishing a hub, but also the need to provide low-threshold information on how to implement CE measures and funding opportunities. Based on the taxonomy and the interview study, we defined the *MobilKreis* Hub's services.

Requirements of a successful Digital Hub: In brief, the DIONA Digital Hub aims to connect small and medium-sized companies and facilitate knowledge transfer, enabling these firms to participate in the automotive CE and gain value from it. The meta-dimension "communication" encompasses the form of communication, the communication within the joint project, and the communication with SMEs and external parties. The website design includes the language and sub-websites and provides a structure for the aspired *MobilKreis* Digital Hub website.

Understanding of relevant sustainability indicators in the automotive sector: Besides the fundamental scientific research on sustainability and transformation features of a Digital Hub, we conducted a literature review to identify current evaluation methods and indicators for a sustainable CE within the automotive industry. Consequently, we created a database, detailing the content of these indicators and evaluation methods whilst also assessing their applicability for use in the automotive sector, and their alignment with the three dimensions of sustainability. To determine practical examples in the automotive industry regarding the use of evaluation methods for a sustainable CE, we carried out semi-structured interviews with external partners. These interviews offer insight into the best practices for the evaluation methods used for the transition to a CE and the challenges in collecting relevant data, including challenges for SMEs. Challenges are, for example, the unwillingness of upstream suppliers to pass on data, customers' inconsistent wishes concerning the reporting of sustainability and circularity indicators, and particularly in the case of SMEs, the fact that they often collect little or no data on their production and could be overstretched in terms of personnel in the future due to the desired legal requirements for environmental reporting.

Data Space potentials fostering a CE: From a systematic literature review, we revealed the potential of data spaces for sustainability and circular economy. Despite the fact, that data spaces are a relatively new technological phenomenon, we derived a range of data space projects hitting the market. Based on our findings, an overview of data space potentials in line with the three dimensions of sustainability was obtained through an empirical examination of real-world data spaces. We arranged and processed our data by the methodology of Kundisch et al. [25] and presented them as a potentials map, describing the discovered dimensions and data space potentials for sustainability as well as CE.

5 Discussion

By drawing on the results, four major observations emerged, which we will discuss in more detail.

First, our findings underscore the promising potential of Digital Hubs as a viable option for fostering sustainable change across organizations by serving as dynamic platforms for collaboration, knowledge sharing, and innovation. However, our findings also emphasize the various hurdles that organizations face when implementing Digital Hubs, which need careful consideration to fully realize their potential in practice.

Second, one of the significant challenges is the varied understanding of sustainability and CE among organizations and practitioners. This lack of a common definition can result in misalignments of goals and strategies within organizations. To address this, organizations must invest in developing a shared understanding of sustainability and align it with the objectives of the CE. This emphasizes the importance of a conceptual groundwork for digital initiatives, ensuring that participants have a unified vision for sustainable practices.

Third, our findings also draw attention to the time constraints faced, particularly in Small and Medium Enterprises (SMEs), which hinder in-person interviews. This practical challenge underscores the need for researchers and organizations to find innovative ways to conduct thorough interviews efficiently. Whether through leveraging technology for virtual interviews or creating flexible scheduling options, overcoming this barrier is essential for gathering valuable insights from participants, especially in SMEs where time is often a limited resource.

Fourth, another critical aspect that emerged during the interview studies are barriers to sharing data. Despite the potential benefits of collaboration in a CE, participants are very reluctant to share data. This hesitation may be due to concerns about data privacy, security, or ownership. Additionally, multiple technical hurdles prevent the sharing of data. In particular, there is a lack of IT infrastructure, standardized interfaces, and data spaces.

Nevertheless, the challenges identified underline the importance of addressing both conceptual and practical issues as part of digital initiatives for sustainable change. Organizations need to invest in developing a shared understanding of sustainability and the CE and align this with solving technical challenges. The results and activities to date enable the development of an ecosystem that brings together different actors to enable the participation of organizations and especially SMEs in the CE - the DIONA Digital Hub.

6 Contributions and Outlook

Climate change and resource scarcity pose an existential threat to humanity. To counter these problems, CE is a viable concept for reducing greenhouse gas emissions and conserving resources by reusing products and materials. Following this, we set ourselves the goal of supporting the efforts of the German automotive industry in terms of CE by developing and operating the DIONA Digital Hub, facilitating cross-company collaboration to promote and share sustainable CE practices. Within this paper, we highlighted our strategic roadmap toward the realization of the DIONA Digital Hub. Thus, the roadmap represents a pioneering effort to solve pressing economic, social and environmental problems and offers a vision for a more sustainable and circular automotive industry.

Thereby, we make important contributions to both **research** and **practice**. For practical purposes, we conceptualized the strategic roadmap by planning and developing the Digital Hub to foster CE in the German automotive industry. Thus, we support a collaborative culture and encourage companies to share best practices, pool resources, and strengthen collective investments in sustainable initiatives. In addition, our roadmap aims to support stakeholders beyond automotive companies. Government agencies, research institutions, and non-governmental organizations will also be involved in the Digital Hub, forming a holistic approach to sustainable CE. From a more academic perspective, our paper provides theoretical knowledge about Digital Hubs fostering CE in the automotive industry as well as insights into a framework for their development. Additionally, various insights were gained about the understanding of sustainability, evaluation metrics, data, and technologies in the context of the CE. This guides academia in adopting and building upon our knowledge.

Naturally, this paper also has some **limitations**, opening the discussion for further investigation and development. First, our results are limited as the concept of Digital Hubs is relatively unknown in the scientific literature. While efforts were made to provide background information about Digital Hubs, the unfamiliarity with the concept introduces a potential source of bias in the interpretation of results. Second, there is a relatively low representation of SMEs within the *MobilKreis*. The findings may not fully capture the diversity of perspectives and practices within the SME landscape, potentially limiting the generalizability of the results. Third, it was challenging to schedule feedback interviews with external partners due to time constraints faced by the partners. This limitation may have influenced the depth and breadth of information gathered, potentially limiting the overall richness of the qualitative data. Despite these limitations, the insights gained from this paper contribute to broader implications of the Digital Hub concept within sustainability and CE.

Our results pave the ground for the implementation of the DIONA Digital Hub in practice. In **future research**, we will expand our research through interviews and workshops with external experts from academia and industry. This provides deeper insights into the challenges and potentials of establishing the Digital Hub in practice. To present the Digital Hub to organizations and the society, various communication channels will be present. For instance, we will use an independent website and social media channel in the future to publish information about the *MobilKreis* Hub. The *MobilKreis* Hub is at the heart of our efforts and acts as a central platform to optimally combine networking and knowledge transfer. The launch of the DIONA Digital Hub could be a catalyst for the transition to a CE in Germany's automotive industry and provide important knowledge, especially for SMEs, to tackle the most pressing challenges regarding CE. After the launch, we plan to collect additional relevant data that will help us target our outreach to stakeholders.

Acknowledgements. This work was partially funded by the "Digital Ecosystem for a Sustainable Circular Economy in the Automotive Industry" (DIONA) project. The project DIONA is funded by the German Federal Ministry of Education and Research (FKZ: 02J21E202).

References

1. IPCC, „Synthesebericht zum Sechsten IPCC-Sachstandsbericht (AR6): Hauptaussagen aus der Zusammenfassung für die politische Entscheidungsfindung," 2023.

2. United Nations, *Transforming our world: The 2030 agenda for sustainable development.* [Online]. Available: https://sdgs.un.org/2030agenda, last accessed 2023/11/10.

3. UNFCCC, *Paris Agreement.* [Online]. Available: https://unfccc.int/files/meetings/paris_nov_2015/application/pdf/paris_agreement_english_.pdf, last accessed 2023/11/29.

4. German Federal Government, "German Sustainable Development Strategy" 2021.

5. M. Fuchs and J. Westermeyer, „Automobilindustrie" in *Wirtschaftsgeographie Deutschlands,* E. Kulke, Ed., 3rd ed., Berlin, Germany, Heidelberg: Springer Spektrum, 2023, pp. 215–225.

6. Zukunft der Wertschöpfung, *Projekte Archive - Zukunft der Wertschöpfung.* [Online]. Available: https://www.zukunft-der-wertschoepfung.de/projekte/?filter-funding-measure=3533, last accessed 2023/11/29.

7. Federal Ministry of Education and Research, *DIONA – Digital ecosystem for a sustainable circular economy in the automotive industry.* [Online]. Available: https://diona-hub.de/, last accessed 2023/11/29.

8. H. Brundtland and M. Khalid, *Our Common Future: Report of the World Commission on Environment and Development,* 1st ed. USA: Oxford University Press, 1987.

9. W. J. Ripple *et al.,* "World Scientists' Warning of a Climate Emergency, 2021" *BioScience,* vol. 71, no. 9, pp. 894–898, 2021, https://doi.org/10.1093/biosci/biab079.

10. W. B. Gallie, "Essentially Contested Concepts," *Proceedings of the Aristotelian Society,* vol. 56, pp. 167–198, 1955. [Online]. Available: http://www.jstor.org/stable/4544562, last accessed 2023/10/05

11. D. Gottschlich, S. Hackfort, T. Schmitt, and U. Winterfeld, *Handbuch Politische Ökologie: Theorien, Konflikte, Begriffe, Methoden.* Bielefeld: transcript, 2022. [Online]. Available: https://www.researchgate.net/publication/365180505_Handbuch_Politische_Okologie_Theorien_Konflikte_Begriffe_Methoden, last accessed 2024/01/15

12. W. Haas, F. Krausmann, D. Wiedenhofer, and M. Heinz, "How Circular is the Global Economy?: An Assessment of Material Flows, Waste Production, and Recycling in the European Union and the World in 2005," *Journal of Industrial Ecology,* vol. 19, no. 5, pp. 765–777, 2015, https://doi.org/10.1111/jiec.12244.

13. R. F. van Schalkwyk, M. A. Reuter, J. Gutzmer, and M. Stelter, "Challenges of digitalizing the circular economy: Assessment of the state-of-the-art of metallurgical carrier metal platform for lead and its associated technology elements," *Journal of Cleaner Production,* vol. 186, pp. 585–601, 2018, https://doi.org/10.1016/j.jclepro.2018.03.111.

14. M. C. den Hollander, C. A. Bakker, and E. J. Hultink, "Product Design in a Circular Economy: Development of a Typology of Key Concepts and Terms," *Journal of Industrial Ecology,* vol. 21, no. 3, pp. 517–525, 2017, https://doi.org/10.1111/jiec.12610.

15. Ellen Macarthur Foundation, *Growth within: A circular economy vision for a competitive Europe.* [Online]. Available: https://www.ellenmacarthurfoundation.org/growth-within-a-circular-economy-vision-for-a-competitive-europe, last accessed 2023/10/11.

16. J. Kirchherr, D. Reike, and M. Hekkert, "Conceptualizing the circular economy: An analysis of 114 definitions," *Resources, Conservation and Recycling,* vol. 127, pp. 221–232, 2017, https://doi.org/10.1016/j.resconrec.2017.09.005.

17. Ellen Macarthur Foundation, *The circular economy in detail - Deep dive.* [Online]. Available: https://www.ellenmacarthurfoundation.org/the-circular-economy-in-detail-deep-dive, last accessed 2023/10/20.

18. P. Schroeder, K. Anggraeni, and U. Weber, "The Relevance of Circular Economy Practices to the Sustainable Development Goals," *Journal of Industrial Ecology,* vol. 23, no. 1, pp. 77–95, 2019, https://doi.org/10.1111/jiec.12732.

19. European Commission, *A Green Deal Industrial Plan for the Net-Zero Age.* [Online]. Available: https://commission.europa.eu/document/41514677-9598-4d89-a572-abe21cb037f4_en, last accessed 2023/03/21

20. P. Ghisellini, C. Cialani, and S. Ulgiati, "A review on circular economy: the expected transition to a balanced interplay of environmental and economic systems," *Journal of Cleaner Production*, vol. 114, pp. 11–32, 2016, https://doi.org/10.1016/j.jclepro.2015.09.007.

21. R. Zeiss, A. Ixmeier, J. Recker, and J. Kranz, "Mobilising information systems scholarship for a circular economy: Review, synthesis, and directions for future research," *Information Systems Journal*, vol. 31, no. 1, pp. 148–183, 2021, https://doi.org/10.1111/isj.12305.

22. A. Georgescu, S. Avasilcai, and M. K. Peter, "Digital Innovation Hubs—The Present Future of Collaborative Research, Business and Marketing Development Opportunities," in *Smart Innovation, Systems and Technologies*, vol. 205, *Marketing and Smart Technologies: Proceedings of ICMarkTech 2020*, Á. Rocha, J. L. Reis, M. K. Peter, R. Cayolla, S. Loureiro, and Z. Bogdanović, Eds., 1st ed., Singapore: Springer Singapore; Imprint Springer, 2021, pp. 363–374. [Online]. Available: https://www.researchgate.net/publication/349939460_Digital_Innovation_Hubs-The_Present_Future_of_Collaborative_Research_Business_and_Marketing_Development_Opportunities, last accessed 2023/12/08.

23. A. Crupi *et al.*, "The digital transformation of SMEs – a new knowledge broker called the digital innovation hub," *JKM*, vol. 24, no. 6, pp. 1263–1288, 2020, https://doi.org/10.1108/JKM-11-2019-0623.

24. J.-L. Hervás-Oliver, M. D. Parrilli, A. Rodríguez-Pose, and F. Sempere-Ripoll, "The drivers of SME innovation in the regions of the EU," *Research Policy*, vol. 50, no. 9, p. 104316, 2021, https://doi.org/10.1016/j.respol.2021.104316.

25. D. Kundisch *et al.*, "An Update for Taxonomy Designers," *Bus Inf Syst Eng*, vol. 64, no. 4, pp. 421–439, 2022, https://doi.org/10.1007/s12599-021-00723-x.

Multi-Criteria Evaluation of Potential Circular Economy Paths for Automotive Components Considering Targets

Anneke Schleusener[1](✉), Max Juraschek[2], Marko Gernuks[1],
and Christoph Herrmann[2,3]

[1] Volkswagen Aktiengesellschaft, Berliner Ring 2, 38440 Wolfsburg, Germany
anneke.schleusener@volkswagen.de
[2] Institute of Machine Tools and Production Technology, Chair of Sustainable Manufacturing and Life Cycle Engineering, Technische Universität Braunschweig, Langer Kamp 19B, 38106 Braunschweig, Germany
[3] Fraunhofer Institute of Surface Engineering and Thin Films, Riedenkamp 2, 38108 Braunschweig, Germany

Abstract. Increasing circularity of products, components and materials can provide an important approach in the automotive industry to actively conserve resources and achieve sustainability targets. In the near future, regulations will strengthen the requirements on the utilization of recycled materials for the manufacturing of new parts. It is a complex task to identify the potentially best performing circularity pathway for specific components against the different objectives of a company, as a practically applicable method for determining the circular economy potential for automotive components is not yet available.

Against this background, the most significant motivations and goals for the implementation of circular economy strategies are summarized and compared regarding their scope, established performance indicators, available assessment methods as well as current and future relevance. Subsequently, a framework is presented, which enables the evaluation of potential circular economy paths for automotive components. It considers a multi-criteria goal-setting and provides the basis for the implementation of an assessment tool.

Keywords: Circular economy · Automotive components · Evaluation framework

1 Introduction

The automotive industry is at a turning point as the transition from traditional combustion engines to electric vehicles is taking place [1]. This transition is a response to climate change and aims to ensure sustainable mobility in the future. However, the sometimes claimed zero emissions of electric cars only refers to the use stage and neglects the emissions in the production and end-of-life stage [2–5]. Consequently, car manufacturers face the challenge of introducing changes not only in vehicle use but also in production and end-of-life.

K. Dröder and T. Vietor (Eds.): CD 2024, Proceedings, pp. 254–266, 2025.
https://doi.org/10.1007/978-3-658-45889-8_20

The implementation of a circular economy has various advantages [6]. Companies can use it to achieve high-quality recycling of end-of-life vehicles (ELVs) on the one hand and to replace primary raw materials with recycled materials on the other. This offers the opportunity to reduce emissions in both the procurement and recycling stage, thus promoting a sustainability with a life cycle perspective. Another important mitigation strategy is the extension of the use stage in order to maximize the service life of products [7]. The circular economy can offer many benefits for companies and the environment. However, scenarios must be identified in which this is also advantageous from a business perspective.

2 Background

With regard to the circular economy, there are various push and pull factors for automotive companies that can be grouped in four categories for consideration: Economic, environmental and social challenges as well as compliance with legal requirements [8, 9]. Economic challenges can be exemplified by competitive pressure, rising energy and raw material costs. The contribution of emissions to climate change is an example for environmental challenges. In addition, companies also have a responsibility to society, which means that falling groundwater levels and greenhouse gas emissions can result in social challenges [9]. Drivers from legislation are, for example, the End-of-Life Vehicle Directive, the regulation concerning batteries and waste batteries, and the Supply Chain Duty of Care Act [10–12].

One approach to address the challenges is the implementation of circular economy. The aim of the circular economy is to preserve the value of the materials used for as long as possible in order to use them for the longest possible time and thus conserve resources taking into account cost and environmental aspects [13]. New technologies may be more efficient, making continued material use impractical. In such cases, material recycling offers a solution for producing new products. In the automotive industry, this could mean that vehicles are re-manufactured at the end of their life cycle instead of being scrapped. A second life can then be achieved by refurbishing an ELV. This would be followed by the dismantling of components and subsequent recycling/ upcycling of the materials [14].

The recycling of vehicles gives manufacturers the opportunity to fulfill the product responsibility required by the End-of-Life Vehicles Directive [10]. It can also create a business case for automotive companies: Recycling makes it possible to recover strategically important raw materials and thus generate added value. As already mentioned, there are compliance requirements that stipulate that new components, such as batteries, must contain a certain proportion of recycled materials [11]. This forces companies to recover their raw materials and secure them for production. Batteries in particular contain a variety of strategically important and rare raw materials, which is why a circular economy process is particularly suitable for electric vehicles. In addition to scarcity, the recycling of materials would also mean that less mining of materials would have an impact on the promotion of human rights.

In a scenario, in which manufacturers take back their vehicles at the end of their useful life in order to obtain strategically important raw materials of the battery, this also gives them the opportunity to use further parts of the vehicle for the circular economy. For this purpose, it is necessary to find the best recycling path for the respective components, as there are many different circular economy strategies. The "R-strategies", for instance, show that there are numerous ways to apply the circular economy by prioritizing the targets Refuse, Rethink, Reduce, Reuse, Repair, Refurbish, Remanufacture, Repurpose, Recycle and Recovery [13, 15].

Each strategy contains a variety of technologies for achieving the targets, with ongoing advancements, such as the continued development of industrial recycling pathways for lithium-ion traction batteries [16]. A pathway as a circular economy pathway describes the process by which components or materials are recycled, reprocessed or reused to achieve the goals of the circular economy. By establishing such pathways, waste can be reduced and resources used efficiently to close the loop and minimize environmental impact. The wide range of options shows that it is essential to have a decision-making basis for circular economy strategies in order to consider a variety of factors for the analysis.

3   Multi Criteria Targets for Circular Economy

The establishment of a circular economy presents an opportunity to pursue diverse objectives for automotive companies, facilitating the transition towards sustainable practices. One challenge lies in clearly defining these objectives and incorporating various goals with specific weightings to identify the most suitable strategy for specific components and products. Key factors in this context are identified in Bruzs' paper, 'A Review on the Outlook of the Circular Economy in the Automotive Industry', which serves as a foundation for the collection of the elements for the targets in this paper [17]. These factors were augmented through additional insights gleaned from discussions and supplementary literature, thereby enriching the foundational framework for the research. In this context, Table 1 outlines selected objectives that can be pursued by automotive companies in the context of circular economy. Each of the examined objective is analyzed across different sustainability dimensions economic, environmental, and social [18, 19].

The sustainability dimensions provide a comprehensive framework for evaluating the impact and relevance of each objective. A rating scale from one to five is employed to quantify the current and future significance of each goal. This assessment considers both existing relevance and forecasts for future importance A high rating indicates a high relevance for automotive companies, which has a significant impact on the potential success of products and business models. Targets with high ratings pose a high risk that a business model will fail if ignored, while 5 as the highest rating represents the most significant risk. The assessment of relevance does not claim to be a statistically valid survey, but rather a current industrial assessment based on various opinions of experts in the field of circular economy.

Table 1. Multi-criteria targets of circular economy in the automotive industry

Target	Dimension	Exemplary evaluation method	Example	Current relevance	Prospective relevance
Economic performance	Economic	Annual report, total cost of ownership, ESG Ratings	[8, 17, 18, 20, 21]	●●●●●	●●●●●
Technical quality performance	Economic	Requirement specification	[12, 22]	●●●●●	●●●●●
Access to resources	Economic	Risk management, criticality ranking	[8, 17]	●●	●●●●
Marketing	Economic	Customer perception, company image	[23]	●	●●●
Product function customization	Economic, environmental	Product specifications and comparison	[24]	●●●●●	●●●●
Lifecycle data information transparency	Economic, environmental	Systematic data collection	[8, 25, 26]	●●●●	●●●●
Product lifetime extension	Economic, environmental	Life Cycle Analysis, average vehicle age	[13, 27]	●●	●●●●
Material origin	Economic, social, environmental	Quality eg. Processing properties, price fluctuations	[12, 17]	●●	●●●●
Performance end of life	Economic, social, environmental	Life Cycle Cost, proof of activities	[10, 21, 26]	●●	●●●
Environmental impact reduction	Environmental	Life Cycle Analysis, "Impact Points"-method, ecological scarcity method	[8, 17, 18, 28]	●●●●	●●●●●

(continued)

Table 1. (*continued*)

Target	Dimension	Exemplary evaluation method	Example	Current relevance	Prospective relevance
Common good	Social	Common good matrix	[29]	●●	●●●

The transformation to a circular economy aims to keep products and goods in circulation for as long as possible and thus preserve their value. The **economic** aspects of a circular economy also include the required recycled content in new products. Due to the legal requirements, industrially manufactured materials with recycled content are expected to become more cost-efficient, so it is economically advantageous for companies to incorporate parts of their used products into their production [8, 17, 18, 20, 21].

In order to maintain the **technical quality performance** during the application of circular economy, it is essential to reduce the requirements to a necessary minimum. In certain respects, the requirements set out in the specifications may exceed the specifications required by the market, even for high-end vehicles. The use of recycled materials will result in fluctuations in quality, but these have to remain within the limits of customer satisfaction performance [12, 22]. By adjusting, raw materials can be procured in response to requirements and economical production might be achieved.

Access to resources is one of the objectives of the circular economy. By recycling materials, availability is guaranteed and independence from market fluctuations is created. The risk of resource independence will also be reduced [8, 17].

The focus on the circular economy in **marketing** is an economic factor. This can enhance the company's image and customers can also make purchasing decisions based on the sustainable actions of companies. The increasing urge to act sustainably due to advancing climate change influences customer behavior and strengthens the importance of marketing in the area of the circular economy [23].

The **product function customization** represents a challenge for the circular economy. The demand for universal use of vehicles around the world and individual customer requirements leads to numerous product configurations [24]. By reducing the possibility of creating variants during vehicle configuration, inhomogeneity decreases. The reduction of individuality creates a greater potential for circular economy, which brings environmental benefits. A reduction also has economic advantages, as the reduction in variants reduces production costs.

Lifecycle data information transparency enables a customized and comprehensive maintenance and repair strategy through precise data analysis of the product and component condition. Consequently, this can increase the product lifetime, resulting in reduced life cycle costs and a lower environmental impact [8, 25, 26].

Product lifetime can be significantly **extended** if the accessibility of parts is ensured. Extending the functional service life of resources means that the extraction and use of new resources, e.g. raw materials, can be avoided. The accessibility and separability of parts increases the guarantee that vehicles can be repaired and maintained [13, 27]. This can reduce the number of working hours required and make repairs more economical for the customer.

The **material origin** has multi-criteria dimensions. The cost of procurement is a decisive criterion. However, the focus is also shifting to the social dimension. A review of the supply chain for compliance with legal requirements and protection of human rights is essential. The ecological factor also plays a significant role. Not only transportation but also the extraction of raw materials can have a major impact on the environment [12, 17].

End-of-life performance is essential for a circular economy. Vehicles offer great potential as the basis for a reuse business. A cascading approach is used to determine which scenario is selected for the vehicle components in order to maximize recycling as opposed to downcycling. Recycling has economic advantages, as sales of recycled parts and materials can generate additional earnings. Valuable raw materials and materials can also be recovered. This can reduce manufacturing costs. In addition, it reduces the need for raw material extraction. This also improves social conditions. Overall, the reuse of components both as spare parts and as raw materials reduces environmental impact [10, 21, 26].

One of the targets of automotive companies with regard to the circular economy is to **reduce** their **environmental impact.** The reuse of components, application-oriented planning and the reduction of waste contribute to the reduction [8, 17, 18, 28].

Sustainability also plays a role in the **common good** assessment of a company. In this sense, circular economy activities can also be included in the assessment. The common good orientation looks at the extent to which products and services also bring social benefits. This means that the support of the circular economy can also be included in the common good assessment [29].

The evaluation of the objectives illustrates the different approaches and drivers that companies have in relation to the circular economy [8]. The reduction of environmental impact is poised to emerge as a paramount goal in the future, driven by the imperative to address and mitigate climate change. Numerous companies have established internal targets aimed at reducing their environmental impact.

The targets for economic performance and technical quality performance will consistently maintain a high level. Economic performance holds paramount importance for companies, as it directly influences their profitability. Technical performance, distinguishing the company, serves as one of the primary reasons why customers opt for specific products or services.

The targets of end-of-life performance and material origin are projected to attain a moderate level of importance. Similarly, product lifetime extension and lifetime data transparency will garner a moderate level of prospective significance. These, alongside the objectives related to product function inhomogeneity, serve as the juncture between economic and environmental considerations.

The economic objectives encompass resource accessibility, marketing, technical quality performance, and economic performance, targets will increase in relevance in the future.

The illustration indicates a diminishing significance of an objective tied to product requirements. The functional consistency in products highlights the imperative need for a more application-centric approach in product planning and design, which would encourage reusability. Extending the product life of vehicles, supported by certain services and maintenance by manufacturers, would simultaneously reduce the production of new vehicles and consequently reduce the impact on the environment.

Implementing circular economy principles into company processes involves a range of different organization-specific methods. There's no one-size-fits-all solution in this domain. Each company must evaluate and prioritize its objectives in order to find the most suitable model for its activity. Nevertheless, it is necessary that the top down prioritized strategies should be coherent with the strategies of other players in the circular economy market. This individualized approach acknowledges the necessity for bespoke strategies, recognizing that what proves effective for one firm may not be applicable to another. Ultimately, achieving a circular economy necessitates a thorough evaluation of internal objectives and the customization of implementation strategies to ensure positive outcomes.

Figure 1 provides a summarized overview of the objectives and how they are related to the three dimensions of sustainability: Economic, Environmental and Social. The relevance of the individual dimensions is indicated by the size and direction of the trends, with both increases and decreases shown. It is noticeable that the objectives of the circular economy are currently primarily linked to the ecological and environmental dimensions. However, many objectives are also linked to the social dimension, which indicates interfaces and connections between these areas. This visualization highlights the complexity of the circular economy objectives and emphasizes the strong focus on ecological and environmental aspects while at the same time recognizing their interconnectedness with social aspects [29, 30].

The investigation and classification of these objectives can serve as a tool to aid companies in prioritizing and planning their strategies to promote a circular economy. This approach allows for the comprehensive assessment of multifaceted objectives regarding their significance and contribution to sustainability. Consequently, it enables companies to implement more targeted and effective measures in advancing a sustainable and environmentally friendly economy. To leverage the identified circular economy objectives, their assimilation into an assessment methodology is crucial. This necessitates the development of a framework for an evaluation method.

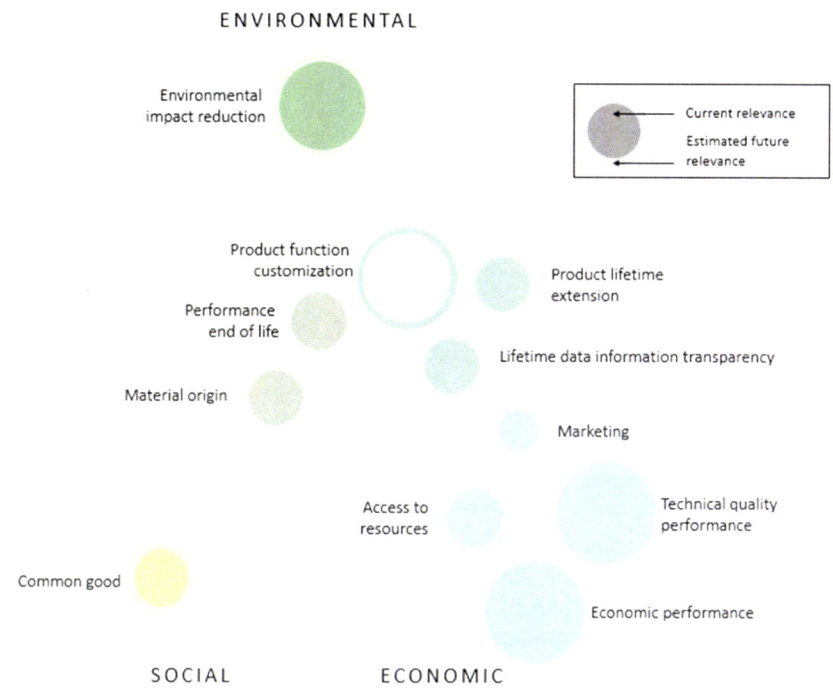

Fig. 1. Variation relevance of the targets in the dimension of sustainability

4   Framework for Evaluation of Potential Circular Economy Paths

The framework aims to assess potential pathways within the circular economy to identify viable circular economy routes for components from ELVs. This design provides a way to achieve results by considering specific key performance indicators (KPIs) and their variable weighting as part of a circular economy methodology to find solutions. This framework builds on basic inputs such as defined targets, product features and manufacturing process specifications.

It should be applied during the product development phase. By anchoring the circularity of the components in the specifications, it can be ensured that the developers are obliged to find possible reuse scenarios for the components. The tool should be operated by an environmental expert who works in the development department in an operational context. This activity forms an interface between life cycle analysis and development (Fig. 2).

The process outlined is divided into three main process steps: The phase in which input is generated, the evaluation of potential circular economy opportunities and the output including the implementation phase.

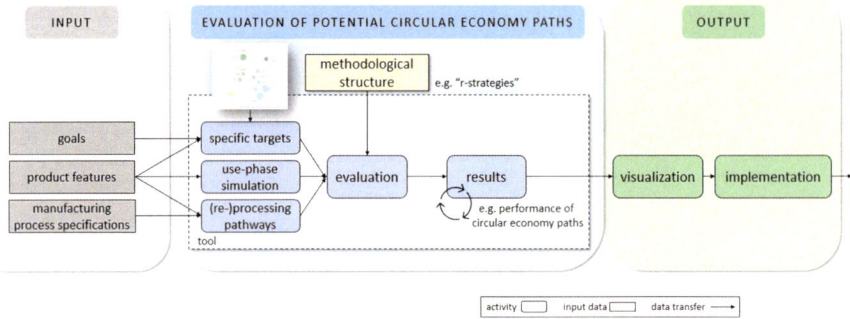

Fig. 2. Evaluation Framework of potential circular economy paths

In the input phase, the objective is to carry out comprehensive data collection. Parameters and data for the defined objectives, the product features and the manufacturing process specifications are collected. This step is necessary in order to obtain the data basis for the subsequent evaluation and calculation processes.

The core of this framework is the evaluation phase, in which the entire evaluation of the circular economy pathways takes place. The evaluation will take place in the shape of a tool. The input for this phase is the data already generated. The phase is divided into five process steps, three of which take place simultaneously: The selection of specific targets, the use-phase simulation and the mapping of possible (re-)processing pathways. The specific targets are selected individually for the use case by the user on the basis of the targets developed in Fig. 1 as well as information from the goals and the product features from the input phase. The goals refer to the general intentions or objectives of a company that are often anchored in the corporate policy or corporate governance. An evaluation method, e.g. pairwise comparison, is used to select which focus should be pursued in the evaluation. This step makes it possible to adjust the focus of the evaluation and thus to consider multiple criteria. The use-phase-simulation aims to simulate the ELV state of the vehicles during the development of the vehicle, the use phase, in order to carry out the evaluation on the basis of these results. For this purpose, a variety of use phases are cumulated to form the overall optimum and thus the probable average state of the ELVs. The simulation is based on information from the product features. Mapping the (Re-)processing pathways provides an overview of the possible circular economy pathways for the components. For this purpose, a process chain is created from current and future accessible technologies. Depending on the process step used in the assembly of the vehicle, different process steps are possible for disassembly. For example, a component that has been joined using the screwing process can only be separated again by screwing. The step is supplied with information from the product features and the manufacturing process specifications. Within the evaluation step, the previously generated data is used to carry out the assessment. The methodological structure also flows into this step, in this case the R-strategies, for example. This determines the procedure during the evaluation. Using the R-strategies as an example, a cascading check is carried out to determine whether the respective strategy can be considered as a path. If yes, the result is available for the respective component. If not, the same procedure is carried out with the next strategy in the list. This step forms the main step of the framework, as the output is the results in the form of performance of possible circular economy paths.

This is followed by the output phase. At the beginning, the results of the evaluation will be visualized in the form of possible circular economy paths to give the developers a visual overview. The developers have the opportunity to design the future circular economy of the components during their work. This is done in the next step, implementation. This is where the results are integrated into the company processes. These identified pathways provide the company with concrete opportunities to promote sustainable practices through the reuse of end-of-life vehicle components and thus contribute to circular economy efforts. Planned circular economy pathways must also be technically prepared, for example in the form of a dismantling area.

In the implementation of the framework, a specific vehicle development project will represent a use case. For example, an electric vehicle could be considered as a specific use case. The application of this framework is carried out by a person in the development department who is an environmental expert. During the development phase of a vehicle, possible circular economy paths for vehicle components in the end-of-life phase are to be aligned. Based on the targets, an goal needs to be selected, e.g. the aim of reducing the environmental impact. An evaluation can be carried out on the data and specifications of this electric vehicle. The results are circular economy paths for individual components. One potential outcome could be that a component is replaces on a cyclical basis and therefore does not have the same age as the vehicle. Consequently the specific component could be returned for refurbishment and then be available as a spare part for repairs. Subsequent the outputs have to be integrated into company processes and combined. In order to anchor the application of the framework and its implementation in the company processes in the specifications in the long term, a decision must be made at divisional/product responsibility level.

5 Conclusions and Outlook

Implementing circular economy for automotive companies is connected to a multitude of objectives on a multidimensional level, while relevance of these objectives is currently changing. This will give companies the opportunity to act more sustainably by establishing the principles of circular economy in their processes.

This paper presents a framework for identifying possible circular economy pathways for automotive components with an individual prioritization of targets. In order to be able to use the method, it is necessary to set up an interface during development that defines exactly which data must be used. In order to check the application of the method, it is foreseen to carry out an exemplary application on a component. In this way, the defined required data can be checked on the component. The method will also be tested in the application and calculation. The results obtained can then be interpreted and it becomes clear how they can be integrated into the industry.

The integration of circular economy principles in the automotive industry will become a necessity in the future against the background of escalating climate change and the tightening legal framework. Therefore, the efforts of car manufacturers to develop innovative solutions will be central to shaping a sustainable and adaptable sector.

Disclaimer. *The results, opinions and conclusions expressed in this publication are not necessarily those of Volkswagen Aktiengesellschaft.*

References

1. Valladares Montemayor, H. M., Chanda, R. H.: Automotive industry's circularity applications and industry 4.0. Environmental Challenges 12, (2023).
2. Hawkins, T. R., Singh, B., Majeau-Bettez, G., Strømman, A. H.: Comparative Environmental Life Cycle Assessment of Conventional and Electric Vehicles. Journal of Industrial Ecology 1 (17), 63–73 (2012).
3. Sharma, R., Manzie, C., Bessede, M., Crawford, R. H., Brear, M. J.: Conventional, hybrid and electric vehicles for Australian driving conditions: part 2: Life cycle CO_2-e emissions. Transportation Research Part C: Emerging Technologies 25, 238–249 (2012).
4. Tagliaferri, C., Evangelisti, S., Acconcia, F., Domenech, T., Ekins, P., Barletta, D., Lettieri, P.: Life cycle assessment of future electric and hybrid vehicles: A cradle-to-grave systems engineering approach. Chemical Engineering Research and Design 112, 298–309 (2016).
5. Hoa, H., Qioa, Q., Liu, Z., Zhao, F.: Impact of recycling on energy consumption and greenhouse gas emissions from electric vehicle production: The China 2025 case. Resources, Conservation and Recycling 122, 114–125 (2017).
6. Ferronato, N., Rada, E. C., Gorritty Portillo, M. A., Cioca, L. I., Ragazzi, M., Torretta, V.: Introduction of the circular economy within developing regions: A comparative analysis of advantages and opportunities for waste valorization. Journal of Environmental Management 230, 366–378 (2019).
7. Prox, M.: Circular Economy. In Schwager, B. (ed):CSR und Nachhaltigkeitsstandards, pp. 261–274. Springer Gabler, Berlin (2022).
8. Schwager, B.: CSR und Nachhaltigkeitsstandards: Normung und Standards im Nachhaltigkeits-kontext. Springer Gabler, Berlin (2022).
9. Herrmann, C.: Ganzheitliches Life Cycle Management. Springer Berlin Heidelberg (2010).
10. Ellen MacArthur Foundation. Completing the picture – How the circular economy tackles climate change (2021).
11. European Commission: Proposal for a regulation of the European parliament and of the council on circularity requirements for vehicle design and on management of end-of-life vehicles amending Regulations (EU) 2018/858 and 2019/1020 and repealing Directives 2000/53/EC and 2005/64/EC. 13.07.2023.
12. European Parliament and the Council of the European Union: Regulation (EU) 2023/1542 of the European Parliament and of the Council of 12 July 2023 concerning batteries and waste batteries, amending Directive 2008/98/EC and Regulation (EU) 2019/1020 and repealing Directive 2006/66/EC. 28.07.2023.
13. Federal Ministry of Justice: Gesetz über die unternehmerischen Sorgfaltspflichten zur Vermeidung von Menschenrechtsverletzungen in Lieferketten. 16.07.2021.
14. Morseletto, P.: Targets for a circular economy. Resources, Conservation & Recycling 153, (2020).
15. Gernuks, M.: Circular Economy. Engel Mobility Days. (2023).
16. Potting, J., Hekkert, M., Worell, E., Hanemaaijer, A.: Circular Economy: Measuring innovation in the product chain. PBL Netherlands Environmental Assessment Agency, Den Haag (2017).
17. Blömeke, S:, Scheller, C, Cerdas, F., Thies, C., Hachenberger, R., Gonter, M., Herrmann, C., Spengler, T. S.: Material and energy flow analysis for environmental and economic impact assessment of industrial recycling routes for lithium-ion traction batteries. Journal of Cleaner Production 377 (2022).

18. Buruzs, A., Torma, A.: A Review on the Outlook of the Circular Economy in the Automotive Industry. International Journal of Environmental and Ecologic Engineering 6 (11), 576–580 (2017).

19. Kara, S., Hauschild, M., Sutherland, J., McAloone, T.: Closed-Loop Systems to Circular Economy: A Pathway to Environmental Sustainability? In CIRP Annals – Manufacturing Technology 71, vol. 2, pp. 505–528 (2022).

20. Kropp, A.: Grundlagen der Nachhaltigen Entwicklung. Springer Gabler Wiesbaden (2018).

21. Neske, A., Schulz, J., Scheer GmbH: How to: Wie sich Unternehmen für die Kreislaufwirtschaft aufstellen müssen. IM+io Best & Next Practices aus Digitalisierung | Management | Wissenschaft 1 (2023).

22. Kalvenkamp, M., Raabe, T.: Automotive Remanufacturing in the Circular Economy in Europe: Marketing System Challenges. Journal of Macromarketing, Vol 38 (I) 112–130 (2018).

23. Berg, E., Dahlmann, R., Schön, M.: Verunreinigungen auf der Spur – Die Eigenschaften von Kunststoffen ermitteln. Kunststoffe: Werkstoffe, Verarbeitung, Anwendung 11, 60–63 (2022).

24. Vassileva, B., Ivanov, Y.: "3G" Business Model for Marketing 4.0: Implications for Circular Economy. Journal of Engineering Trends in Marketing and Management, 1(1), 124–134 (2017).

25. Daffner, R.; Juckenack, S.; Schurrer, S.: The good, the bad, and the complicated. Dealing with car complexity. OliverWyman (2021).

26. Wilde, A.-S., Wanielik, F., Rolinck, M., Mennenga, M., Abraham, T., Cerdas, F., Herrmann, C.: Ontology-based approach to support life cycle engineering: Development of a data and knowledge structure. In: 29th CIRP Life Cycle Engineering Conference, pp. 398–403 (2022).

27. Saidani, M., Yannou, B., Leroy, Y., Cluzel, F.: Heavy vehicles on the road towards the circular economy: Analysis and comparison with the automotive industry. Resources, Conservation & Recycling 135, 108–122 (2018).

28. Krause, C., Wei, L., Ortiz, V., Lobeck, F.: Digitale Transformation und Integration im Automotive Aftersales. In Proof. H. (ed.): Towards the New Normal in Mobility, pp. 489–505. Springer Gabler, Wiesbaden (2023).

29. Gebler, M., Witte, S., Blume, S., Muhl, M., Finkbeiner, M.: The "Impact Points"-method: A distance-to-target weighted approach to measure the absolute environmental impact of Volkswagen's global manufacturing system. Journal of Cleaner Production 386 (2023).

30. Bookhagen, A.; Seymour, G.: Sustainable-oriented Entrepreneurship: Sozial und ökologisch verantwortliches Handeln als Teil des unternehmerischen Selbstverständnisses, PraxisWISSEN Marketing, Arbeitsgemeinschaft für Marketing (AfM) 4, 112–126 (2019).

31. Nikolaou, I. E., Jones, N., Stefanakis, A.: Circular economy and sustainability: the past, the present and the future directions. Circular Economy and Sustainability, 1–20 (2021).

Modelling Usage-Specific Circularity Measueres for Individual Life Cycle Paths in Life Cycle Engineering

Marie Schwahn[1]([✉]), Lukas Block[2], Thomas Potinecke[2], and Maximilian Werner[2]

[1] Institute of Human Factors and Technology Management (IAT), University of Stuttgart, Nobelstraße 12, 70569 Stuttgart, Germany
marie.schwahn@iat.uni-stuttgart.de

[2] Institute for Industrial Engineering (IAO), Fraunhofer IAO, Fraunhofer, Nobelstraße 12, 70569 Stuttgart, Germany
{Lukas.Block,Thomas.Potinecke}@iao.fraunhofer.de
https://www.iat.uni-stuttgart.de/, https://www.iao.fraunhofer.de

Abstract. Circularity measures like repair, remanufacture and refurbish provide options to enhance the life cycle and thus the sustainability of a product besides measures, which primarily concentrate on the end-of-life of a product. To integrate the concept of circular value creation engendered by such measures in the development process of a product, we extended the idea of modelling standard life cycles as introduced by Block et al. [3]. This extension introduces the novel notion of undesired states within a product's life cycle. Undesired states can be dissolved with the help of circularity measures. A Model Based Systems Engineering (MBSE) approach was developed to implement this strategy. The resultant modelling approach is implemented in form of a software tool. The system is evaluated within an automotive development project revealing a significant advantage in contrast to previous methods.

Keywords: Circularity · Product development · Life cycle engineering · Meta model · Model-based systems engineering

1   Introduction

Circular Economy is a promising approach to enhance the sustainability of products. Resources of raw materials are limited, and resource scarcity will become more and more noticeable within the next years. Moving from linear to circular value creation drives the optimization of resource utilization by reducing the consumption of raw materials and enabling second life of products or subcomponents. Circularity measures like repair, remanufacture and refurbish provide options to enhance the life cycle and thus the sustainability of a product besides measures, which focus on the end-of-life of a product (e.g., recycling, see [18]). Such circularity measures are usually employed in case an undesired state of the product occurs in its production or usage phase. Examples for such undesired states are defects or obsolete technologies in a certain component. This

K. Dröder and T. Vietor (Eds.): CD 2024, Proceedings, pp. 267–278, 2025.
https://doi.org/10.1007/978-3-658-45889-8_21

is where the application of circularity measures can obviate the transfer to an end-of-life state. To maximize the positive impact of circularity measures their application during a product's life cycle must be considered and assessed in an early development stage. A promising and widely used methodology of systems engineering in the field of Circular Economy is Model Based Systems Engineering (MBSE). MBSE allows for precise and comprehensive modelling of the system under consideration, enabling the developer to cope with complex system dependencies. MBSE is a suitable methodology to comprise the full range of necessary factors to be considered in Circular Economy based product development. To integrate the concept of circular value creation in the product development process, we extended the MBSE based approach of modelling standard life cycles as introduced by Block et al. [3].

To adequately integrate circularity measures into the estimated standard life cycles of a product, it is important to identify points during the life cycle where undesired states are likely to occur. Currently, the probabilities of such undesired states are only considered statistically in product development. However, the statistical numbers are lacking information about the circumstances under which products are entering undesired states. Thus, options to return the component into a useful state cannot be designed systematically. Further, the impact of the application of a circularity measure in case of an undesired state cannot be assessed. Therefore, circularity measures as "return paths" into a useful state cannot be described and evaluated in terms of their quality and quantity.

Consequently, the challenge lies in the early qualitative and quantitative analysis of the impact of circularity measures in a product's life cycles. Within this paper we utilize MBSE to address this challenge. We present a meta model to integrate undesired states into product life cycle. The novel modelling approach enables the developer to identify suitable circularity measures and immediately assess their overall impact.

2   Literature Review

The field of product development is confronted with new sustainability challenges, particularly in relation to circular-oriented concepts such as the Circular Economy [1, 6, 16]. The Circular Economy is an economic model that seeks to minimize waste and resource consumption by emphasizing continuous product and material use, recycling, and regeneration [18]. Within this context, Potting et al. [18] have identified three key measures for achieving circular design: material efficiency (recover and recycle), extended product life cycles (repurpose, remanufacture, refurbish, repair, and re-use), and intelligent product development (reduce, rethink, refuse). The German Institute of Norms (DIN) has also introduced a similar approach aimed at aligning business processes with circular principles [9]. While these strategies provide concrete definitions for various circularity approaches, they lack specific design and analysis guidance. It is crucial to extend product life cycles and maximize the utilization of products, components, and materials for as long as possible, requiring consideration of additional phases during the product development stage, such as product usage [8, 13]. Consequently, a life cycle model for circularity assessments during the development phase is required. It should describe the potential life cycle trajectories and defects that a product may encounter throughout its lifespan. The trajectories and defects then serve as the foundation for determining circularity measures that enhance the longevity of the product [12, 14].

More than to 6,000,000 publications can be identified, when looking at scientific literature containing the keyword "Life Cycle Model" [15]. Most of this literature describes life cycle models in other scientific fields, e.g., biology, social research, or informatics. However, when condensing the literature pertaining to product development, a significant portion is dedicated to the definition of abstract life cycle descriptions. These descriptions primarily revolve around a single unified life cycle that encompasses abstract phases like development, production, logistics, usage, maintenance, and recycling (see e.g., [2, 19, 21]). However, representing the life cycle of a product with a single model may not accurately depict the distribution among and the variations within the individual products. These variations stem from diverse factors such as differing product usage, personalization, and software or hardware updates. Bougain et al. [4], Yvars et al. [22], Cerdas et al. [5], Dér et al. [7], and Tao et al. [20] propose approaches that consider such individual life cycle circumstances, for example through Bayesian Networks. However, these approaches are primarily focused on data and simulation, making them less suitable for the development phase where detailed data and simulations may not be readily available, and outcomes must be understandable to reason about the design decisions. Consequently, Kong and Wang et al. [14] states that a life cycle model that integrates and explicitly represents the overall life cycle as well as the different paths and their hierarchical dependencies is still missing.

To address this, Block et al. [3] proposed a new approach to model individual life cycle paths during product development for Life Cycle Engineering. They define a life cycle as a series of states that a product can transition between, with assigned probabilities indicating the likelihood of such transitions. This enables a comprehensive description of both the overall life cycle of the product and its individual life cycles. However, it should be noted that the circularity aspect is not explicitly incorporated into Block et al.'s [3] life cycle model. It lacks to describe the defects that a product may encounter throughout its lifespan, which are necessary to identify appropriate circularity measures and promptly assess their overall impact. As such, the research gap is that an integrated life cycle model is missing, which can describe the overall life cycle, individual life cycle trajectories as well as defects that a product may encounter throughout its lifespan. Yet, such a model is necessary to incorporate circularity measures into the product's design.

3 Approach

We present an MBSE based approach to enable an early qualitative and quantitative analysis of the impact of circularity measures in product life cycles. To integrate the concept of circular value creation in the product development process, we extended the MBSE based approach of modelling standard life cycles as introduced by Block et al. [3]. The extended meta model enables the user to model extensions composed of undesired states and return paths via circularity measures. The extension therefore introduces the concept of circular value creation. Based on the modelled Circular Economy related life cycle extension, the impact of circularity measures can be immediately assessed. The developer can improve the sustainability of the product iteratively by adapting the strategy to apply circularity measures based on the feedback on their impact.

3.1 The Concept of Undesired States and Return Paths via Circularity Measures

Circular aware product development aims for a product design that allows for the application of circularity measures and in general for circular value creation during its life cycle. Our approach for circular aware product development, is based on the concept that each product has its individual life cycle [3]. Block et al. present a model to define product life cycles in system architecture models providing the application structure of the system and its components. Building upon a life cycle model allows us to incorporate undesired states and circularity measures by extending the regular model. The modeling of an extension enables us to set the appearance of an undesired state into relation and give context to the abstract concept of undesired states. The basis for the MBSE based approach towards Circular aware product development is composed of two basic modelling steps. The first one is the modelling of the product architecture as a foundation to integrate all functional relations and life cycle modelling. The second step is the modelling of life cycles which can be modelled for the product as a whole as well as for a subcomponent as provided by the physical product architecture. Life cycles are modeled in a probabilistic manner. The modeling approach enables the definition and analysis of different life cycle paths of components. A life cycle is modelled as a composition of states and transitions with probabilities [3]. During the development of a product, the developer can estimate such an expected life cycle based on experience and expertise in the field. Transitions feature probabilities. The probabilities describe how likely it is that a product will enter the related subsequent state defined as end point of the transition. The probabilistic description of life cycle paths enables the modeling of uncertainty in an early development state. The life cycle of a product is not deterministic, and all possible options need to be considered for a holistic evaluation of life cycle variants. The modeling of estimated life cycles is based on the assumption that the product stays in a desired condition. This means the product has the intended quality and functionality throughout the entire life cycle. The introduced modeling approach for regular life cycles enables a holistic analysis and evaluation of product variants. It does not include a methodology to deal with unforeseen defect product conditions. Utilizing this modeling approach would lead to a product that most likely will be transferred to an end-of-life state when leaving the desired condition. This work aims to enable circular value creation and avoid immediate end-of-life transfers.

For this purpose, we extend the idea of modeling faultless product life cycles by the concept of undesired states. Undesired states are commonly either defects, obsolete technologies or product conditions which are unprofitable or will become unprofitable in the foreseeable future. The life cycle extensions including undesired states are based on an extension of the metamodel for regular life cycle modeling. The concept of modelling life cycle paths composed of states and transitions is reused and undesired states are integrated into this methodology as a special type of state. We do not only consider undesired states statistically. Instead, the underlying definition of undesired states in the extended metamodel includes a description of the product condition at that time. When entering an undesired state, knowledge of the product condition enables an educated decision on the products further handling. This is where circularity measures can take effect. Instead of devaluating the products at this point, a circular value creation approach aims to return the component to a useful state. The return to use can take place either in the same system or across system borders. The undesired state does not immediately result in an end-of-life state. Potting et al. [18] identify ten possible strategies for closing energy, material, and usage cycles. We refer to them as circularity measures. They are categorized as follows: "recycle", "repurpose", "remanufacture", "refurbish", "repair", "reuse", "rethink", "reduce" and "recover". The extension of the modelling approach provides a library of circularity measures which are functional measures based on those strategies. These circularity measures can be integrated into the life cycle models. Fig. 1 visualizes the concept of modeling the expected product life cycle and the extension composed of undesired states and return paths via circularity measures.

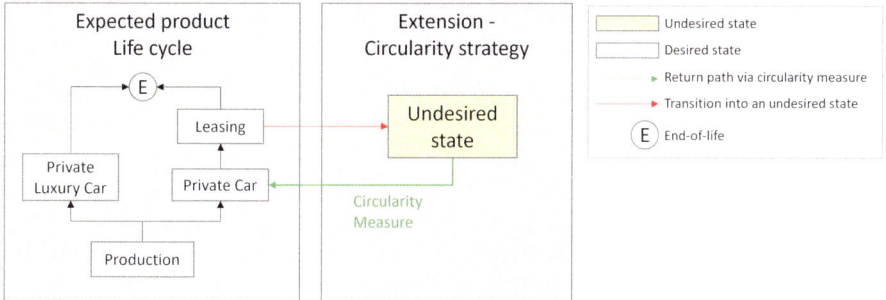

Fig. 1. A state desired state within the expected life cycle depicted on the left hand side can lead to an undesired state which is part of the extension to model a circularity strategy as visualized on the right hand side. An undesired state can be dissolved by applying circularity measures and return the product into a desired state.

The transition from a regular state to an undesired state is defined as normal transition with a probability. For the reverse direction, a special type of transition is introduced. A connection from an undesired state to a regular state can only be established by applying a circularity measure. Meaning once a product or component reaches an undesired state this state can only be dissolved with the help of the application of a circularity measure. In case an undesired state does not feature any circularity measure transitions as return paths, the undesired state is assumed to mark the end-of-life of the related physical component.

3.2   Assessment of Circularity Measures

In the previous section we introduced the developed approach to model usage-specific circularity measures in life cycle engineering. The new model helps to deal with undesired product behavior and introduce the concept of Circular Economy and circular value creation in product development. To enable the developer to optimize the application of circularity measures for his project, a strategy for a holistic assessment of the generated models must be developed. In the following we introduce such an assessment methodology.

In conventional product development a generated product version is evaluated based on metrics such as costs, material consumption, energy consumption, lifetime, and other parameters. These metrics can be calculated based on specifications of the product characteristics and an estimated product life cycle. To allow for a comparability between traditional product development and Circular Economy focused development, the metrics, the same metrics can be applied to the circular approach including sustainability metrics. The developer itself can choose to specifically focus on certain metrics. To assess the consideration of undesired states and especially the impact of circularity measures, changes in the evaluation metrics are monitored. The updated values of metrics can help to perform an informed decision. To enable these calculations, the individual impact of a single circularity measure and the impact of the appearance of an undesired state on each metric must be specified. The introduced meta model provides a concrete description of the products condition when entering an undesired state. Further, the circularity measures imply a description of their impact on the product condition. Moreover, each of the elements of a circular life cycle extension features information on the time, energy consumption and other qualitative and quantitative measures defining the extensions impact on the product. This knowledge enables the calculation of evaluation metrics for extended life cycles. To immediately realize changes in the metrics a dynamic recalculation and a visualization of the metrics are required.

Within this work the conceptional integration of metrics is carried out. However, the computation of metrics is only implemented for the lifetime (see Sect. 4.2.) as an example and prove of concept for the methodology to iteratively improve the circular strategy based on changes in the metrics. The developed approach does conceptionally include performance metrics with a special focus on sustainability. However, there are no metrics introduced which can only be evaluated for circularity measures. The utilization of standard metrics with a sustainability focus enables a comparability of approaches which follow a linear value creation and those that incorporate the circular approach.

4   Implementation and Evaluation

The developed modeling approach is evaluated within an automotive development project. A sustainable center console was developed by means of a Circular Economy approach. The following chapter illustrates how the developed approach is implemented in form of a software tool. The tool enables modeling of circular life cycles as described in Subsect. 4.1. To evaluate modeled product variants, the possible paths of life cycles and the evolving product conditions throughout the process need to be analyzed. To allow for such an analysis, we introduce a Monte Carlo simulation approach in Sect. 4.2. Lastly, we discuss the evaluation of the developed approach on the example of an automotive center console in Sect. 4.3.

4.1   Implementation of the Meta Model

Chapter 3 presents the developed MBSE based approach to enable an early qualitative and quantitative analysis of the impact of circularity measures in product life cycles. A software tool is developed to provide modeling capabilities and allow for a comprehensive conceptualization of circular product life cycles. Further, the tool enables an automated assessment of the quality of the developed circular strategy.

The open-source systems modelling tool Capella [17] builds the basis for the implementation. The implementation of the developed metamodel is in turn based on the *Eclipse Modeling Framework* (EMF) [11]. An existing metamodel and visualization of Capella is used to model the physical product architecture as a basis for any life cycle evaluation and integration of circular strategies. The custom orthogonal metamodel was implemented to extend the basis of a product architecture with the concept of product life cycles including circular related aspects.

The orthogonal metamodel is composed of two parts. The modeling of regular life cycles and the modeling of extensions to enable circular value creation.

The metamodel defines a structure to model life cycles with circular value aspects, as introduced in Sect. 3.1. For the regular life cycle those are states and transitions from one state to another. For the circular extension the metamodel defines undesired states and return transitions. A return transition, as defined in the meta model, contains a circularity measure. The tool provides a library of those measures. Based on the meta model any possible product life cycle with alternative paths can be modelled in a probabilistic manner. The model allows to document the products evolvement during the process from production to end-of-life.

4.2   Model Based Lifetime Estimation

Section 3.2 introduced the concept of evaluating different product variants and the impact of introduced circularity measures based on the changes in values of evaluation metrics. Automated dynamic recalculations would allow for an immediate assessment of changes in the product model. Within this work we implemented the automated calculation of the estimated lifetime of the product under development. The lifetime is an important performance metrics especially in Circular Economy. Enabling circular value creation aims to increase the lifetime of a product. Costs of production and other factors can be distributed over the years of usage and the relative indicator of costs per year of usage decreases. Further, a longer time of life lowers the overall demand for raw materials. There are many more advantages in an increased lifetime.

To evaluate the impact of introduced circularity measures on the lifetime of a product, we need to take into consideration all possible life cycle paths. We use a Monte Carlo Simulation to address this challenge. A Monte Carlo simulation creates a model of possible outcomes by taking a probability distribution for any variable with inherent uncertainty. The results are recalculated multiple times each time using a different probability. The results can be used to assess a system despite inherent uncertainties within it.

In the case of life cycle simulations, uncertainty is introduced through the possibility of multiple valid life cycle paths for the same product. We implemented a monte carlo simulation with a sample size of 1000 products to cope with the uncertainty. The alternative transitions from one state to multiple possible subsequent states are assigned with a probability. During the monte carlo simulation decisions on the path of a single entity are based on the assigned probabilities. These rules hold for the regular life cycle as well as for the extensions composed of undesired states and return paths via circularity measures. Each state, regular or undesired, and each transition features an estimated average duration. Based on the individual durations along a path, the lifetime can be computed for a single deterministic run of the monte carlo simulation.

The results of the monte carlo simulation are used to assess the impact of changes in the life cycle and specifically the addition of new circularity measures on the expected lifetime. Based on the values for the estimated lifetime of a single run, an average, minimal or maximal lifetime can be computed. This approach allows for educated estimates of the benefit of circularity measures. The implementation is designed to automatically update a displayed evaluation metric whenever the life cycle with the circularity measures is updated or altered in any way.

4.3 Evaluation at the Example of an Automotive Center Console

The presented approach was evaluated within the research project *Cyclometric*. An automotive center console was redesigned and digitally optimized for Circular Economy and sustainability aspects [10] Based on the digital design a physical product architecture was modelled in *Capella*. The characteristics of physical components are defined in more detail. Materials, weights, and further details are provided. For a selection of those physical components, a life cycle was modelled. In a first iteration, the modeling is performed in a conventional way without a Circular Economy focus. Based on this first model, desired states which likely result in an undesired state where identified. Consequently, we added the most likely undesired states to test the implementation approach. For each case several circularity measures were tested to dissolve the undesired state. The impact of each circularity measure could be directly assessed by monitoring the updated lifetime value.

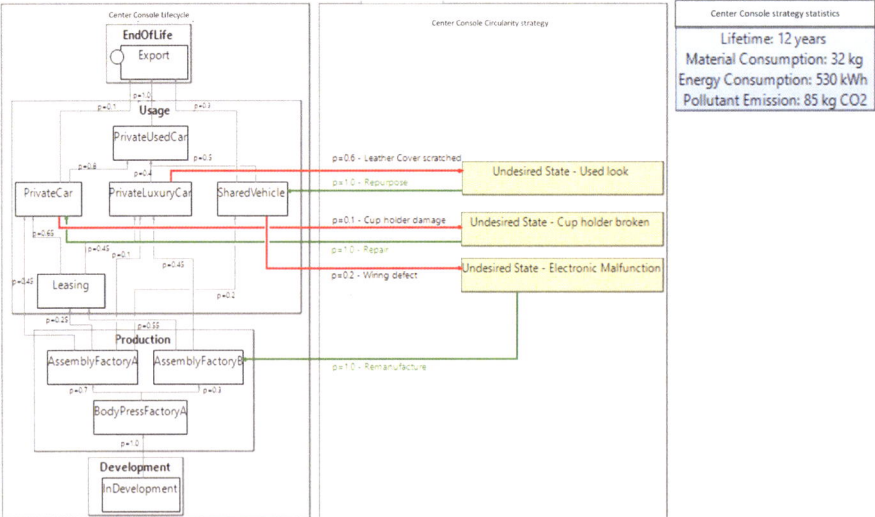

Fig. 2. Implementation of the center console's life cycle with the extension composed of undesired states and circularity measures. Probabilities p are assigned to each transition to indicate the likelihood of each life cycle path. Dynamically updating strategy statistics are displayed for immediate assessment.

The visualization of such an extended life cycle is depicted in Fig. 2 On the left-hand side, a regular life cycle is modelled. The states are cumulated in phases – Production, Usage and EndOfLife. The red arrows indicate a transition from a desired condition of the product to some undesired state. They feature a short description of the error or functional limitation causing the product to escape the regular life cycle. The resulting undesired states in turn are depicted in yellow. They are part of the right-hand side of the visualization. This part comprises all Circular Economy related components. To return from an undesired state on the right-hand side to a regular state, the green arrows are added to the model. These represent a return path featuring a circularity measure. The last component of Fig. 2 is the blue section displaying the evaluation metrics. The lifetime is already dynamically updated based on the monte carlo simulation. With the current state of development, the remaining values are assumptions.

During the application, the developed approach has proven to enable the identification of suitable circularity measures to increase circularity of the product under development. By modeling the estimated regular life cycle first, the identification of critical states is made easy. Undesired states can be directly added to the model. They are no longer only considered statistically, but instead they are integrated in a meaningful way providing the necessary context and information to trigger the application of a suitable circularity measure. The qualitative and quantitative assessment of the impact of the applied measure is enabled by the recalculation of evaluation metrics. Providing those values in the same view as the life cycle allows for a very user friendly and immediate optimization of the developed circular strategy. The immediate assessment of developed circular strategies is a significant advantage in contrast to previous methods. These findings are based on the first implementation for testing the approach. Further analysis including a comprehensive user study is required to evaluate the approach in more detail and extract additional results.

5   Discussion and Conclusion

The challenge addressed in this work lies in the qualitative and quantitative analysis of the impact of circularity measures in product life cycles. The aim was to develop an approach which enables the developer to identify product states where undesired behavior or conditions are likely to appear. Moreover, the goal was to provide a method to integrate those states into the life cycle model and dissolve them with the help of circularity measures to increase sustainability of the product. To complete the methodology and create a holistic modelling approach, direct evaluation of the measures and the circular strategy should be made possible. This allows for an approach to iteratively improve the strategy and directly give feedback to the developer.

For this purpose, an MBSE based approach was developed exploiting the idea that each product has an individual life cycle. The concept of modeling regular life cycles was extended by Circular Economy aspects. The idea of undesired states and the method to dissolve those with the help of circularity measures was introduced in an extension of the metamodel.

The approach allows to easily identify potential unwanted conditions throughout a product's lifecycle and permits comparison of possible circular strategies to return to a desired state. Hence, the approach facilitates immediate and qualitative assessment of the impact of circularity measures. The ability to assess the impact of the developed strategies leads to more informed decision-making and thus a better product quality or sustainability of the product.

The implementation utilizes the open-source system modelling tool Capella (Eclipse 2023a; Roques 2018). The implementation of the developed modeling approach was successful. The tool allows the user to obtain a good understanding of the complex system and the interdependencies between changes of the product life cycle at different states. The dynamic recalculation of evaluation metrics based on changes in the model was implemented at the example of the product lifetime. The example succeeded to show that an immediate assessment of the impact of a circularity measure on evaluation metrics is possible and indeed very useful for the optimization of the circularity strategy.

Nevertheless, the developed approach has potential for further improvement and extensions. The tool can be used for the development of different types of products from various industry sectors. However, to adequately assess the impact of circularity measures for a product from the support tool the product architecture must include all parameters that will be changed by the circularity measures. The complexity of the system arises from the large number of possibilities for circular strategies and the many variants of individual life cycle paths. Identifying all critical states and finding an optimal strategy for a return of the product to a useful state asks for a skilled and experienced developer. The current implementation enables a product developer to easily optimize the system, but an automation of the process could lead to more optimal solutions in the future. Furthermore, the hierarchical architecture of products increases the complexity of the task. The presented approach allows for a modeling of life cycles with circular aspects at each level of the product architecture. However, the developed models are not transferred to subcomponents or parent components which would simplify the process for the developer. Implementing this functionality can be part of future work. The implementation of the recalculation of the lifetime of a product lies the foundation for further development regarding other evaluation metrics. The computation of some metrics might require an extension of the model or a more detailed description of component characteristics. For some metrics it can also be relevant to discuss the topic of allocation. In case a product is not returned to a state within the same system, but instead finds a second life in another system, the impact of its productions and further processes needs to be distributed across both systems.

Acknowledgments. The research of this paper has been carried out within the research and development project Cyclometric. Cyclometric is funded by the German Federal Ministry of Education and Research (BMBF) (funding no. L1FHG42421) and implemented by Project Management Agency Karlsruhe (PTKA). The authors are responsible for the content of this publication.

References

1. Aguiar, M. F., Mesa, J. A., Jugend, D., Pinheiro, M. A. P., and Fiorini, P. D. C. 2022. Circular product design: strategies, challenges and relationships with new product development. *MEQ* 33, 2, 300–329.

2. Bargende, M., Reuss, H.-C., Wagner, A., and Wiedemann, J., Eds. 2019. *19. Internationales Stuttgarter Symposium*. Proceedings. Springer Fachmedien Wiesbaden, Wiesbaden.
3. Block, L., Werner, M., Spindler, H., and Schneider, B. 2023. A Variability Model for Individual Life Cycle Paths in Life Cycle Engineering. *Future Automotive Production Conference 2022*, 73–85.
4. Bougain, S. and Gerhard, D. 2017. Integrating Environmental Impacts with SysML in MBSE Methods. *2212-8271* 61, 715–720.
5. Cerdas, F., Thiede, S., and Herrmann, C. 2018. Integrated Computational Life Cycle Engineering — Application to the case of electric vehicles. *0007–8506* 67, 1, 25–28.
6. 2017. *Circular by design: Products in the circular economy.*
7. Dér, A., Kaluza, A., Reimer, L., Herrmann, C., and Thiede, S. 2022. Integration of Energy Oriented Manufacturing Simulation into the Life Cycle Evaluation of Lightweight Body Parts. *Int. J. of Precis. Eng. and Manuf.-Green Tech.* 9, 3, 899–918.
8. Diaz, A., Schöggl, J.-P., Reyes, T., and Baumgartner, R. J. 2021. Sustainable product development in a circular economy: Implications for products, actors, decision-making support and lifecycle information management. *2352–5509* 26, 1031–1045.
9. DIN Deutsches Institut für Normung e. V. *CIRCULAR THINKING in Standards. Wie Normung eine Circular Economy unterstützen kann.* https://www.din.de/resource/blob/952460/817 ac05b868cad6959d3358b58127050/flyer-circular-economy-data.pdf. Accessed 18 January 2023.
10. Fraunhofer IAO. *Lebenszyklusorientierte Entwicklung von Fahrzeug-Komponenten.* https://www.iao.fraunhofer.de/de/presse-und-medien/aktuelles/lebenszyklusorientierte-entwic klung-von-fahrzeug-komponenten.html. Accessed 18 January 2023.
11. Gronback, R. 2024. *Eclipse Modeling Project | The Eclipse Foundation.* https://eclipse.dev/modeling/emf/. Accessed 18 January 2024.
12. Herrmann, C. 2010. *Ganzheitliches Life Cycle Management. Nachhaltigkeit Und Lebenszyklusorientierung in Unternehmen.* VDI-Buch. Springer, Dordrecht.
13. Hollander, M. C. den, Bakker, C. A., and Hultink, E. J. 2017. Product Design in a Circular Economy: Development of a Typology of Key Concepts and Terms. *J of Industrial Ecology* 21, 3, 517–525.
14. Kong, L., Wang, L., Li, F., Tian, G., Li, J., Cai, Z., Zhou, J., and Fu, Y. 2022. A life-cycle integrated model for product eco-design in the conceptual design phase. *Journal of Cleaner Production* 363, 132516.
15. 2024. *life cycle model—Google Scholar.* https://scholar.google.de/scholar?hl=de&as_sdt=0%2C5&q=life+cycle+model&btnG=. Accessed 13 March 2024.
16. Mestre, A. and Cooper, T. 2017. Circular Product Design. A Multiple Loops Life Cycle Design Approach for the Circular Economy. *The Design Journal* 20, sup1, S1620–S1635.
17. 2024. *Model Based Systems Engineering | Capella MBSE Tool.* https://mbse-capella.org/. Accessed 18 January 2024.
18. Potting, J., Hekkert, M. P., Worrell, E., and Hanemaaijer, A. 2017. Circular Economy: Measuring Innovation in the Product Chain. *Planbureau voor de Leefomgeving,* 2544.
19. Stark, J. 2015. *Product Lifecycle Management.* Springer International Publishing, Cham.
20. Tao, J. and Yu, S. 2018. A Meta-model based Approach for LCA-oriented Product Data Management. *2212–8271* 69, 423–428.
21. Wiesner, S., Freitag, M., Westphal, I., and Thoben, K.-D. 2015. Interactions between Service and Product Lifecycle Management. *2212–8271* 30, 36–41.
22. Yvars, P.-A. and Zimmer, L. 2021. A Model-based Synthesis approach to system design correct by construction under environmental impact requirements. *2212–8271* 103, 85–90.

Information as Key for Strategic Stakeholder Opportunities Within the Value Chain of an Advanced Circular Economy

Phillip Wallat[1]([envelope]), Sebastian Lawrenz[2], and Mathias Nippraschk[3]

[1] DfACE Strategy Systems, Gerhard-Rauschenbach-Straße 14, 38678 Clausthal-Zellerfeld, Deutschland
`wallat@dface.eu`
[2] ReDo, Sebastian Bach Straße 29, 31141 Hildesheim, Deutschland
`sebastian@redo-please.com`
[3] Recycling und Kreislaufwirtschaftssysteme, Institut Für Aufbereitung, Walther-Nernst-Str. 9, 38678 Clausthal-Zellerfeld, Deutschland
`mathias.nippraschk@tu-clausthal.de`
`https://www.dface.eu`, `https://www.redo-please.com`,
`https://www.ifad.tu-clausthal.de/`

Abstract. Designing resource-efficient products, aided by digitalization, offers additional potential to consider improved circularity of the product for the whole value chain. Distribution of data is crucial to provide key information for relevant stakeholders, to guide and support their decisions regarding the product of interest. Stakeholders have different interests in certain sets of data. Streamlining the data exchange will lead to improved efficiency for all involved parties. Therefore, the lifecycle of products has to be examined; the most relevant stakeholders, with their ability to extract and distribute data, are identified; and who has an interest in combined datasets from all stakeholders at what point in time were investigated. Hence, combining datasets has to take into account that time is a factor. Loss of information over time, as well as the necessary data-velocity, are important to provide valid datasets.

The result is a schematic overview of the data and information exchange between stakeholders over time. Additional stakeholder data sets are defined and who is involved in providing data to X. Tailored information will be generated out of those data sets, depending on individual interests. The Stakeholder-Datasets are structured in static and dynamic information, potential loss of information, and information velocity between stakeholders. The mobility value chain is used to schematically show which stakeholders need which data set as information, in order to utilize potential that has not yet been tapped. To integrate these results along recent research, the product passport model is applied. The developed schematics are still on a meta-level and shall provide a procedure for further research based on case studies to enhance the advanced circular economy (Definition: Advanced Circular Economy: A circular economy supported with digitized information systems to aid optimization of the whole value chain in the economic system).

K. Dröder and T. Vietor (Eds.): CD 2024, Proceedings, pp. 279–290, 2025.
https://doi.org/10.1007/978-3-658-45889-8_22

Keywords: Advanced Circular Economy · Combustion vehicle · Electric vehicle · Data exchange · Data trading · Innovation platform

1 Introduction

Insufficient information about the environment leads to negative decision-making for stakeholders in almost every aspects in hindsight. The current necessary changes for business practices concerning Europe [1] will be aided by the right distribution of information. The legal or illegal export of end-of-life products represent a loss of materials, furthermore lost business opportunities for resource companies [2] and may lead to unnecessary sourcing of material for manufacturers with the upcoming legislations from the European Union [3] and new opportunities for individual authorities in the union [4]. Which is not only valid for technical driven companies but also for companies in the food and fashion sectors [5, 6] This will lead to the question:

Which Stakeholders inside the Circular Economy need which information from whom at what point during the product life cycle, and how should the necessary information be distributed while keeping unnecessary information?

The gathering of necessary data to generate the information seems to start with the design phase of a product. Most relevant data for many stakeholders is created by decisions within the design process. For well-balanced decisions data from downstream processes of the product life cycle is necessary. Without data exchange, previous to the design phase, the product cannot be developed along with the needs of other stakeholders. The same facts are valid for stakeholders positioned later in the product life cycle, to set up efficient processes for the treatment of the product. For those reasons, the whole information system is a mixture of data-driven hindsight, insight, and foresight. While constantly evaluating the data regarding different externalities e.g. energy costs, availability of material, TRL of new technology, and individual economic circumstances. Furthermore, the generated and collected data over the product lifespan can even be used by researchers, governments, and innovators, beyond the lifecycle of products for valid secondary research.

2 Previous Work

An overview of the authors' previous work in the field of information and data in connection with the "advanced circular economy" is provided first; see the base paper [7]. This is then expanded to include further work on this topic. The referenced paper also tackle the related work part.

One of the main barriers towards an optimal circular economy or an advanced circular economy is the information gap between the stakeholders along the value chain [7, 8]. Regardless of which product stream is considered: whether it is, for example, the irreparability of smartphones or the inappropriate dismantling processes of traction batteries. In 2021, the authors showed that there is a distinction between static and dynamic information and that these develop differently over time. Static information, such as the production date, does not change over time and its dataset is constant, while the set of dynamic information such as the remaining capacity of a battery evolves and increases. Moreover, a change of ownership of a certain product is always followed by a flow of information, whereby the gap between real[1] and ideal information[2] increases over time. This gap should ideally be zero, or at least kept as small as possible [7].

These research findings were taken up or further developed by these authors in joint and own publications or even by other scientists. In addition, certain research findings are also taken up by other authors without finding a citation to these authors. It should be noted here that the authors do not claim to be the only scientists researching this topic. However, it shows that the ideas are also being pursued by others and that the topic is relevant; furthermore the parallel research generates benefit within the whole research field.

The newer publications mentioned in the former paragraph, which referred to the main paper from 2021 [7] can be divided into three areas: firstly, as evidence that lack of information and data is part of the problem of why optimal CE has not yet been or even will not be achieved. This part will not be discussed further in this paper.

The second area deals with ways in which this information can be made available to stakeholders. Berger et al. [9] takes this up in order to define the data requirements and availability for a battery passport. Product passports can ultimately be applied to all products that today follow a take-make-waste principle and should strive for a circular economy from an economic, ecological and social point of view. This ranges from electrical and electronic equipment [10] or the fashion industry [5] to approaches that can be applied in general [9, 11–14]. The best-known passport for batteries is discussed in more detail in Sect. 4.

The last area deals with models for technical implementation in order to evaluate information and make it available beyond their primary sources and use cases. Many data streams are not designed to be (re)-used, filtered and tailored for other stakeholders. Or the required information are not even stored, mostly dynamic information. To overcome these obstacles technical solutions are required, such as an information marketplace. The concept of a data and information marketplace was already introduced and discussed in different publications such as [7, 15–17].

3 Stakeholders and Data Sets

Many different Stakeholders are involved in the value chain of the advanced circular economy. Accordingly, different data sets are required (due to the specific needs of the stakeholders), and different stakeholders are involved. This section delves into the types of information critical for these processes and examines the dynamics of stakeholder engagement, shedding light on how these interactions can either propel or hinder the advancement of sustainable practices.

---

[1] Definition: Real Information: Information that is known, documented and accessible for a stakeholder.

[2] Definition: Ideal Information: Every bit of information that could have been documented about something.

Static Information

Definition: Static information refers to data that remains constant over time. In the sustainability sector, this could include material compositions, design specifications, and baseline environmental impact assessments.

Challenges: Despite its importance, the sharing of static information often encounters obstacles. The lack of incentive for stakeholders to share such data, coupled with the fear of exposing proprietary information or trade secrets, contributes to a gap in the collective knowledge base, impeding collaborative efforts for sustainability.

Dynamic Information

Definition: Contrasting static data, dynamic information is characterized by its evolving nature. This includes data on consumer usage patterns, recycling rates, and the environmental impact of products over their lifecycle.

Challenges: The collection and updating of dynamic information faces significant hurdles. In many instances, there's a lack of consistent and systematic efforts to gather this data. The variability in formats due to different technological standards or stakeholder requirements further complicates this landscape, posing challenges for effective data utilization.

Main Stakeholders in the Advanced Circular Economy

Supplier: Suppliers provide the foundational materials or products, making their contribution of both static and dynamic data crucial for mapping out sustainable supply chains.

OEMs (Original Equipment Manufacturers): OEMs play a central role in the product lifecycle, influencing the generation, consumption, and relevance of sustainability data. Their engagement is vital in aligning industry practices with sustainability goals.

Customer: Customers, as end-users, significantly impact and benefit from sustainability data. Their informed choices, driven by accurate and transparent information, can steer market trends towards more sustainable products.

Manufacturer: Manufacturers who remanufacture products are key players in the sustainability cycle. The data they use and generate can optimize the reuse and repurposing of materials, reducing waste and environmental impact.

Second Life Supplier: These stakeholders are instrumental in extending the life of products through reuse. Access to comprehensive data sets enables them to identify and utilize viable parts effectively.

Recycler: Recyclers are at the forefront of transforming waste into resources. The exchange of both static and dynamic data is essential for them to adapt their processes to changing material streams and sustainability standards.

Innovator: Innovators drive the evolution of products towards greater sustainability. Their reliance on a mix of static and dynamic information is critical in developing solutions that are both innovative and environmentally responsible.

To make the assignment of data easier for all involved stakeholders, datasets shall be used. Those are either composed depending on actions in the value chain or for stakeholder groups. By structuring those sets in this particular manner, the classification of their own provided data as well as the required data is more accessible. Therefor the datasets must be composed out of data form different stakeholders, furthermore they are evolving depending on the ration of static and dynamic information. Later, stakeholders can request datasets to utilize them for their corresponding decision process. Hence they are able to make decisions regarding circularity based on "insight" or "foresight" (see Fig. 1) instead of "hindsight". This will provide possibilities to increase the overall value within the circular value chain for the whole society in the business environment.

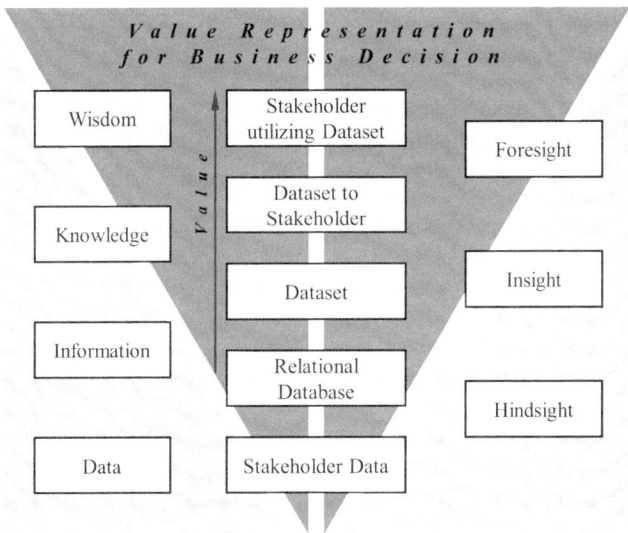

Fig. 1. 'Data-Value-Funnel' as schematic for the transfer of data to wisdom and added value by enhancing (business) decisions for stakeholder due to the distribution of various data condensed into specific datasets for stakeholders.

In Table 1 there are three Data sets with two circles (C1 and C2) which represents not only the product-life-cycle but also the material-life-cycle. The first cycle represents the design of the product without any former data about the input materials. While the OEM needs data from the material supplier and the manufacturer of certain parts that will be included in the whole product, they also need information regarding the specific use of the material or part to determine more suitable alternatives. Both data flows aiming for the most valuable use and increased efficiency of resources. Depending on the product, the customer might also be interested in all those information, to influence their purchase decision based on the dataset insights. If the information about the product will positively influence the customers decision, stakeholder specific value can be assigned to the information. The former data input of the upstream stakeholders, additionally enables the tailoring of individual datasets to provide second life supplier and recycler with information they can structure their processes with; based on foresight by knowledge, which again will add value to the individual data due to the more efficient processes at the products end of life.

The next cycle will increases the value of the data even more. The data used as wisdom from former processes of all stakeholders can be utilized for every stakeholder in future decision making (see Fig. 1 and Table 1). The information market place is

Table 1. Development of three datasets over two circles with exemplary stakeholders and where they provide data to a dataset or use the information from a dataset. Enhancement to the circular economy are marked in green.

Stakeholder	Pre-Life Dataset		Life-Cycle Dataset		Post-Life Dataset	
	Provide	*Use*	*Provide*	*Use*	*Provide*	*Use*
Supplier	C1 C2	[C2]				[C2]
OEM	C1 C2	C1 [C2]		[C2]		[C2]
Customer		C1 C2	C1 C2			
Recycler		C1 C2		C1 C2	C1 C2	
Information Marketplace	[C2]	C1 C2	[C2]	C1 C2	[C2]	C1 C2

able to take the role as distributer and an memory of data over different lifecycles (see Sect. 5). For each cycle the data is collected and transformed into information. This information will provide insights for stakeholders who before had no information to base their decisions on.

> *Example: The material supplier for the OEM of breaking equipment will provide static data regarding the material properties and manufacturing process, which in return the OEM needs for quality control. The OEM again generates data for the 'Pre-Life' dataset. This is particular interesting to use for the customer to purchase the right part for the mark and model, afterwards the customer will produce dynamic data for the 'Life-Cycle' dataset. After the use phase of the part it is beneficial for the recycler to know what kind of material it is, hence the use of the 'Pre-Life' dataset. For wear parts like in this case, the dynamic information about the operation hours are not that important, this differs for more complex assemblies. In both cases the recycler will generate static and dynamic data for the 'Post-Life' Dataset. In case ever stakeholder in this example provides ideal data to the information marketplace, this data can be transformed to information and be provided for various purposes to other stakeholders. Furthermore, the new available information will offer benefits for various decisions in the second cycle of the material for stakeholders which were not available or existent before.*

4   Transfer to Other Products: Product Passport

The information exchange concept can be transferred to other products. By using a product passport, benefits for all stakeholders along the value chain can be generated. Within the automotive value chain, different circular approaches (CA) are involved.

Parts and assemblies have to be repairable to prolong the life of the whole car, this is especially valid for wearing parts. This method was driven by economic factors; a car should not be disposed of, just because one part will no longer function as with the intended performance. The overall monetary value is too high to not consider repair. This positive circumstance represents a problem for products with smaller monetary value.

With a product passport for those products, it is possible to set the monetary and material value of products into a context. Therefor assisting the decision-making process during the product lifetime. Implementing data driven product passports will enable engineers to decide with a certain degree of foresight during the design phase while developing the product. With less overall product value, repair operations have to be more efficient for parts that are exposed to wear and tear, otherwise the breakdown of one designated wear part can lead to a malfunction of the associated assembly and as a result carries over to economic obsolescence of the whole product.

One of the best-known product passport is the one for batteries, which will be introduced in the EU in 2027. This passport, defined in Article 77 of the EU Battery Regulation, will be mandatory for all batteries in electric vehicles (EV), light means of transport (LMT) and industrial batteries with a capacity greater than 2 kWh that are placed on the EU market or put into service [18].

A wide range of data and information is collected along the value chain and made available to certain stakeholder groups on a mandatory basis. The value chain of batteries is the same as that of other products with different names for their stakeholders or value creation steps (see Fig. 2) preliminary products or components are manufactured from raw materials, which in turn are manufactured by an OEM into a complete product, sold to a customer and, after a different product life cycle, are fed into the various circular economy strategies (e.g. repair, remanufacturing, repurpose/2nd-life or material recycling).

With the introduction of the Battery Passport, three additional stakeholder groups are now introduced at the data and information level, which have different access to the data and information that is generated and collected during the product life cycle: "General public", "Notified bodies, market surveillance authorities and the Commission" and "Any natural or legal person with a legitimate interest in accessing and processing that information". The information that must be made available to these respective groups is defined in Annex XIII of the Battery Regulation [18].

The Battery Pass Consortium, in collaboration with the Battery Alliance, has summarized the definitions and attributes of the Battery Pass based on the Battery Regulation. It is important to note that in addition to listing the data and information and which of the three stakeholder groups it is made available to, this document also makes a distinction between static and dynamic information. This again shows that the distinction between the two types of information made by the authors in their previous publications is taken up in real projects such as the Battery Pass [19].

The published document contains 90 attributes that are mandatory under the Battery Regulation or other regulations. In addition, there are a further 17 attributes that are voluntary. Of these 107 attributes, 79 are static data or information and 28 are dynamic in nature [19].

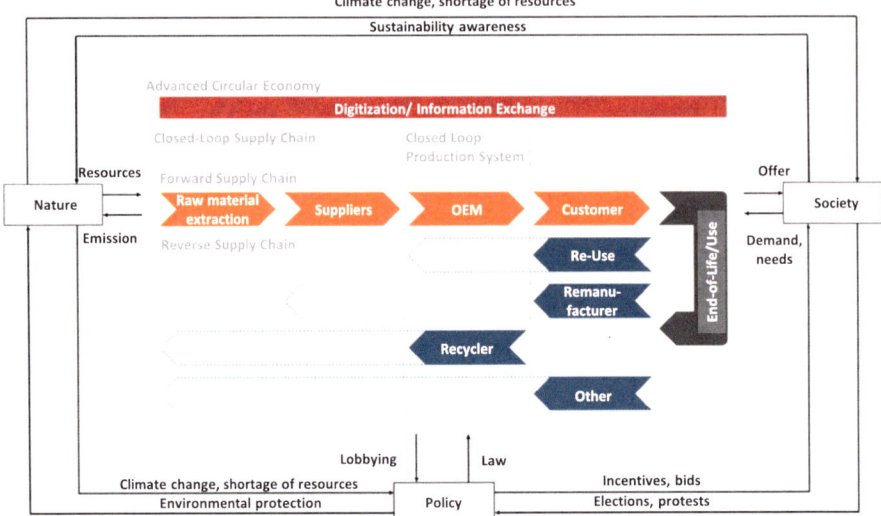

Fig. 2. Framework of the advanced circular economy based on the Recycling 4.0 approach (translated) [8]

5 Platform Based Approach

In the quest for an advanced circular economy, the concept of a platform-based approach emerges as a critical enabler. This approach leverages digital platforms to facilitate open data and information exchange, vital for the functionality of products like the battery passport. Such platforms act as centralized hubs, where various stakeholders can access, contribute, and utilize data efficiently and transparently. These platforms are for example data marketplaces, such as described in [1–3], matching platforms for different stakeholders along the circular economy [20] or support circular processes such as repair [21]. Platforms for tracing the sustainable impact [22], or just to exchange data [23, 24].

This section explores the nuances of the platform-based approach, its implications for the circular economy, and how it underpins products like the battery passport.

5.1 The Role of Digital Platforms in Circular Economy

Digital platforms in the circular economy serve a dual purpose. Firstly, they function as repositories for both static and dynamic information, ensuring that data is not only stored but is also accessible and usable. Secondly, these platforms act as facilitators for communication and collaboration among stakeholders, bridging information gaps and fostering a more integrated approach to sustainability. The functionality of these platforms can be enhanced with advanced technologies like hashing, which ensures data integrity and traceability. Additionally, the integration of AI and machine learning can aid in analyzing vast data sets, providing insights that can drive efficient decision-making and innovative solutions. Based on large language models data consumers get support in exploring the data, foster new use cases or support a data purchase decision.

5.2   Case Study: The Battery Passport Platform

The proposed battery passport is an exemplary instance of a platform-based app-roach. This digital platform will encompass a comprehensive set of information, including both static and dynamic attributes of batteries used in electric vehicles and other applications. By centralizing this information, the platform not only sim-plifies access for stakeholders but also ensures that data is updated and accurate, thus supporting sustainability efforts throughout the product's lifecycle.

5.3   Advantages of the Platform-Based Approach

The platform-based approach offers several advantages:

1. **Enhanced Transparency and Traceability:** By providing a centralized data reposi-tory, these platforms enhance the transparency of information across the value chain. This transparency is crucial for stakeholders to make informed decisions and for consumers to understand the sustainability aspects of their choices.
2. **Improved Stakeholder Collaboration:** Platforms provide a common ground for stakeholders to collaborate, share insights, and develop joint strategies for sustain-ability. This collaboration is essential for addressing the complex challenges of the circular economy.
3. **Data-Driven Decision Making:** The wealth of information available on these platforms enables stakeholders to make data-driven decisions. This is particularly important in designing products and processes that align with sustainability goals.
4. **Scalability and Adaptability:** Digital platforms can be scaled and adapted to accommodate different products and industries, making the approach versatile and applicable across various sectors of the circular economy.

5.4   Challenges and Considerations

Despite its advantages, the platform-based approach also presents challenges. These include ensuring data security and privacy, managing the complexities of data standard-ization, and fostering widespread adoption among stakeholders. Addressing these chal-lenges will require collaborative efforts, clear regulatory frameworks, and continuous innovation in platform design and technology.

6   Conclusion

One of the main barriers to achieving an optimal circular economy is the information gap that still exists between individual stakeholders. The concept of information loss when the product changes users or owners, which was introduced by the authors in previous publications, has since been taken up and further developed in other publications to reduce the information gap. Various types of data passports have also been developed that can be utilised by the future introduction of the battery passport.

In conclusion, the platform-based approach represents a significant step forward in achieving an advanced circular economy. By facilitating open data and informa-tion exchange, these platforms can enhance transparency, collaboration, and innovation across the value chain. As the case of the battery passport illustrates, such platforms are not just theoretical concepts but practical tools that can drive real change in the pursuit of sustainability. Furthermore, 'Large Language Models' are a game changer for new platform based approaches. A fine tuned model in combination with the battery passport can increase the value of such a platform highly.

Furthermore, it is crucial to identify and address potential challenges in adopting these platforms, such as ensuring data security and quality, managing stakeholder interests, and overcoming technical barriers. Future research should focus on these aspects to optimize platform efficacy and broaden their applicability. Additionally, exploring the scalability of such platforms in various economic sectors can provide insights into their adaptability and impact at a larger scale. This holistic approach will be instrumental in fully realizing the potential of platform-based solutions in contributing to a more sustainable and circular economy.

Literaturs

1. Bobba, S.; Carrara, S.; Huisman, J. et al.: Critical raw materials for strategic technologies and sectors in the EU. A foresight study. Luxembourg: Publications Office of the European Union 2020.
2. Mathur, N.; Last, N.; Morris, K. C.: A process model representation of the end-of-life phase of a product in a circular economy to identify standards needs. Frontiers in Manufacturing Technology 3 (2023).
3. Europäisches Parlament und Rat: REGULATION (EU) 2023/956 OF THE EUROPEAN PARLIAMENT AND OF THE COUNCIL of 10 May 2023 establishing a carbon border adjustment mechanism 2023.
4. Domingo, L.; Melchor, D.: Designing out waste: which levelers for local authorities? Procedia CIRP 105 (2022), S. 535–540.
5. Esbeih, K. N.; Molina-Moreno, V.; Núñez-Cacho, P. et al.: Transition to the Circular Economy in the Fashion Industry: The Case of the Inditex Family Business. Sustainability 13 (2021), Nr. 18, S. 10202.
6. Bag, S.; Srivastava, G.; Cherrafi, A. et al.: Data-driven insights for circular and sustainable food supply chains: An empirical exploration of big data and predictive analytics in enhancing social sustainability performance. Business Strategy and the Environment 33 (2024), Nr. 2, S. 1369–1396.
7. Lawrenz, S.; Nippraschk, M.; Wallat, P. et al.: Is it all about Information? The Role of the Information Gap between Stakeholders in the Context of the Circular Economy. Procedia CIRP 98 (2021), S. 364–369.
8. Blömeke, S.; Mennenga, M.; Herrmann, C. et al.: Recycling 4.0. An Integrated Approach Towards an Advanced Circular Economy. ICT4S'20: 7th International Conference on ICT for Sustainability, Bristol 2020.
9. Berger, K.; Rusch, M.; Pohlmann, A. et al.: Confidentiality-preserving data exchange to enable sustainable product management via digital product passports—a conceptualization. Procedia CIRP 116 (2023), S. 354–359.
10. Andersen, T.; Jæger, B.: Circularity for Electric and Electronic Equipment (EEE), the Edge and Distributed Ledger (Edge&DL) Model. Sustainability 13 (2021), Nr. 17, S. 9924.
11. Berger, K.; Baumgartner, R. J.; Weinzerl, M. et al.: Data requirements and availabilities for a digital battery passport—A value chain actor perspective. Cleaner Production Letters 4 (2023), S. 100032.

12. Berger, K.; Schöggl, J. P.; Baumgartner, R. J.: Digital battery passports to enable circular and sustainable value chains: conceptualization and use cases 2021.

13. Reich, R. H.; Alaerts, L.; van Acker, K.: The Circular Economy as a Requirement for Smart Information System Designs. In: 19th International Conference on Distributed Computing in Smart Systems and the Internet of Things. DCOSS-IoT 2023 : 19–21 June 2023, Pafos, Cyprus : proceedings. 2023 19th International Conference on Distributed Computing in Smart Systems and the Internet of Things (DCOSS-IoT), Pafos, Cyprus, 6/19/2023–6/21/2023, S. 359–366. Piscataway, NJ: IEEE 2023.

14. Reich, R. H.; Ayan, J.; Alaerts, L. et al.: Defining the goals of Product Passports by circular product strategies. Procedia CIRP 116 (2023), S. 257–262.

15. Kintscher, L.; Lawrenz, S.; Poschmann, H.: A Life Cycle Oriented Data-Driven Architecture for an Advanced Circular Economy. Procedia CIRP 98 (2021), S. 318–323.

16. Lawrenz, S.; Poschmann, H.; Rausch, A. et al.: Data Trading Similarity Signature An Extended Data Trading Framework for Human and Non-Human Actors. In: Bui, T. (Hrsg.): Proceedings of the 55th Hawaii International Conference on System Sciences. Hawaii International Conference on System Sciences. Hawaii International Conference on System Sciences 2022.

17. Lawrenz, S.; Rausch, A.: Don't Buy A Pig In A Poke A Framework for Checking Consumer Requirements In A Data Marketplace. In: Bui, T. (Hrsg.): Proceedings of the 54th Hawaii International Conference on System Sciences. Hawaii International Conference on System Sciences. Hawaii International Conference on System Sciences 2021.

18. Europäisches Parlament und Rat: Verordnung (EU) 2023/1542 des Europäischen Parlaments und des Rates vom 12. Juli 2023 über Batterien und Altbatterien, zur Änderung der Richtlinie 2008/98/EG und der Verordnung (EU) 2019/1020 und zur Aufhebung der Richtlinie 2006/66/EG 2023.

19. The Battery Pass consortium: Battery Passport—Data Attribute Longlist 2023. Internetadresse: https://thebatterypass.eu/assets/images/content-guidance/pdf/2023_Battery_Passport_Data_Attributes.xlsx. Zuletzt aufgerufen am 21.01.2024.

20. Rudolf, S.; Lawrenz, S.; Blömeke, S. et al.: Matching the Supply and Demand within the Circular Economy for Used Electrical and Electronic Equipment applying Condition Assessment. Conference: THE 10TH INTERNATIONAL CONFERENCE ON LIFE CYCLE MANAGEMENT 2021. Internetadresse: https://www.researchgate.net/publication/358149722_Matching_the_Supply_and_Demand_within_the_Circular_Economy_for_Used_Electrical_and_Electronic_Equipment_applying_Condition_Assessment. Zuletzt aufgerufen am 14.03.2024.

21. Rudolf, S.; Blömeke, S.; Niemeyer, J. F. et al.: Extending the Life Cycle of EEE—Findings from a Repair Study in Germany: Repair Challenges and Recommendations for Action. Sustainability 14 (2022), Nr. 5, S. 2993.

22. Lawrenz, S.; Leiding, B.; Mathiszig, M. E. A. et al.: Implementing the Circular Economy by Tracing the Sustainable Impact. International journal of environmental research and public health 18 (2021), Nr. 21.

23. Sikarwar, R.; Yadav, P.; Dubey, A.: A Survey on IOT enabled cloud platforms. In: 2020 IEEE 9th International Conference on Communication Systems and Network Technologies (CSNT). Proceedings. 2020 IEEE 9th International Conference on Communication Systems and Network Technologies (CSNT), Gwalior, India, 4/10/2020–4/12/2020, S. 120–124. Piscataway, NJ: IEEE 2020.

24. Iwasa, D.; Hayashi, T.; Ohsawa, Y.: Development and Evaluation of a New Platform for Accelerating Cross-Domain Data Exchange and Cooperation. New Generation Computing 38 (2020), Nr. 1, S. 65–96.

AI-based Optimisation

A Markovian Approach to Automated Detection of Product Traceability Anomalies and Gaps in Complex Discrete Manufacturing

Roudi Bakir[✉] and Christine Blume

Robert Bosch Elektronik GmbH, John-F.-Kennedy-Straße 43, 38228 Salzgitter, Germany
roudi.bakir@de.bosch.com
https://www.bosch.com

Abstract. Product traceability is essential in modern manufacturing to ensure quality standards and regulatory compliance. It also provides transparency that resonates with customers' growing emphasis on sustainability. Traditional methods of traceability verification often involve manual inspections, which are time-consuming and prone to human error. This paper proposes an automated approach leveraging machine learning to detect anomalies and gaps in product traceability within discrete production lines. The production line is modeled as discrete sequences, with process parameter names for each station hashed and represented as states in Markov Chains. Through continuous analysis, the system updates the Markov Chain model, adapting to changing production scenarios without manual intervention. Anomalies are detected based on deviations in transition probabilities, allowing for the automatic identification of traceability anomalies and gaps. This not only ensures the integrity of the production process but also optimizes operational efficiency, significantly reducing the need for manual oversight. The proof-of-concept is demonstrated in a case study at the automotive electronic manufacturing plant of Robert Bosch Elektronik GmbH, showcasing the effectiveness and scalability of the automated approach in real-world manufacturing environments.

Keywords: Traceability · Discrete manufacturing · Machine learning · Anomaly detection · Markov chain

1 Introduction

In the current landscape of discrete manufacturing, ensuring product traceability has become a crucial requirement. The need for reliable traceability systems is motivated by several factors, such as regulatory compliance, quality assurance, and the growing intricacy of global supply chains. Traceability is the systematic tracking and recording of product and component movement throughout the entire manufacturing process, from raw materials to finished goods. This capability provides manufacturers with valuable insights, enabling them to enhance product quality, streamline operations, and respond promptly to issues such as recalls or defects. Traceability is crucial in industries such as

K. Dröder and T. Vietor (Eds.): CD 2024, Proceedings, pp. 293–304, 2025.
https://doi.org/10.1007/978-3-658-45889-8_23

automotive, aerospace, pharmaceuticals, and electronics, where strict quality standards and regulatory requirements demand a thorough comprehension of the product lifecycle. In the automotive sector, for example, traceability enables the tracking of faulty components to their source, making targeted recalls possible and reducing the impact on consumer safety. Achieving effective traceability in complex discrete manufacturing environments presents numerous challenges due to the volume of data generated and the intricacies of modern supply chains. Manual traceability management is often error-prone and inefficient, and traditional methods may not address the dynamic nature of manufacturing processes, resulting in gaps and anomalies in traceability records. Moreover, the complexity of traceability efforts is exacerbated by the integration of legacy systems and disparate technologies. To overcome these challenges, there is an increasing recognition of the transformative potential of enhanced digitalisation and artificial intelligence (AI) in manufacturing traceability. The combination of advanced data analytics, machine learning, and AI-based algorithms shows potential in automating traceability processes, offering real-time insights, and detecting anomalies or gaps in the traceability chain. These technologies enhance the accuracy and efficiency of traceability systems, empowering manufacturers to proactively address issues and prevent potential disruptions to the production workflow. Our paper introduces a novel Markovian approach to automated detection of product traceability anomalies and gaps in complex discrete manufacturing. Leveraging the inherent stochastic nature of manufacturing processes, Markov models offer a powerful framework for capturing the dynamic dependencies among events in the production sequence. By integrating Markovian principles with AI-driven analytics, our proposed methodology aims to enhance the precision and reliability of traceability systems, providing manufacturers with a potent tool for ensuring the integrity of their product lifecycles. Through the exploration of this innovative approach, we contribute to the ongoing discourse on advancing traceability in complex manufacturing environments, offering a viable solution to the persistent challenges faced by industry practitioners.

2   Background

2.1   State of the Art

Traceability in discrete manufacturing systems is the ability to track and trace the production process, as well as the flow of materials, components, and products throughout the entire manufacturing lifecycle. This process is crucial for ensuring quality, compliance, and efficiency in manufacturing operations. Traceability involves identifying, recording, and tracking the history, location, and usage of each component, subassembly, and final product in the manufacturing process. Traceability is a crucial quality management principle highlighted in ISO 9000 standards [1] It stresses the importance of establishing clear and documented connections between product-related information to enable effective control, analysis, and improvement of the quality management system. Traceability can refer to products, such as tracing the amount, quality, or type of products, or activities, such as the type or duration of life cycle stages [2].

Traceability in manufacturing systems serves several critical objectives, each contributing to the overall effectiveness and quality of the production process. One of the primary goals of traceability is quality assurance. By establishing a comprehensive production history, traceability facilitates the swift identification and correction of defects, ensuring the consistent production of high-quality products. Secondly, traceability is crucial in ensuring compliance with the often stringent regulations and standards prevalent in various industries. It is important to note that traceability is not only important for compliance but also for quality control and process improvement. Sectors such as automotive, aerospace, and healthcare have specific requirements that must be met to guarantee the safety, reliability, and legality of products [3]. Traceability provides a documented trail of materials and processes, aiding manufacturers in demonstrating adherence to regulatory frameworks and standards. Additionally, traceability significantly contributes to the overarching goal of efficiency and optimization in manufacturing operations. By systematically tracking the movement of materials from their origin through various stages of production, manufacturers gain valuable insights into the dynamics of their supply chain. This insight allows for the identification of inefficiencies, bottlenecks, or areas of waste. As a result, manufacturers can optimize their supply chain processes, reduce unnecessary resource consumption, minimize waste generation, and enhance the overall efficiency of the production system (Fig. 1).

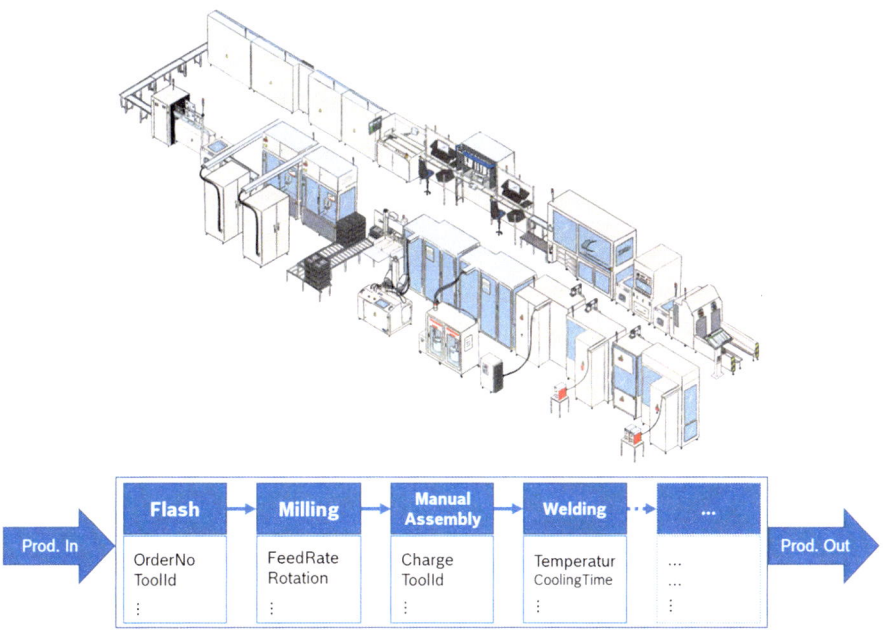

Fig. 1. Final assembly line for engine control units (schematic) [4]

For instance, in manufacturing industry like automotive electronics, traceability is crucial for ensuring quality, safety, and compliance, which is achieved through meticulous documentation and monitoring. Integrating data processing in cloud environments or blockchain technology enhances the robustness and security of traceability systems. This allows for accurate identification and tracking of components, addressing challenges such as optimizing production efficiency. Real-time data acquisition, facilitated by technologies such as RFID and IoT within cloud or blockchain infrastructures, enables manufacturers to promptly detect anomalies. The analytical framework can be extended beyond traditional statistics to incorporate advanced methodologies like Markov chains and machine learning. This integration mitigates risks associated with defects and promotes sustainability and innovation in energy storage technologies.

2.2   State of Research

Traceability is often used to analyse and document trace links. Some applications also allow for progress monitoring or system synthesis. When analysing, there are various functionalities available across different tools. [5]. In specific application areas, such as industrial manufacturing, enhanced traceability systems can provide additional benefits, such as optimizing production and improving product quality. Wessel et al. present concepts for traceability in battery production in research pilot production scale [6, 7]. Leng et al. propose data methodologies for enhanced traceability systems based on literature, which they compare to their technical feasibility. [8]. In general, it was found that as data analysis methods advance towards AI-enhanced process optimization, technical feasibility decreases. For example, data visibility and traceability systems have already been demonstrated in industrial production cases. [9], whereas a ledger based trusted framework towards zero-defect manufacturing has yet to be tested on a case study [10]. While there are frameworks using Markov Chain, they either do not include the manufacturing focus [11] or the traceability aspect [12]. The focus of proposed traceability systems in manufacturing is the scheduling of process steps. As indicated in Guo et al., these approaches realized data analysation with other methods [13]. Therefore, this papers' aim is to apply a new traceability system based on Markov Chain for anomaly detection in the manufacturing realm with an accompanying industrial case study, which is not yet represented in the state of research (Fig. 2).

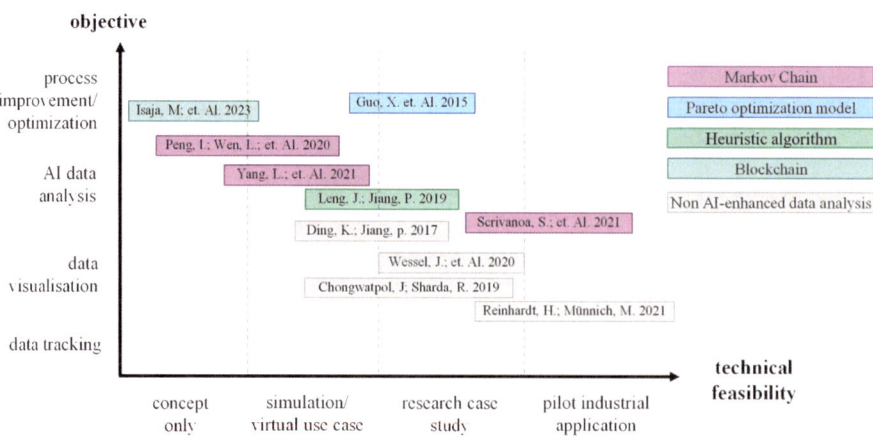

Fig. 2. Research based concepts of enhanced traceability systems for industrial manufacturing

3 Concept

The core idea of this paper is to employ Markov Chains to dynamically monitor the flow of products Traceability through the manufacturing line, thereby identifying deviations from expected paths. Anomalies or gaps in traceability are detected using a threshold-based approach: if the transition probability between two processes exceeds a predefined threshold (indicating an unexpected or unrecorded transition), it signals a potential anomaly. This method allows for real-time detection of irregularities, such as skipped processes or unauthorized modifications, providing a mechanism for ensuring product quality and integrity in discrete manufacturing settings.

The subsequent sections will elaborate on the definition of Markov Chains, the concept of a Discrete Production Line, and the role of hash functions. These components are integral to understanding how they collectively contribute to modeling Markov Chains in Production and, importantly, how they enable the detection of traceability gaps and anomalies.

3.1 Markov Chains

A Markov Chain is a mathematical model that encapsulates a stochastic system with a discrete state space. Formally, a Markov Chain is defined by a set of states, $S = \{S_1, S_2, \ldots, S_n\}$, where n is the number of distinct states in the system. A Transition Matrix $\mathcal{M}^{n \times n}$ represents the number of occurrences of transition from status $\mathcal{M}_i$ to $\mathcal{M}_j$. The transition probabilities between these states are described by a transition probabilities matrix $P^{n \times n}$, where p_{ij} represents the probability of transitioning from state S_i to state S_j. Mathematically, the transition probability p_{ij} is calculated as $p_{ij} = \frac{\text{Number of transitions from } S_i \text{ to } S_j}{\text{Total transitions from } S_i}$. The matrix P satisfies the following properties: non-negativity: $p_{ij} \geq 0 \forall i, j$. Row Sum Property: $\sum_{j=1}^{n} p_{ij} = 1 \forall i$. Memoryless Property: The probability of transitioning to a future state depends solely on the current state, not on the history of states.

3.2 Discrete Production Line

In the context of discrete manufacturing, a production line can be represented as a sequence of stations or processes, each responsible for specific manufacturing tasks. In this paper, we represent it as a sequence of processes because a station could have multiple processes. Formally, $proc = \{proc_1, proc_2, \ldots, proc_m\}$, where m is the number of processes in the line.

Each process $proc_i$ is associated with a set of process parameters, $Param_i = \{Param_{i1}, Param_{i2}, \ldots, Param_{ik}\}$ where i the index of the process the k the number of parameters. The process parameters are the data transmitted from the production station to the Manufacturing Execution System (MES) for storage. These parameters include crucial information, such as the temperature of an oven during the processing of a part or the force applied in an operation. The traceability significance of these parameters lies in their ability to provide a detailed record of the manufacturing process. The presence or absence of these parameters is pivotal, as it directly impacts the traceability of the production, ensuring accurate monitoring and analysis.

3.3 Hash Function

A hash function, denoted as H, is a deterministic function that converts an input string of arbitrary length into a fixed-size output, typically presented as a sequence of characters. In the context of our model, the hash function H takes the process parameters associated with each process $proc_i$ as input. It then generates a unique hash value H_i for the given parameters. The hash function can be expressed as $H : proc_i \rightarrow H_i$, where H_i represents the resulting hash value derived from the input process parameters $proc_i$.

3.4 Markov Chains in Production Modeling

By utilizing Markov Chains, we can formally depict the transitions of product process states as they progress through various manufacturing stages in a production line. In this context, states are defined as unique combinations of station identifiers, such as names or IDs, along with their corresponding hashed process parameters. Each state $S_i \in S$ is defined as $S_i = proc_i \cdot H_k$. The following illustration provides a concise overview of these concepts through a practical example. Let's consider a production line consisting of 5 processes {101, 102, 103, 104, 105}, which has produced a total of 340 parts. The data parameters for each part in each process undergo hashing. Consequently, the Markov chain states transform into {101_10, 102_11, 103_1A, 104_2B, 105_3A} (where the hash codes are provided as an illustration), as all 340 processed parts have followed the same route (Fig. 3).

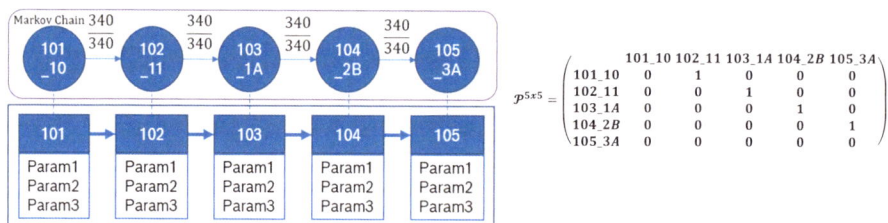

Fig. 3. Example of discrete production line and its Markov chain.

3.5 Traceability Anomalies and Gaps

Our approach revolves around the application of Markov models to represent and analyze the dynamics of discrete manufacturing. The transition probabilities, denoted as p_{ij}, between states in the Markov Chain signify the likelihood of transitioning from state S_i to state S_j. In the model, these transition probabilities are essential for capturing the dynamic behavior of the production line and detecting anomalies or gaps in product traceability.

There are two categories of anomalies or gaps which can be detected, the first type occurs when a product bypasses a step in the production line, which may result from manual intervention by a worker or a lack of configuration in MES following an update. If a process is skipped, the transition from the preceding process to the subsequent process, as illustrated in the figure below, will no longer be zero. A threshold, such as

1%, can be established, and based on this threshold, the presence of an anomaly or gap can be identified (Fig. 4).

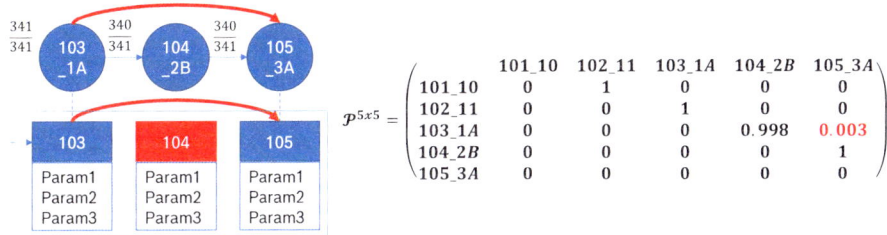

$$\mathcal{P}^{5x5} = \begin{pmatrix} & 101\_10 & 102\_11 & 103\_1A & 104\_2B & 105\_3A \\ 101\_10 & 0 & 1 & 0 & 0 & 0 \\ 102\_11 & 0 & 0 & 1 & 0 & 0 \\ 103\_1A & 0 & 0 & 0 & 0.998 & 0.003 \\ 104\_2B & 0 & 0 & 0 & 0 & 1 \\ 105\_3A & 0 & 0 & 0 & 0 & 0 \end{pmatrix}$$

Fig. 4. Concept of the Markov Chain enhanced traceability

The second anomaly involves alterations to the parameters of a process, such as deletions, additions, or modifications. Changes to the names of process parameters result in a modification of the hash function mentioned earlier. Consequently, this modification leads to a new state in the Markov chain, given that the combination of the process ID and the hash function is not present. Like the approach used in the first anomaly/gap category, employing a threshold enables the identification of this anomaly (Fig. 5).

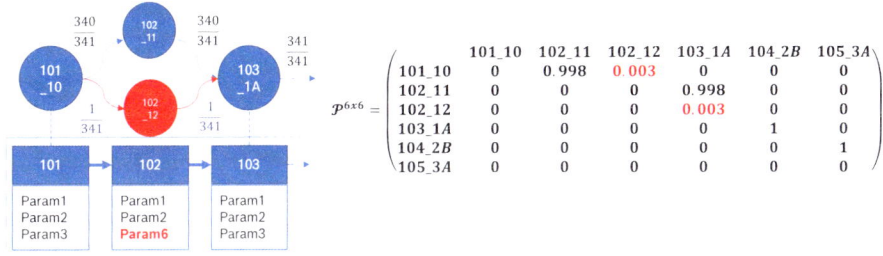

$$\mathcal{P}^{6x6} = \begin{pmatrix} & 101\_10 & 102\_11 & 102\_12 & 103\_1A & 104\_2B & 105\_3A \\ 101\_10 & 0 & 0.998 & 0.003 & 0 & 0 & 0 \\ 102\_11 & 0 & 0 & 0 & 0.998 & 0 & 0 \\ 102\_12 & 0 & 0 & 0 & 0.003 & 0 & 0 \\ 103\_1A & 0 & 0 & 0 & 0 & 1 & 0 \\ 104\_2B & 0 & 0 & 0 & 0 & 0 & 1 \\ 105\_3A & 0 & 0 & 0 & 0 & 0 & 0 \end{pmatrix}$$

Fig. 5. Concept of the Markov Chain enhanced traceability

4 Case Study

This case study demonstrates the practical application of our proposed methodology, utilizing Markov Chains for the identification of product traceability gaps and anomalies in discrete manufacturing settings. In this case study, we have chosen one of our Engine Control Unit (ECU) production lines as the focus. For each part, the data collected from each station within this line are processed, utilized, and fed into the transition matrix to update it. Subsequently, the probabilities within the matrix are scrutinized for transitions below the defined threshold to detect anomalies.

We applied the concepts from Sect. 3 to develop a functional model aimed at improving traceability. Given the versatility of our methodology, it is adaptable across various computing platforms and programming languages. For this implementation, we selected a Hadoop Ecosystem cluster due to its superior distributed computing capabilities, essential for efficiently processing the large datasets characteristic of manufacturing environments (Fig. 6).

Principle product data tracing

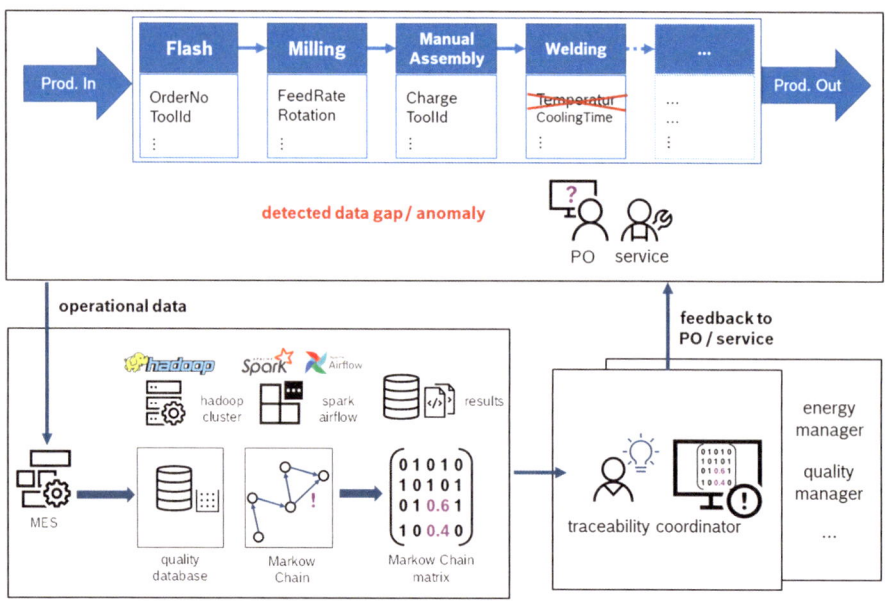

IT architecture with Markov Chain algorithm

Fig. 6. Prototypical implementation and information flow in the proposed concept

4.1 Data Collection

After each manufacturing process, a station sends a message (telegram) to MES. This message contains the part identifier and the corresponding process parameter data. To avoid disruption to the production MES database, a Data Pipeline retrieves data from MES and deposits it in the Hadoop file system. The data is structured into tables using the Parquet file format, offering an efficient and scalable storage solution for handling large datasets. The processing of the process data can be executed in either a batch or micro-batch processing mode (time triggered), or in real time mode (event triggered). For our proof of concept, we opted for the batch method.

Then at regular intervals, Apache Airflow™[1], a workflow orchestration platform, triggers an Apache Spark job. Apache Spark™[2], a distributed data processing engine, executes the job, loading data from a table, constructing the Markov chain matrix, and calculating the probabilities matrix.

---

[1] Https://airflow.apache.org/
[2] Https://spark.apache.org/

The results of the Spark job are stored for subsequent anomaly and gap detections. Moreover, these outcomes are used for visualization. The implementation also includes an alerting mechanism, notifying stakeholders in case of anomaly detection during the analysis.

4.2   Markov Chain Construction and Transition Probability Calculation

Considering the sparse nature of the Markov chain matrix in our case, the implementation opts for a dictionary-based representation. Each process ID serves as a key and the corresponding values indicate connections to other processes. This approach reduces storage requirements and enhances computational efficiency.

In the initialization phase, both the transition matrix p and the Markov States Matrix $\mathcal{M}$ begin as empty matrices. As the batch processing starts, data is loaded for each unique product that has traversed the production line A at a specific time. Subsequently, states are calculated for each station $\mathcal{A}_i$ with $S_i = \mathcal{A}_i \cdot H_k$ as defined earlier.

This product states sequence is then inputted into a function $update\_matrix : \mathcal{M} \times S_i \rightarrow \mathcal{M}$ responsible for updating the states matrix. The function iterates through the sequence to update the matrix, counting transitions between states. For each pair of consecutive states in the sequence, it checks if the previous state exists in the matrix. If not, a new entry is created. Then, it verifies if the current state exists as a transition from the previous state in the matrix. If not, a new entry is initialized with a count of zero. Finally, the function increments the count of transitions from the previous state to the current state by one. In essence, the function populates a matrix with counts of transitions between states based on the given sequence.

Following the update of the matrix $\mathcal{M}$, the function *calculate_transition_probabilities calculate_transition_probabilities* $: \mathcal{M}_I \times p \rightarrow p_I$, which computes the transition probabilities in the matrix p. It populates it with the transition counts, calculates row sums for normalization, and finally computes transition probabilities by dividing the transition matrix by the row sums. The resulting transition probabilities matrix is returned, providing a normalized representation of the likelihood of transitioning from one state to another.

4.3   Anomaly Detection

To identify potential traceability anomalies or gaps, we set the threshold to 1% for transition probabilities. That means when the probability of transitioning between states falls equal or below this threshold, it acts as a trigger for further investigation. This threshold is carefully chosen, taking into consideration the voluminous data, the multitude of production lines, stations, and the intricacies of the data generated in our manufacturing environment.

The probability matrix can be converted into a graph for visualizing and analyzing anomalies using the following pseudocode:

```
1:    for source, targets in probabilities_matrix.items():
2:        for target, probability in targets.items():
3:            G.add_edge(source, target, weight = probability)
4:            If probability < = threshold:
5:                print("Anomaly or gap detected")
```

This code traverses the entries of the probability matrix, retrieving source nodes and their respective target nodes along with their associated weights (probabilities). It integrates this information by adding edges to the graph (denoted as G), facilitating a visualization of the data. Additionally, the code checks if the probability is equal to or below a specified threshold, and if so, it signals the detection of an anomaly.

4.4 Results

The implementation of the Markov Chain based traceability system in a manufacturing setting has yielded insightful results. The construction of the Markov chain matrix and the calculation of transition probabilities were successfully executed on a Hadoop Ecosystems cluster, demonstrating the feasibility and of the approach. The anomaly detection mechanism, guided by a transition probability threshold, flagged inconsistencies in the manufacturing process. A thorough examination of these inconsistencies was conducted to determine their root causes, facilitating improvements in both the manufacturing process and the traceability system. This, in turn, contributed to the swift identification and resolution of traceability gaps.

4.5 Discussion

While the proposed methodology for automated traceability in manufacturing offers numerous advantages, it's important to consider potential drawbacks for a comprehensive discussion. A potential drawback is the choice of the transition probability threshold for anomaly detection requires careful consideration. Sensitivity to this threshold may impact the balance between false positives and false negatives, influencing the system's overall accuracy. Other potential drawback could be the continuous Model Maintenance. Ensuring the ongoing accuracy and relevance of the Markov Chain model demands continuous maintenance. Changes in the manufacturing process, introduction of new products, or modifications to the production line may necessitate frequent updates to the model.

5 Conclusion and Outlook

This paper introduced a method to improve product traceability in discrete manufacturing using Markov Chains. The implemented system effectively identifies irregularities and shortcomings within the manufacturing process. Future endeavors in this research involve refining the anomaly detection algorithm to adapt to diverse manufacturing environments and exploring real-time data analysis for prompt corrective actions. Additionally, potential research avenues may explore applying this methodology in other areas where traceability and process integrity are crucial. This could include integrating additional data, such as energy consumption, to calculate product-specific energy and carbon emission footprints. Furthermore, the development of tailored digital tools for operational improvement and daily business processes could be considered. The goal is to extend the application of this approach to various plants within the international Bosch factory network.

References

1. DIN EN ISO 9000:2015, "Quality management systems—Fundamentals and vocabulary."
2. T. Moe, "Perspectives on traceability in food manufacture," *Trends Food Sci. Technol.*, vol. 9, no. 9, pp. 211–214, 1998, S0924-2244(98)00037-5.
3. C. Kuhrt and M. Neußmann, "Produktrückruf und Produkthaftung, Qualitätssicherung," in *Betriebliches Risikomanagement und Industrieversicherung*, R. (Hg. . Mahnke, Ed., 2020, pp. 557–580. https://doi.org/10.1007/978-3-658-30421-8.
4. Robert Bosch Elektronik GmbH, "No Title." 2023.
5. S. F. Königs, G. Beier, A. Figge, and R. Stark, "Traceability in Systems Engineering—Review of industrial practices, state-of-the-art technologies and new research solutions," *Adv. Eng. Informatics*, vol. 26, no. 4, pp. 924–940, 2012, https://doi.org/10.1016/j.aei.2012.08.002.
6. J. Wessel, A. Turetskyy, O. Wojahn, C. Herrmann, and S. Thiede, "Tracking and tracing for data mining application in the lithium-ion battery production," *Procedia CIRP*, vol. 93, no. March, pp. 162–167, 2020, https://doi.org/10.1016/j.procir.2020.03.071.
7. J. Wessel, A. Turetskyy, O. Wojahn, C. Herrmann, and S. Thiede, "Tracking and Tracing for Data Mining Application in the Lithium-ion Battery Production," in *Procedia CIRP*, 2019, pp. 162–167. https://doi.org/10.1016/j.procir.2020.03.071.
8. J. Leng and P. Jiang, "Dynamic scheduling in RFID-driven discrete manufacturing system by using multi-layer network metrics as heuristic information," *J. Intell. Manuf.*, vol. 30, no. 3, pp. 979–994, 2019, https://doi.org/10.1007/s10845-017-1301-y.
9. H. Reinhardt *et al.*, "Retrieving properties of manufacturing systems from traceability data for performance evaluation and material flow simulation," *Procedia CIRP*, vol. 104, no. March, pp. 20–25, 2021, https://doi.org/10.1016/j.procir.2021.11.002.
10. M. Isaja *et al.*, "A blockchain-based framework for trusted quality data sharing towards zero-defect manufacturing," *Comput. Ind.*, vol. 146, no. July 2022, 2023, https://doi.org/10.1016/j.compind.2023.103853.
11. L. Peng, L. Wen, L. Qiang, D. Yue, D. Min, and N. Y. Ying, "Research on complexity model of important product traceability efficiency based on Markov chain," *Procedia Comput. Sci.*, vol. 166, pp. 456–462, 2020, https://doi.org/10.1016/j.procs.2020.02.065.
12. S. Scrivano and T. Tolio, "A Markov Chain model for the performance evaluation of manufacturing lines with general processing times," *Procedia CIRP*, vol. 103, no. March, pp. 20–25, 2021, https://doi.org/10.1016/j.procir.2021.10.002.
13. Z. X. Guo, E. W. T. Ngai, C. Yang, and X. Liang, "An RFID-based intelligent decision support system architecture for production monitoring and scheduling in a distributed manufacturing environment," *Int. J. Prod. Econ.*, vol. 159, pp. 16–28, 2015, https://doi.org/10.1016/j.ijpe.2014.09.004.

Smart Characterization of Secondary Materials to Enable Resilient Material Processing

Tom Hoppe$^{(\boxtimes)}$, Louisa Türke, and Thomas Vietor

Institute for Engineering Design, Technische Universität Braunschweig, Hermann-Blenk-Str. 42, 38108 Braunschweig, Germany
t.hoppe@tu-braunschweig.de

Abstract. The focus of this research is on supporting a resource-efficient economy to minimize post-consumer plastic waste by incorporating secondary materials into existing product design and thereby closing material cycles in plastic usage. In addition to lower primary resource demand and lower energy consumption in the production phase, the potential opportunities include lower overall environmental impacts. An important challenge arises from batch-dependent variations in the quality of recycled plastics, influencing product development in the design phase. Finite element analysis (FEA) plays a crucial role in the early stages of product development, demonstrating a concept's potential for further development at a low cost and with minimal production effort. Detailed modeling of material properties within FEA is essential to ensure the utmost validity of calculation results and to pinpoint any weaknesses in the early design stages. The present paper discusses various approaches and applications for using neural networks (NNs) in constitutive modeling, particularly when dealing with heterogeneous material behavior resulting from the recycling process. The use of virtual training data derived from a phenomenological constitutive model and its advantages and potential applications to recyclates are highlighted. This leads to the proposal of using inverse surrogates of the phenomenological constitutive model as one method for obtaining suitable constitutive laws from experimental data in the future.

Keywords: Sustainability · Digital tools · Recyclates

1 Introduction

To achieve the goal of a sustainable, low-carbon and resource-efficient economy, not only must the amount of waste be minimized, but resources must be kept in a cycle for as long as possible. Therefore, new efficient production and consumption methods must be developed. At the same time, irreversible damage caused by air, water or soil pollution should be avoided and energy should be saved. The EU's "Closing the Loop" strategy therefore proposes measures that should be implemented to achieve a circular economy. In the long term, the aim is to reduce landfilling and prepare waste streams for recycling, for example through appropriate waste management. [1].

The production of plastics not only generates large amounts of waste, but also large amounts of CO_2. For example, one ton of plastic produces two tonnes of CO_2. At the end

© The Author(s) 2025
K. Dröder and T. Vietor (Eds.): CD 2024, Proceedings, pp. 305–316, 2025.
https://doi.org/10.1007/978-3-658-45889-8_24

of a plastic product's life, only 20% of this mass is currently recycled. Around two-thirds of this plastic mass is thermally recycled, where an additional 2.7 tonnes of CO_2 are produced. [2].

In 2019, a total of 50.7 megatons of plastics were processed as primary resources in the European Union, of which an average of 10% went into the automotive industry [3]. The areas of application for plastics in cars are diverse. For the interior, plastics can be used not only for decorative parts, but also for seats and headrests, seals, instrument panels and supports. In vehicle exteriors, in addition to the possibility of using plastics for headlights and windows, they can also be used for bumpers, sunroofs, sills and underbody protection. Additionally, to standard plastics such as Polypropylene (PP) and polyvinyl chloride, engineering plastics such as Polyoxymethylene and Polyamide 6, high-performance plastics for example Polyetheretherketone are also used. [4].

For example, Audi uses up to 27 parts in the entire vehicle with recycled content in the Q4 model. Recycled plastics are used in assembly carriers, headlight shells and covers in the exterior. In the interior, recyclates are mainly used in insulation and damping materials. PET bottles are also used to make fibers for sports seat covers. [5].

In Germany, two megatons of recyclates were produced and reintroduced to the market in 2019, which corresponds to a recyclate use rate of just 13.4% [3]. The low proportion results from the challenges in the area of recyclates. Challenges occur in the areas of technology, economy, regulation and culture. Technical challenges exist primarily in the area of separating plastics, as these can be contaminated. In addition, post-consumer plastic waste can be aged. Both of these factors lead to an uncertain composition of plastic recyclates. [6].

The author Schatz (2023) characterized different batches of a post-consumer recyclate using standard test methods. Variations in the different material properties of polypropylene recyclates were identified. The results show that not only is a doubling of the melting rate possible with several batches, but the content of inorganic substances can also triple. In direct comparison with primary raw materials, the service life of the samples is reduced under cyclic loads. [7].

The batch variations described above have a significant influence on part design. To be able to design a part with a recycled material, the variations in material properties in each existing batch must be determined and taken into account. The part's functional performance has to be guaranteed all the time and the part is designed in a risk-appropriate manner.

NNs allow the representation of any continuous functional relationship and therefore the modeling of arbitrarily complex material properties [8]. The learned functional relationship is stored in the parameters of the NN and can be retrieved within milliseconds. NN-based approaches are much more time and resource-efficient than the iterative optimization of conventional material models, where computationally intensive simulations are performed repeatedly. Thus, at different levels of constitutive modeling, NNs are used as surrogate models [9]. This paper presents some of these applications and discusses their suitability for modeling the properties of recycled plastics. With a focus on simplicity and efficiency, a preferred methodology is presented that can facilitate the transition to a circular economy by providing a resilient response to varying material properties during the design phase.

This publication describes the role of plastics in the automotive industry and their potential for moving towards a circular approach. It highlights the challenges that arise in the production and use of post-consumer recyclates, with a focus on the varying material properties that can adversely affect the product development process. The knowledge gained is used to discuss the suitability of different NN-based approaches for the generation and calibration of constitutive models for recycled plastics in FEA. For this purpose, several publications are reviewed and a preferred method is proposed.

2 Plastics in the Circular Economy

The increasing consumption of raw materials and the resulting higher demand will lead to irreversible damage to ecosystems in the long term. At present, finite primary resources are mainly used to meet the high demand for raw materials. In Germany, these are mainly imported; in the case of metallic raw materials, imports account for almost 100%. This linear production method generates a constant flow of waste. To make economy less dependent on availability, innovations in technology and design must be promoted in order to make products more durable and to view waste as a valuable resource. [10].

There are various strategies in the circular economy that aim to reduce the consumption of primary resources and minimize waste production. The recovery of materials through recycling requires only minor innovations in product design. However, recycling is close to the linear economy. To ensure that secondary resources recovered through recycling can be used meaningfully, recycling must be carried out to a high standard, which means that the quality of the materials must be guaranteed. [11].

Plastics are used in almost all areas of life, with the packaging industry accounting for the largest market share in the EU at 40%. In Germany, 6.3 megatons of plastic waste were generated in 2019. There are various approaches to counteracting the amount of waste. One solution in Germany has been established with the introduction of the deposit system. In 2019, 93% of these were recycled. Plastic parts from an end-of-life vehicle can also be recycled. There are approaches to reusing car parts, such as a bumper, by using a mixture of primary and secondary raw materials in the same application. [3].

The advantage of recycling plastic is that existing resources can be used. This not only increases the security of raw material supply, but also reduces energy requirements. [12] For example, the environmental services provider ALBA Group was able to avoid 4.2 million tons of greenhouse gases in 2019 by recycling 6 million tons of recyclable materials. [13] Plastics have a short life cycle and are not intended for further life cycles. One example of this is plastic film for PP packaging. Mechanical stress and exposure to the environment, such as heat, oxygen and UV light, can lead to the degradation of plastics. Reprocessing of plastics can also lead to degradation, caused by a change in the polymer structure. The resulting effects can be divided into three categories. On the one hand, a change in viscosity can occur due to the change in molecular weight, cross and chain branching can form and, on the other hand, oxygen-saturated compounds can occur. [14].

One challenge in the field of plastics is proper separation by type, as this is crucial for the quality of the recyclates [12]. In particular, materials such as PP are difficult to recycle if the materials are not in a pure form. Mechanical recycling of plastics takes place by remelting the shredded material and is therefore limited to thermoplastic materials. Re-melting is associated with degradation of the polymer chains. For this reason, primary resources are often added to the secondary resources obtained. A distinction is made between primary and secondary recycling. Materials that originate from production waste and are available in pure form are used for primary recycling. In contrast to sources that originate from post-consumer waste, the composition is known and they are not contaminated. Secondary recycling involves further steps such as collecting, sorting and cleaning to remove impurities. Raw material recycling is used for thermoplastics, for example, if the quality achieved through mechanical recycling is too low. Processes can be used to convert the polymer chains into compounds with a low molecular weight. Chemical processes can be used to produce original polymers again. [3, 14].

Experimental results show that despite the choice of the same recyclate as the starting material for post-consumer waste, the identical material properties do not necessarily have to be achieved. High variations in material properties have been observed in different tests. In addition to the proportion of inorganic content, the Melt Flow Rate, for example, also fluctuates by a factor of two. The use of fillers can improve the mechanical properties, but leads to corresponding differences in material behavior. [7] As already mentioned it is possible to use recycled materials as an additive to primary resources in addition to fillers. For example, the authors Barbosa et al. (2017) in [15] found that PP can be used as a secondary material if it is exposed to tensile loads, but not if impact loads occur. The tensile properties of primary, secondary and mixed PP are similar, but the notched impact strength of the secondary materials is significantly lower. The author Frounchi (1999) also shows that the molecular weight, impact strength and tensile strength of PET decrease with the number of reprocessing cycles [16].

In summary, it should be mentioned that both degradation and contamination, the material composition of primary and secondary resources and the number of reprocessing cycles influence the material properties of plastics.

3   Constitutive Modeling for Recyclates

The previous chapter illustrates the complexity of the factors that influence the material behavior of a secondary plastic material. At present, the transition to a circular economy is only possible through a resilient approach to these uncertainties during the design phase due to the lack of standards and the correspondingly varying quality of recycled materials. The characterization and modeling of the constitutive behavior of the recycled material play a central role in part design. The constitutive behavior, i.e. the mathematical relationship between stress, strain, time and, for example, temperature, forms the basis for subsequent FEA [17]. By accurately modeling this relationship, the properties of the material in question can be taken into account when designing parts and adjustments can be made if necessary. The use of FEA also allows the functional suitability of parts to be continuously evaluated without the need numerous prototypes [18].

In addition to physics-based and phenomenological models, the field of constitutive modeling has been extended in recent decades by the use of machine learning, in particular using NNs [9]. NNs make it possible to represent any continuous functional relationship and thus to model arbitrarily complex material properties [8]. They describe a mathematical function that maps a set of input values to output values. The function is created by composing many simpler functions and allows complex relationships to be expressed that are complex or impossible to express using conventional mathematical methods [19, 20]. The parameters of the mathematical function are adapted during the learning phase, based on the information flowing through the network, so that the differences between the network output and the true output of the data set are minimized for all training examples [19]. In this way, the network learns to predict output values to input values from unseen data. The learned functional relationship is stored in the parameters of the NN and can be retrieved within milliseconds. NNs are therefore used as surrogate models at various levels because they are much more time and resource efficient [9].

4   Nn-Based Approaches for Constitutive Modeling

In the context of secondary materials, the use of surrogate models to predict the 28-day compressive strength of concrete containing recycled aggregates is a common example. The authors Duan et al. (2013) in [21] and Naderpour et al. (2018) in [22] use several physical input parameters, including the amount of recycled aggregate and the percentage of impurities, to predict the 28-day compressive strength. The appeal of this model formulation is that it takes into account the physical influences of the recyclate, and in this way provides compressive strength without experimental testing for a wide range of design decisions. To generate or calibrate a complex constitutive model as a basis for FEA, more information about the material behavior is required. This includes information on how the material will behave under different load cases and when it will fail.

In one of the first papers on constitutive modeling using NNs, Ghaboussi et al. (1991) in [23] also model the mechanical behavior of concrete. One of the models employed is strain-guided and utilizes experimental datasets for biaxial monotonic loading. In this method, inputs include the prior stress-strain states and the current strain increment, with the aim of predicting the resultant stress increment. In their later work [24], Ghaboussi et al. (1998) describe a method of training NNs called autoprogressive training. This method enables NNs to learn the complex stress-strain behavior of materials by using global load-deflection response. Jordan et al. (2020) in [25] use a NN-based surrogate model to describe the temperature and strain rate-dependent response of PP. A hybrid approach of mechanism-based and data-based modeling is used. The surrogate model describing the viscous behavior acts in series with a temperature-dependent spring describing the instantaneous elastic response of the material. The data base for the NN is captured by a robotic testing system, with uniaxial tensile loading being tested.

In addition to the approaches presented here that learn stress-strain relationships solely from experimental data, some approaches that use stress-strain relationships from existing constitutive models. In this way, it is not necessary to perform extensive series of experiments to generate a meaningful data set. The authors Stoffel et al. (2018) in [26] replace the visco-plastic material law used in the Gaussian quadrature points with a trained NN. The training data of the surrogate model was generated by numerical simulations with a calibrated conventional constitutive model. The input values include stress, back stress and plastic strain tensor components and the output values consist of five plastic strain rate and back stress rate components.

In the approaches mentioned so far, each training example, consisting of input values and corresponding output values, corresponds to an experiment, usually performed under the assumption of uniform loading and homogeneous material behavior, and often limited to simplified scenarios. The approach of Aguir et al. (2008) in [27] consider several inhomogeneous tests to calibrate an existing constitutive model. In this case, instead of stress-strain increments, appropriate model parameters are predicted for the selected material model. Calibration is necessary because the model parameters for reproducing the experimental system response are difficult to measure directly, although the system response can be measured easily. Even for physically motivated material models, the parameters cannot be determined directly from experimental results and must be adjusted in a calibration process [28, 29].

The identification of the model parameters from the measured system response is called inverse analysis [30]. Aguir et al. (2008) in [27] show that the calibration is performed as a multi-objective optimization and an error measure is obtained by comparing the model responses with the experimental system responses. To reduce the computational time for the analysis of the objective functions, NN-based surrogate models are used instead of finite element calculations. These surrogate models were trained with selected material parameters and the resulting model responses. Commercial optimization programs like LS-OPT [31] also use NN-based surrogate models of the existing constitutive models to make the performance of probabilistic analyses more efficient in terms of time and resources. Huber et al. (1999) in [32] and Yagawa et al. (1996) in [33] present surrogate models in inverse formulation, where the model response is the input and the material parameters used to generate it are the output. Here, the experimental system response is still mapped to the simulated output domain. The advantage of the backwards defined surrogate model is that the mapping takes place within a single evaluation of the network, and the associated material parameters are output for the most plausible mapping. In this way, both the selection of suitable initial parameters and their iterative optimization can be avoided. To identify hyperelastic material parameters, Yenigum et al. (2022) in [34] use NNs trained with the resulting force values of a virtual O-ring tensile and compression test as input and the corresponding material parameters as output. Andrade et al. (2022) in [35] present an alternative model where the data set consists of selected elasto-plastic material parameters and the resulting strain fields. The experimental input is image correlation data for which appropriate material parameters are output. Schulte et al. (2023) in [36] uses an inverse surrogate model to generate suitable initial parameters for subsequent conventional parameter identification. The input of the surrogate model are virtual stress-strain curves of the hardening behavior in sheet metal forming processes. The output-suitable material parameters serve as initial parameters for a conventional optimization based on image correlation data.

Koch and Haufe (2019) in [37] present an equivalent model that learns the parameters for defining yield curves based on force-deformation curves from virtual tensile tests. By considering the full realistic range of parameters, the model should be able to predict an appropriate set of parameters for a given force-deformation curve. The perspective of the paper is to learn the full spectrum of the material model in advance and then predict the material properties from a few experiments with minimal effort. Meißner et al. (2022) in [38, 39] use this approach to calibrate the LS-DYNA material model MAT_SAMP-1 and the failure model GISSMO for a virtual data set derived from acrylonitrile-butadiene-styrene. In this framework, the parameters defining the yield curves and the variable plastic Poisson's ratio under tension and compression were predicted by an inverse surrogate model. In addition, the points of the curve describing the equivalent plastic strain at failure for different triaxialities were learned. The deviation of the yield curves defined by the predicted parameters from the yield curves initially used to generate the force-deformation curves was used as an error measure during training.

5   Proposal of Inverse Material Parameter Identification Method for Enabling Smart Characterization of Secondary Materials

Building on the results of Meißner et al. [38, 39], this paper proposes the use of NNs to learn a comprehensive spectrum of a material model by varying the material parameters of the model in a design of experiments. FEA is performed for each unique configuration of material parameters, corresponding to the required experimental test for characterizing the recyclates. The resulting force-deformation curves from the simulation are used as input labels to an inverse NN, with the material parameters used to generate the curves serving as output labels. The NN thus learns the phenomenological material model in an inverse manner by statistically approximating the relationship between the material parameters and the resulting behavior. After being trained on a large virtual dataset, a NN can process different batches of recycled material. Each batch possesses unique material properties due to varying degrees of degradation and contamination. Therefore, a specific stress-strain relationship is required, which in turn necessitates a newly parameterized material model. Figure 1 b) shows an abstract representation of this approach. Experimental force-displacement curves are required for each material under consideration. The NN extracts patterns from these curves and predicts appropriate material parameters using the learned inverse relationship between FEA results and the corresponding material parameters. The calibrated phenomenological constitutive model is then used for stress calculations. This allows an adaptive response to fluctuating material properties without the need for repeated simulation efforts or re-training.

In contrast, the direct approach depicted in Fig. 1 a) eliminates the need to define a yield surface, a flow rule, and a numerical algorithm to update the stresses. The NN outputs stress increments in the Gaussian quadrature points directly, allowing for a faster simulation process. The approaches employed by Ghaboussi et al. in [24] and Jordan et al. in [25] solely necessitate the input of strain increments into the network. This method assumes homogeneous material behavior and requires a comprehensive experimental database to properly train the NN. The stress-strain relation for the material in question is generalized from the stress-strain pairs in the training data. The stress-strain relationship for the material is generalized from the stress-strain pairs in the training data. If the material's behavior is not homogeneous due to degradation or contamination, it is necessary to derive experimental training data for each new material or provide additional information about the material to the strain-dependent model. Otherwise, data from a strain-controlled test, such as the uniaxial tension test, contains samples with identical input labels but varying output labels.

a)   b)

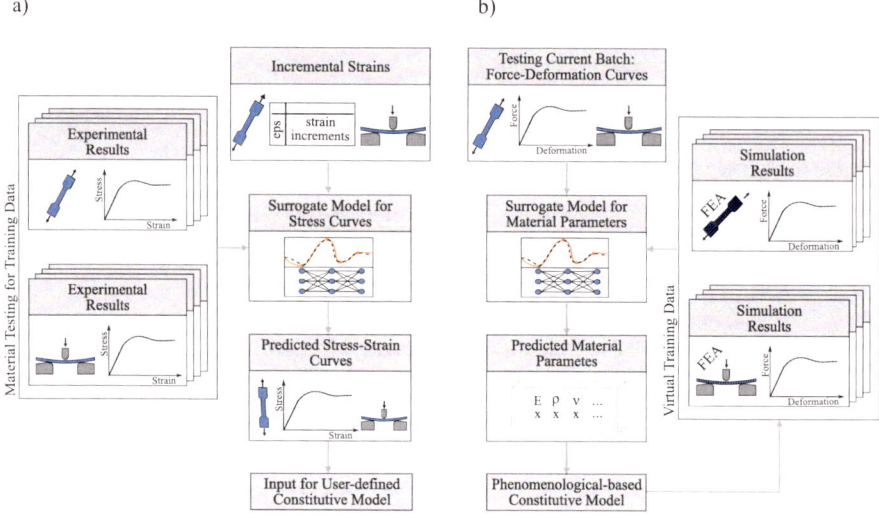

Fig. 1. Comparison of NNs for a) direct constitutive modeling and b) inverse calibration

To efficiently capture the diverse constitutive behaviors of recycled materials within a unified NN framework, it is suggested to train the network using a virtual dataset. This enables the representation of a broad spectrum of material properties and the generalization of an appropriate correlation for the experimental data at hand. Employing the method of inverse parameter identification enables the derivation of a stress-strain relation by fitting the phenomenological model based on the experimental observations. Subsequently, advancing towards establishing a NN-based constitutive law can be pursued through autoprogressive NNs [24] or intelligent finite elements [26]. This process may be implemented for each outcome derived from FEA within the virtual dataset, aiming to establish a NN framework capable of predicting stress increments at Gaussian quadrature points based on empirical observations. Alternatively, a pre-trained NN can be retrained and serve as a surrogate model for each calibrated phenomenological model, which has been fitted to the experimental observations.

Although such an approach does not require extensive experimental data collection, it still requires the resource-intensive generation of labeled data using numerical simulations of the material to be characterized. This data, or pre-trained NNs, could be provided by finite element software companies or even material suppliers. This would allow users to generate suitable material maps for subsequent finite element analysis without any calibration effort. The necessary expertise in continuum mechanics and materials engineering would be provided in the form of complex material models, while the expertise required to calibrate the models would be replaced by statistical decisions learned from sufficiently large data sets using NNs. Such an approach can make it easier for small and medium-sized companies in particular to use finite element analysis in the product development process, as only the experimental force-deformation curves need to be determined and imported into the NNs. [40].

With this in mind, it is desirable to determine the minimum number of experimental tests required as input to the inverse surrogate model that still provides a sufficiently accurate representation of the material properties [37].

6 Conclusion & Outlook

The inadequate and variable quality of recycled plastics represents one of the main barriers to moving towards a circular approach to plastics [6]. Quality variations stem from various factors, including mixtures with other polymers, degradation during use and mechanical recycling, and impurities that can't be removed during the recycling process [7]. The variations in material properties within each existing batch must be identified and considered in part design when incorporating recycled content. The ongoing development of machine learning methods opens up new opportunities to make traditional product development processes more resilient to these uncertainties. The present paper proposes a possible NN-based inverse surrogate model to enable the use of accelerated and accurate FEA. With this approach, a virtual data set can be generated by varying the input parameters of an existing constitutive model. The dataset is intended to cover a wide range of possible material properties and captures all potential variations in the quality of the recycled material [37]. A calibrated constitutive model can be output within milliseconds for each input of experimental force-deformation curves [39]. This enables the prompt availability of recycled material properties for integration into FEA. Furthermore, the dataset holds potential for configuring a more sophisticated NN capable of directly predicting stress increments.

This potential will be experimentally validated in the future by deriving virtual data sets for primary PP in different mixing ratios with secondary PP of different quality from different phenomenological material models. It will then be investigated whether the phenomenological material laws of the material models can be replaced using the ability of NNs for universal function approximation. The stress-strain values of the FEA results in the virtual data will be used as a basis for training. In this context, different test methods, load paths and NN architectures will be analysed. Finally, it will be investigated whether the combination of approaches allows the prediction of stress increments in the Gaussian quadrature points from coupon test data.

References

1. European Commission, "Closing the loop - An EU action plan for the Circular Economy" Communication from the Commission to the European Parliament, the Council, the European Economic and Social Committee and the Committee of the Regions, Brussels, 2015.
2. DIW Berlin, "Auf dem Weg zur Klimaneutralität: Plastikrecycling muss stärker in den Fokus rücken", 2021. [Online]. Available: https://www.diw.de/de/diw_01.c.820601.de/auf_dem_weg_zur_klimaneutralitaet__plasti. [Accessed: Jan. 11, 2024].
3. Orth, P., Bruder, J. and Rink, M.: Kunststoffe im Kreislauf Vom Recycling zur Rohstoffwende. Springer Vieweg, Wiesbaden (2022).
4. Ziegmann, G., Die Leichtbauwerkstoffe im Fahrzeugbau", in Leichtbau in der Fahrzeugtechnik. Springer Fachmedien, Wiesbaden, pp. 362–429, (2013).

5. Audi AG, „Nachhaltigkeit - Rezyklate," [Online]. Available: https://www.audimediacenter. com/de/rezyklat-14681. [Accessed: Jan. 11, 2024].
6. Baldassarre, B., Maury, T., Mathieux, F., Garbarino, E., Antonopoulos, I. and Sala, S., "Drivers and Barriers to the Circular Economy Transition: the Case of Recycled Plastics in the Automotive Sector in the European Union", 29th CIRP Life Cycle Engineering Conference, Elsevier B. V., 2022.
7. Schatz, L., "Batch variations of post-consumer recyclates and their influence on material properties," M. S. thesis, Montan Universität Leoben, 2023.
8. Cybenko, G.: Approximation by superpositions of a sigmoidal function. Math Control Signals Syst 2(4), pp.303–314 (1989).
9. Dornheim, J., Morand, L., Nallani, H.J. et al.: Neural Networks for Constitutive Modeling: From Universal Function Approximators to Advanced Models and the Integration of Physics. Archives of Computation Methods in Engineering (2023).
10. Baron, M., „Ressourcen- und Klimaschutz durch Kreislaufwirtschaft", in Einführung in die Kreislaufwirtschaft. Springer Vieweg, Wiesbaden, pp. 47–64, (2018).
11. Potting, J., Hekkert, M., Worrel, E. and Hanemaaijer, A., "Circular Economy: measuring Innovation in the Product Chain" Netherlands Environmental Assessment Agency, The Hague, 2017.
12. Hunold + Knoop Kunststofftechnik, „Mehr Nachhaltigkeit durch Rezyklate Kunststoff-Recycling schont das Klima und vermeidet Abfälle" [Online]. Available: https://www.hun old-knoop.de/kunststoff-blog/9-fragen-zu-rezyklat/. [Accessed: Jan. 12, 2024].
13. interzero zero waste solutions, „Wissenschaftlich belegt: Recycling wesentlich für Klimaschutz / ALBA Group fordert Mindestquoten für Rezyklat-Einsatz". Available: https:// www.interzero.de/medien/service-fuer-journalisten/pressemitteilungen/detailseite/wissen schaftlich-belegt-recycling-wesentlich-fuer-klimaschutz-alba-group-fordert-mindestquoten-fuer-rezyklat-einsatz/. [Accessed: Jan. 12, 2024].
14. Rudolph, N., Kiesel, R. and Aumnate, C.: Einführung Kunststoffrecycling Ökonomische, ökologische und technische Aspekte der Kunststoffabfallverwertung. Carl Hanser Verlag, München (2020).
15. Barbosa, L., Piaia, M. and Ceni, G.: „ Analysis of Impact and Tensile Properties of Recycled Polypropylene" International Journal of Materials Engineering 2017 pp. 117–120, (2017).
16. Frounchi, M. „Studies on degradation of PET in mechanical recycling. Macromolecular Symposia" pp. 465–469, (1999).
17. Stommel, M., Stojek, M., Korte, W.: FEM zur Berechnung von Kunststoff- und Elastomerbauteilen. 2nd edn. Hanser, München (2018).
18. Kohar, C.P., Greve, L., Eller, T.K et al.: A machine learning framework for accelerating the design process using CAE simulations: An application to finite element analysis in structural crashworthiness. Computer Methods in Applied Mechanics and Engineering 385(4), 114008 (2021).
19. Goodfellow, I., Bengio, Y., Courville, A.: Deep Learning. MIT Press, Massachusetts (2016).
20. Fausett, L.: Fundamentals of Neural Networks: Architectures, Algorithms, and Applications. Prentice-Hall Inc., Englewood Cliffs (1994).
21. Duan, Z.H., Kou, S.C., Poon C.S.: Prediction of compressive strength of recycled aggregate concrete using artificial neural networks, Construction and Building Materials 40, pp. 1200–1206 (2013).
22. Naderpour, H., Rafiean, A.H., Fakharian, P.: Compressive strength prediction of environmentally friendly concrete using artificial neural networks. Journal of Building Engineering 16, pp. 213–219 (2018).
23. Ghaboussi, J., Garrett, J.H., Wu, X.: Knowledge-based modeling of material behavior with neural networks. J. Eng. Mech. 117, pp. 132–153 (1991).

24. Ghaboussi, J., Pecknold, D.A., Zhang, M., Haj-Ali, R.M.: Autoprogressive training of neural network constitutive models. Internat. J. Numer. Methods Engrg.. 42 (1), pp. 105–126 (1998).

25. Jordan, B., Gorji, M.B., Mohr, D.: Neural network model describing the temperature- and rate-dependent stress-strain response of polypropylene. International Journal of Plasticity 135, 102811 (2020).

26. Stoffel, M., Bamer, F., Markert, B.: Artificial neural networks and intelligent finite elements in non-linear structural mechanics. Thin-Walled Structures 131, pp. 102–106 (2018).

27. Aguir, H., Chamekh, A., Belhadjsalah, H. et al.: Identification of Constitutive Parameters using Hybrid ANN multi-objective optimization procedure. International Journal of Material Forming (2008).

28. Mahnken, R.: Identification of Material Parameters for Constitutive Equations. In: in Stein, E., Borst, R. de and Hughes, T.J.R. (eds.), Encyclopedia of Computational Mechanics Second Edition, Vol. 71, pp. 1–21. John Wiley & Sons, Ltd, Chichester (2018).

29. Jones, E.M.C., Carroll, J.D., Karlson, K.N. et al.: Parameter covariance and non-uniqueness in material model calibration using the Virtual Fields Method. Computational Materials Science 152(4), pp. 268–290 (2018).

30. Stavroulakis, G.E., Bolzon, G., Waszczyszyn, Z. et al.: Inverse Analysis. Comprehensive Structural Integrity 24, pp. 685–718 (2003).

31. Stander, N.E.A.:LS OPT User's Manual—A Design Optimization and Probabilistic Analysis Tool for the Engineering Analyst, [Online]. Available: https://www.lsoptsupport.com/docume nts/manuals/ls-opt/lsopt_60_manual.pdf. [Accessed: Jan. 20, 2024].

32. Huber, N., Tsakmakis, C.: Determination of constitutive properties fromspherical indentation data using neural networks. Part i: the case of pure kinematic hardening in plasticity laws. Journals of the Mechanics and Physics of Solids 47(7), pp. 1569–1588 (1999)

33. Yagawa, G., Okuda, H.: Neural networks in computational mechanics. Archives of Computational Methods in Engineering 3(4), pp. 435–512 (1996).

34. Yenigun, B., Gkouti, E., Barbaraci, G. et al.: Identification of Hyperelastic Material Parameters of Elastomers by Reverse Engineering Approach. Materials 15(24), 8810 (2022)

35. Andrade-Campos, A., Bastos, N., Conde, N. et al.: On the inverse identification methods for forming plasticity models using full-field measurements. In: International Deep Drawing Research Group Conference (IDDRG 2022). IOP Conference Series: Materials Science and Engineering, 1238, 012059. IOP Publishing Ltd, Lorient (2022).

36. Schulte, R., Karca, C., Oszwald, R. et al.: Machine learning-assisted parameter identification for constitutive models based on concatenated loading path sequences. European Journal of Mechanics / A Solids 98, 104854 (2023).

37. Koch, D., Haufe, A.: First Steps towards Machine-Learning supported Material Parameter Identification. In: 12th European LS-DYNA Conference, DYNAmore GmbH, Koblenz (2019).

38. Meißner, P., Winter, J., Vietor, T.: Methodology for Neural Network-Based Material Card Calibration Using LS-DYNA MAT_187_SAMP-1 Considering Failure with GISSMO. Materials 15, 643 (2022).

39. Meißner, P., Hoppe, T., Vietor, T.: Comparative Study of Various Neural Network Types for Direct Inverse Material Parameter Identification in Numerical Simulations. Appl. Sci. 12(24), 12793 (2022).

40. Meißner, P., Vietor, T.: New opportunities and benefits in the product development process using the machine learning based direct inverse method for material parameter identification. In: Proceedings of the International Conference on Engineering Design (ICED23), Bordeaux (2023).

Multi-Sensor Data Fusion for Application of Machine Learning in Milling Processes on Press Tool Surfaces

Skender Paturri[1], Lasse Wendland[1] (ID), Marvin Gravert[1], Enea Duka[1], and Maik Mackiewicz[2]([email])

[1] Synergeticon GmbH, Hein-Saß-Weg 22, 21129 Hamburg, Deutschland
{info,m.gravert}@synergeticon.com
[2] Volkswagen AG, Berliner Ring 2, 38440 Wolfsburg, Deutschland
maik.mackiewicz1@volkswagen.de

Abstract. In today's global markets, companies must ensure product quality, transparency, and responsiveness to remain competitive. Efficient and precise milling processes are now more crucial than ever due to the utilization of advanced manufacturing technologies. Deviations of nominal geometry occurring from the milling process can result in time consuming and costly reworking or scrapping. Predicting these deviations before machining has a significant value for optimizing the workflow. Data-based methods have demonstrated their ability to surpass the limitations of classical analysis methods in optimizing machining processes in production. However, their performance depends on an extensive and diverse structured data set. In this paper, we propose a method for fusing all the necessary machine signals collected during the milling process into a large-scale data set suitable for machine learning in milling. This is achieved by recognizing patterns and reference points in various types of sensor systems to transform the signals into a unified data model. Specifically, our data set includes the milling program, internal data of the tool machine, data collected by external accelerometers and spindle sensors, as well as CAD data and 3D scans of the finalized part. Together, this serves as a digital end-to-end representation of the milling process, providing a foundation for machine learning based methods. We apply and evaluate our approach on a test geometry.

Keywords: Data analysis · Multi-sensor data fusion · Milling process

1 Introduction

The increasing digitization of manufacturing presents new opportunities to optimize processes and reduce costs without compromising quality [15]. Process data acquisition is a crucial aspect of this.

optimization, enabling manufacturing companies to analyze complex cause-effect relationships [4]. In Industry 4.0, each manufacturing process generates context-specific data sets. Combining data from multiple sources is necessary for real-time diagnostics

K. Dröder and T. Vietor (Eds.): CD 2024, Proceedings, pp. 317–330, 2025.
https://doi.org/10.1007/978-3-658-45889-8_25

or prediction applications [16]. The ultimate goal of the milling process is to deliver workpieces of the desired quality in a cost effective and timely manner. Deviations from specifications can lead to expensive rework and negatively impact subsequent process steps [11].

In the manufacture of press tools, the geometry of the active surfaces plays a critical role in determining the dimensional and shape accuracy of the resulting components. The milling process comprises roughing, pre-finishing, and fine finishing [2]. Many variables can disrupt the milling process and cause geometric deviations in the active surfaces [11]. These variables can be categorized as machine-related, control-related, cutting tool-related, workpiece-related, and environmental. Lack of rigidity or component wear on the machine side can have a detrimental impact. The control system interprets control commands from the NC program and regulates the position of the cutting tool via an internal. measuring system. Temperature fluctuations or contamination of the measuring system can negatively affect the process. The cutting tool undergoes constant geometry changes due to wear or breakage, while remaining allowances on the workpiece and structural changes can also cause deviations. Environmental factors such as foundation vibrations or temperature influences can also have an. impact on the machining process [5, 10]. Various sensors located within the machine tool, control system, spindle, and external accelerometers can record data in the form of values and time series related to the process and influencing variables. However, different sensor data are typically evaluated separately due to the lack of reference points, which limits their ability to describe and investigate their interrelationships. To expand the investigative capabilities in this area, methods for fusing sensor data are needed that can determine reference points and synchronize data. Moreover, fused data sets form the foundation for more advanced data-based and learning-based methods that are more powerful than classical approaches. Hence, this paper presents a new method for fusing milling-related data to create a digital end-to-end representation of the process. We incorporate data representing the state of the machine components alongside process-related data, enabling the resulting unified data set to be used in serial production.

2   Related Works

Data-based methods, such as predictive modeling and machine learning, have been applied in milling to improve efficiency by predicting tool condition and wear [9, 14], and cutting forces. Among these factors, tool condition is especially critical because it can significantly affect the quality and productivity of the machining process. To predict tool condition, both traditional conditional methods and artificial intelligence approaches, including direct and indirect prediction systems, have been utilized. In [12], the authors provide an extensive overview of the large field of using direct and indirect systems to predict tool conditions in the milling process. The paper reviews different methods and techniques that have been used in this area, including statistical methods, artificial neural networks, and fuzzy logic approaches. The review shows that the data-based approach has been successfully used in predicting tool wear and tool failure in milling processes.

An indirect method for tool wear measurement is investigated in [3, 8]. The authors in [3] are using multi-sensor data fusion to combine vibration, power, temperature and force data, which then acts as input to neural network. The tool wear predictions are evaluated by comparing with manually measured values and it is shown that the presented approach achieves a good accuracy. The fused data reveals valuable correlations and insights into the context of wear. [7] presents an investigation of wear volume of milling tools as a nonlinear approach using multi-sensor data fusion. A neural network processes the cutting force and acoustic data as input. The authors show that the prediction accuracy using fused data is significantly better than using non fused data. The RSME is up to 60% and the MAE up to 75% lower than when using single data.

In addition to classical approaches, many deep learning approaches have also been applied to machine signals to predict tool wear. These deep learning models typically involve CNN and LSTM networks to process the time-series data and predict the tool wear accurately [9, 14]. The authors in [14] compared the performance of various deep learning models, including CNNs, LSTMs, and hybrid models, to predict tool wear in a milling process. The results show that the deep learning models outperformed classical methods, demonstrating the potential of deep learning methods in predicting tool wear and tool failure in milling processes.

Beside tool condition, cutting and process forces affect the resulting quality of the milling process. The potential of fusing simulation and sensor data for machine learning applications is investigated in [1, 6]. [1] proposes a framework for combining real world observations and simulation data and use the cutting force and time information as input to a Random Forest. The utilization of enriched fused data proves useful and significantly improves the performance of the used model compared to single data. Thereby the NRMSE decreases by up to 60%.

From all the mentioned approaches, a type of fusion between different machine signals is attempted for further processing. Many works on surface machining lack a standardized fusion process, despite the necessity of a comprehensive data set merging various sensors proven to impact milling processes. Such a data set is crucial for evaluating performance across different methodologies.

3 Data

In this section we introduce the different kinds of data considered. Namely, we include CAD data, Scan data and NC data as well as data collected by various sensors in the machine control, the spindle, and external accelerometers.

3.1 CAD Data

The CAD Data includes the geometry of the processed part as a volume body. It contains the nominal surface of the part, which is produced during the milling and is used in comparison with the scanned data to determine the actual deviation. Hardening of surface areas serves to increase wear resistance. Depending on the hardening process, dimensional changes may occur, which is why these are carried out before finishing. Hardness has an important influence on the course of the milling operation. Two files are used, they contain the hardened and non-hardened areas of the part respectively.

3.2   NC Milling Program

The machining of the components is planned in a CAM system. The tool traverses, parameters and tool changes are converted into the NC milling program, which represents the control commands for the machine tool. The NC milling program contains block numbers which are processed in ascending order.

3.3   Internal Machine Data (IM)

The internal machine data (IM) describes the parameter values from the machine control. These range from position values of the milling tool, the block number of the current processing step, real-time values of the machining parameters, rotational speed, current, voltage and power values of various drives to control variables and temperatures. The data is collected from the iTNC530 control with the TNC Scope analysis software from Heidenhain inside an CSV file. For recording the maximum frequency of 333.3 Hz is used. The position of the tool center point (TCP) is not directly included in the data. Instead only an internal base reference is recorded which must be transformed into a known coordinate system. Within the machine tool, the organization and technical execution of the movements of the tool are determined by the kinematics. The direction of the machine movement is decomposed into the axes of movement. By transforming the coordinates along the kinematics, points in the machine coordinate system can be transformed to the component coordinate system and vice versa. Four files are extracted from the machine control, including information on swing length, offsets and thermal compensation, reference points of the machine, and the tool table. From this information the kinematic chain can be build up, whose sequence is fixed to transform the coordinates.

3.4   Spindle Data

The spindle of our machine is equipped with a Heidenhain spindle ERM 6000 Dplus sensor, which provides data about the amount and direction of the load in the milling tool. The sensor consists of a bearing in which pressure sensors are installed along the running surface. Recorded data contains the translational displacements in the spatial axes X,Y,Z and the rotation around the respective axes. The data is acquired with a circle time of $20\mu s$ and pre-processed by an multi-channel processing unit (MCP).

3.5   Vibration Data

Vibration of mechanical machine components have an influence on the resulting surface quality. They are measured by a Montronix PulseNG 720011 accelerometer. During the measurement, all axes and the balance and smoothness of the spindle is checked. Therefore the axes are moved individually through the complete working area of the machine and the spindle is measured at a speed of 2000 rotations per minute (RPM). The resulting data contains a time series of the positive and negative acceleration values in three axes at each measuring point.

3.6 Scan Data

After milling, the workpiece surface is manually digitized by an GOM Atos 5 using strip light projection. By this, a digital mesh of the actual surface is generated and again manually compared to the nominal geometry of the CAD model inside the GOM Inspect Professional software. The comparison determines the deviations from the nominal geometry, which are attached to the mesh nodes as numerical value. The enriched mesh is used as an STL file. The alignment of the scan with the CAD model by superposition of the coordinate systems is done manually using the best fit function in the GOM Inspect Professional software.

4 Methods

This section contains the procedures to fuse the outlined data presented. To fuse the data together, the matching steps are considered individually in regards to a base data set. In our case the IM data is the prudent choice for the base data set. An overview of the matching steps and a short form description of the matching involved can be seen in Fig. 1.

The block number is saved as a column within the IM data and links the associated block number to every data point. Therefore, the matching is trivial.

The data acquisition of IM and spindle data is not synchronized and hence the data cannot be trivially linked. To realize a matching the operator drives a known pattern at the start of a manufacturing cycle. The rotational speed of the tool is increased in steps: 0, 3000, 6000, 9000 RPM before settling to the manufacturing speed. As the rotational speed is tracked both in IM and spindle data, they can be matched if at least one common point can be identified in both data sets.

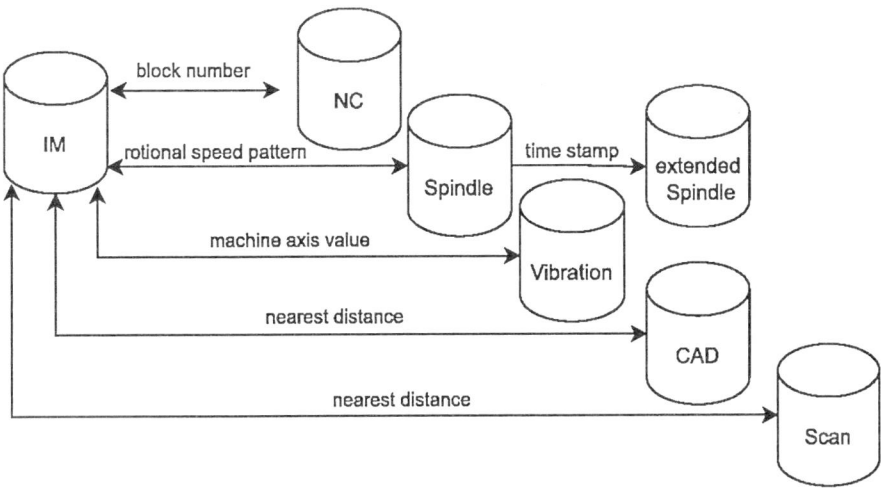

Fig. 1. Overview of the matching

We apply the sliding window change point detection method [13] to find the points where the speed is increased. A window is split in two parts and slid along the signal. Based on the least absolute deviation cost function (Eq. 1) a discrepancy curve (Eq. 2) is calculated for a given window size [13].

$$c_{L_1}(y_{a..b}) = \sum_{t=a+b}^{b} ||y_t - \bar{y}_{a..b}||_1 \tag{1}$$

$$d(y_{a..t}, y_{t..b}) = c_{L_1}(y_{a..b}) - c_{L_1}(y_{a..t}) - c_{L_1}(y_{t..b}) \tag{2}$$

The peaks in the discrepancy curve correspond to prominent changes in the median of the signal. Therefore this method is able to accurately predict the change points in the IM and spindle data. As the Spindle data is sampled at 1250 Hz and the IM at 333.3 Hz, one to one matching is not possible. Considering our use case the most fitting method is the linear interpolation of the IM data. The first change point in the IM and spindle data is used to synchronize and thus serves as t_0. Each IM coordinate can be mapped to its nearest neighbor in the vibration data. To realize this, the data has to be pre-processed and the five dimensional vibration space, consisting of the vibration data of each machine axis, has to be created. In each vibration axis the start position and turning point position are identified by the signal changing. Once the start and turning point positions in the acceleration data are found, the data can be matched to the IM by distributing it over the known workspace dimensions followed by nearest neighbor matching. Since the tool can be moved back and forth several times along a single direction only one forward and one backward movement is considered. Hereafter the methods to isolate the start and turning points are discussed. The signal in these axes show the unique characteristic:

1. stationary phase
2. acceleration along a singular axis
3. consistent motion
4. deceleration along a singular axis

This pattern is repeated multiple times An example signal showing the 3 channel accelerometer data for movement along the linear X-axis can be seen in Fig. 2a. It can be seen that the accelerometer detects acceleration along its three axis. Due to the characteristic motion it is also possible to detect the direction of motion without making any prior assumptions making the algorithm robust against rotations of the accelerometer. To stabilize the change point detection the data is preprocessed by first integrating the acceleration data and then subtracting a linear function to remove the effect of the acceleration due to gravity.

The signal in the rotational A-axis has a different characteristic (see Fig. 2b). The A2 and A3 signals resemble approximate sin and |cos| function respectively.

$$\frac{1}{tan\left(\frac{|cos|}{|sin|}\right)} \tag{3}$$

Hence, the Eq. 3 being piece-wise linear can be used to make the signal piece-wise linear as well. Subsequently, a sliding window change point detection with an adjusted

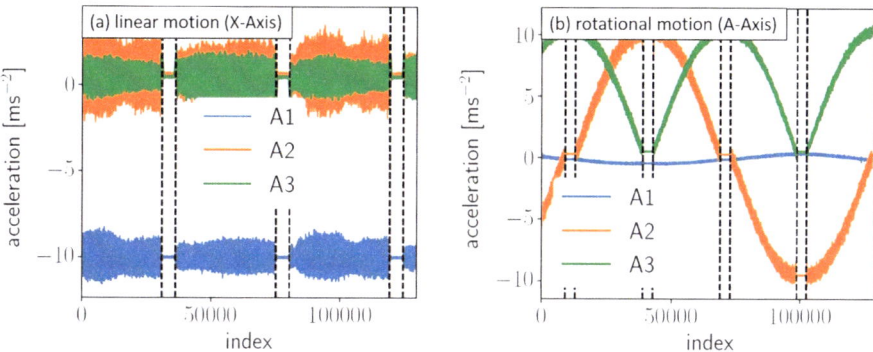

Fig. 2. Segments in the linear and rotational axis vibration data

cost function can be performed. Instead of the absolute value norm, a cost function that measures the error to a piecewise linear signal is used [13]. Although all change points are detected correctly, the location is not exactly at the point where the motion of the tool machine stops.

$$L(y) = \frac{1}{b-a} Var(y_{a..b}) \qquad (4)$$

To address this problem, in a second step the precise location of the change points is optimized using the cost function (Eq. 4) where a and b are the start and end positions of each interval respectively and Var(y) denotes the variance of the signal. The stabilized change points can be seen in Fig. 2b.

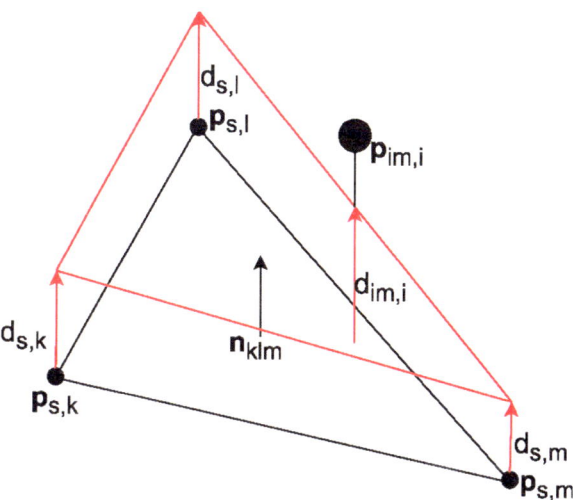

Fig. 3. Sketch of the IM/scan matching process

The IM/scan matching is a two-step process. First, for each IM data point the nearest triangle (within a maximum distance) in the scan is found. Afterwards, an estimated deviation for the IM data point is computed using the values at the three vertices. After finding the closest triangle surface for the machine point $p_{m,i}$, the point is projected onto the surface. In general the projected point will be somewhere on the surface. To estimate the deviation at that point we employ inverse distance weighting. The points in the scan form a triangle k,l,m. Each have a distance d. The linear interpolation yields $d_{m,I}$, as illustrated in Fig. 3.

To include information about hardened areas, for each IM point it is checked if a hardened or non-hardened area is closer. The position of hardened areas is included in the CAD model as a specifically zoned area. The algorithm to find the closest area is the same as in the IM/scan matching.

5 Experiment

The experimental validation of the proposed multi-sensor data fusion methodology was conducted using data acquired during the milling process of a test geometry representative of shapes encountered in press tools. This test geometry, produced from EN-JS2070 through sand casting, encompasses intricate features such as steep flanks, planes, and free-form surfaces with dynamically changing gradients. Its dimensions measure 1240mm (length) x 350mm (width) x 250mm (height). The initial roughing phase reduced a 20mm casting allowance to 1mm, followed by further refinement to 0.15mm through subsequent roughing and fine finishing. The milling procedure was systematically executed using a line-by-line approach with a 0.30mm spacing in the x and z directions. Figure 4a illustrates the resulting part. The resultant data set, thoroughly curated to handle extensive data exceeding RAM limitations, comprised comprehensive parquet or CSV files containing 4098127 rows and 78 columns.

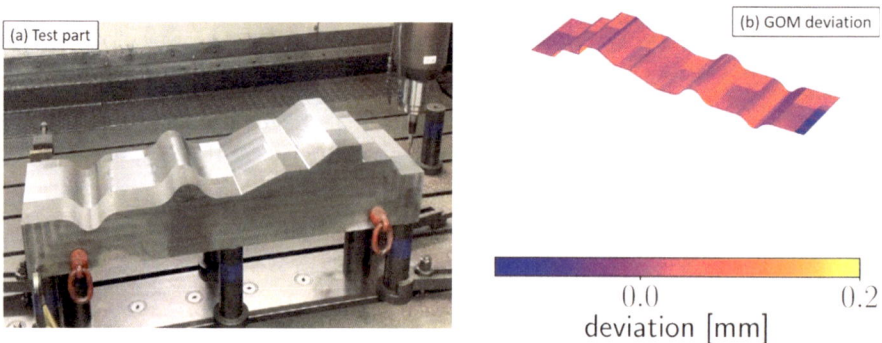

Fig. 4. Image of the test part and visualization of the GOM deviation on the right

5.1   Validation of Matching Algorithm

To assess the precision and efficacy of our matching algorithm, an evaluative task was conducted to predict GOM deviations for a known geometry, as depicted in Fig. 4b. A transformer architecture was tailored for regression analysis to predict mill head deviations at a specific time 't'. This entailed leveraging historical observations within a temporal window spanning from 't-w' to 't', where 'w' represents the window size. We leveraged recent research insights [17] to enhance our method by integrating a dual attention mechanism focusing on both time and channel aspects. These refinements were instrumental in facilitating accurate modeling of GOM regression for a known geometry, incorporating all matched variables and their historical values influencing the deviation at time step 't'. Subsequent preprocessing involved token and channel-wise attention processing of input sequences of observed factors, resulting in average pooling over channels and subsequent application of an MLP layer to derive the GOM deviation value as displayed in Fig. 5.

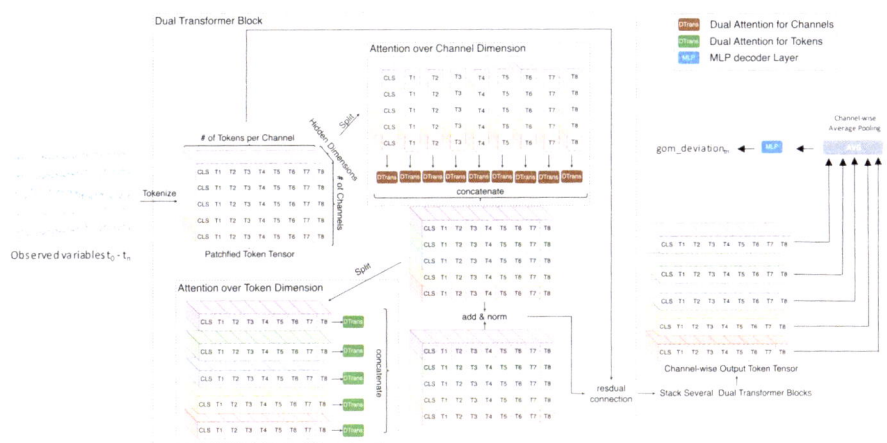

Fig. 5. Contextual regression transformer

5.2   Experimental Setup

In setting up our experiment, we segment the data into sequences of 100 for training and validation purposes. As part of the test split, we simulate a real-world scenario where predicting the next GOM deviation relies on past variables. To achieve this, we extract 1000 time steps from the original data set, splitting them into overlapping sequences of 100 with a stride of 1. This approach mimics the temporal movement of the mill head. Our model utilizes 99 out of 100 steps as input to forecast the complete sequence of 100 values, with the last predicted value representing the future GOM deviation.

We assess our model using the R2 Score as the evaluation metric. We present two scores: the R2 Score, which corresponds to the predicted GOM deviation when the variables of the predicted step are present, and the R2 Future score, indicating the prediction of future GOM deviation when the variables of the predicted step are unavailable.

5.3 Influence of Observed Data Set Size

We investigated the impact of observed factor set sizes on the model's performance. The model trained on a complete set and a subset of observed factors, using 1D Positional Encodings, revealed differences. Tab. 1 presents the performance contrast between these models, indicating the added value of a complete data set. Initial variables we train our model with were:

X, Y, Z, ist/X, s ist/X, s ist/Y, s ist/Z, s ist/A, s ist/C, REF/X, REF/Y, REF/Z.

Tab. 1. GOM deviation prediction performance

Data set	R2 Score	R2 Future Score
Partial Data set	61.2%	58.8%
Matched Data set	**64.5%**	**64.5%**

The observed performance improvement associated with the utilization of a matched data set signifies the validity and effectiveness of our matching method. This substantiates the hypothesis that accurate fusion of diverse sensor data, facilitated by our matching approach, significantly enhances the model's predictive capacity for GOM deviations in milling processes.

5.4 Inclusion of Positional Information

Acknowledging the pivotal role of positional data within the Transformer model, precise 3D Positional Encodings (3DPE) were generated from available 3D mill head position data (X, Y, Z). The 3DPE aimed to explicitly provide the transformer model with precise positional cues pertaining to the mill head's coordinates across different time steps.

$$PE_i = MLP(CAT(Sine(x_i), Sine(y_i), Sine(z_i)))where x_i \in X, y_i \in Y, z_i \in Z \quad (5)$$

The 3D positional encoding (Eq. 5), generated using a sinusoidal function, was integrated into the input sequence to furnish the model with accurate positional information. In this context, "MLP" denotes Multi-Layer Perceptron, "CAT" signifies concatenation, and "SINE" refers to sinusoidal encoding.

Further evaluation focused on the impact of 3D Positional Encodings (3DPE) on overall model performance. Tab. 2 demonstrates the substantial performance improvement when using 3DPE compared to 1DPE.

Tab. 2. Ablation study of position encoding impact on GOM deviation

Data set	Positional Encoding	R2 Score	R2 Future Score
Partial data set	1D	61.2%	58.8%
Matched data set	1D	64.5%	64.5%
Partial data set	3D	92.6%	89.8%
Matched data set	3D	**95.4%**	**95.4%**

The results underscore the significance of precise positional information and complete observed factor sets in enhancing the model's predictive capacity for GOM deviations in milling processes, affirming the efficacy of our matching methodology in leveraging diverse sensor data. In the scenario of predicting the future R2 score with the matched data set, we observe that the performance matches the R2 score attained when all the independent variables are known. This observation underscores the success of our matching process, which provides substantial additional information to the model. Consequently, the model can predict the next step even in the absence of knowledge regarding other predictive variables for that step.

6 Discussion

Our investigation into the accuracy of our method involved a thorough analysis of errors encountered during the fusion of diverse data sources. The precision of our matching varied based on data sources, each presenting distinct characteristics influencing the accuracy of fusion.

The matching between IM and NC data is error-free since these data sets are inherently connected through block numbers. However, pinpointing the precise transition point between two block numbers might be challenging due to the sensor's sampling frequency limitation, which is confined by the machine's internal sampling rate of 3 ms (ms).

In the IM/Spindle matching, two prominent errors were identified: misalignment and interpolation errors. Misalignment, originating from change point detection and sensor sampling noise, led to deviations in identifying corresponding points between IM and spindle data. Additionally, nonlinear data columns introduced interpolation errors, albeit unquantifiable due to data-dependent estimation.

The IM/Vibration matching utilized a change point detection algorithm, entailing an inherent yet unquantifiable error. Moreover, intermittent tests every three months led to limited data capture, necessitating extrapolation to infer the machine's ongoing state and workspace-dependent vibration behavior.

In IM/CAD matching, discrepancies arose from the tool machine's kinematics and NC discretization. Variances between the ideal CAD surface and the actual tool machine path, attributed to motor slippage and mechanical imperfections, manifested as a parameter in the IM, influencing the control loop.

The IM/Scan matching unveiled errors primarily rooted in scan sensor accuracy, manual registration, and interpolation. Manual registration, relying on reference areas and scan scaling to align with CAD geometry, introduced precision limitations despite the machine's accuracy of up to 0.025mm. Additionally, interpolation on deviation data posed potential errors due to nonlinear behaviors.

However, the experimental results presented in the previous section provide substantial evidence supporting the accuracy and effectiveness of our data matching process. The noticeable performance boost observed in our proxy task, employing all matched variables compared to utilizing only the originally matched variables, attests to the success and correctness of our matching approach. This enhancement in performance reinforces the hypothesis that our meticulously devised matching methodology indeed facilitates superior fusion of sensor data, underscoring its pivotal role in improving predictive modeling accuracy for GOM deviations in milling processes.

Evaluating the performance of the matching process poses challenges due to the absence of direct metrics yielding measurable outcomes. Consequently, we relied on downstream tasks like GOM deviation prediction to offer measurable results and validate the efficacy of the matching process. In our experiments, incorporating 3D positional encoding into our transformer yielded a notable performance enhancement. However, this improvement is not solely attributed to this modification but rather to the inherent representation of the transformer model itself, which more effectively models time series data when an explicit positional encoding module is utilized. Thus, the primary advantage remains the matching data set, which drives the performance enhancement.

7   Future Works

Despite the errors that occur, our method allows the fusion of relevant data from different sources. Compared to other known methods, it provides a broader data set in which more complex relationships can be mapped and therefore has the potential to be used in a variety of applications. Beside investigations on tool condition monitoring, process forces, and chatter effects a prediction of the occurring deviations of the press tool surface could lead to significant benefits for companies that rely on the milling process. By mitigating or preventing negative effects on subsequent process steps, milling process efficiency could be improved overall. To achieve this, we will apply various data based and learning-based methods to the fused data and investigate their performance. By using advanced machine learning techniques, our aim is to develop predictive models that can accurately forecast the impact of different milling parameters on the machining process. Exploring the applicability of the developed methods to alternative machine tools or varying milling processes would be a valuable direction for future investigation. Overall, our approach has the potential to significantly improve the efficiency and effectiveness of the milling process by enabling accurate predictions of the optimal milling parameters. The impact of this work could extend beyond milling and have implications for other manufacturing processes as well.

8   Disclaimer

The results, opinions and conclusions expressed in this publication are not necessarily those of the Volkswagen Aktiengesellschaft.

Funded by the European Union - NextGenerationEU. The views and opinions expressed are solely those of the authors and do not necessarily reflect the views of the European Union or the European Commission. Neither the European Union nor the European Commission can be held responsible for them.

Funded by the European Union NextGenerationEU

Supported by:

Federal Ministry
for Economic Affairs
and Climate Action

on the basis of a decision
by the German Bundestag

References

1. Saadallah, A., Finkeldey, F., Buß, J., Morik, K., Wiederkehr, P., Rhode, W.: Simulation and sensor data fusion for machine learning application. In: Advanced Engineering Informatics 52, 101600 (2022), https://doi.org/10.1016/j.aei.2022.101600

2. Birkert, A., Haage, S., Straub, M.: Umformtechnische Herstellung komplexer Karosserieteile, Springer Vieweg Berlin, Heidelberg (2013), https://doi.org/10.1007/978-3-642-34670-5

3. Bagga, P., Makhesana, M., Patel, H., Patel, K.: Indirect method of tool wear measurement and prediction using ann network in machining process. In: Materials Today: Proceedings 44 (2021), https://doi.org/10.1016/j.matpr.2020.11.770

4. Bergs, T., Brecher, C., Schmitt, R.H., Schuh, G.: Internet of Production – Turning Data into Value: Statusberichte aus der Produktionstechnik 2020, Fraunhofer-Institut für Produktionstechnologie IPT, Aachen (2020), https://doi.org/10.24406/ipt-n-589615

5. Brecher, C., Weck, M.: Werkzeugmaschinen Fertigungssysteme 3: Mechatronische Systeme, Steuerungstechnik und Automatisierung, Springer Vieweg Berlin, Heidelberg (2021), https://doi.org/10.1007/978-3-662-46569-1

6. Finkeldey, F., Saadallah, A., Wiederkehr, P., Morik, K.: Real-time prediction of process forces in milling operations using synchronized data fusion of simulation and sensor data. In: Engineering Applications of Artificial Intelligence 94, 103753 (2020), https://doi.org/10.1016/j.engappai.2020.103753

7. Gao, K., Xu, X., Jiao, S.: Measurement and prediction of wear volume of the tool in nonlinear degradation process based on multi-sensor information fusion. In: Engineering Failure Analysis 136, 106164 (2022), https://doi.org/10.1016/j.engfailanal.2022.106164

8. Ghosh, N., Ravi, Y., Patra, A., Mukhopadhyay, S., Paul, S., Mohanty, A., Chattopadhyay, A.: Estimation of tool wear during cnc milling using neural network-based sensor fusion. In: Mechanical Systems and Signal Processing 21 (2007), pp.466-479, https://doi.org/10.1016/j.ymssp.2005.10.010

9. He, Z., Shi, T., Xuan, J.: Milling tool wear prediction using multi-sensor feature fusion based on stacked sparse autoencoders. In: Measurement 190, 110719 (2022), https://doi.org/10.1016/j.measurement.2022.110719

10. Hirsch, A.: Werkzeugmaschinen: Anforderungen, Auslegung, Ausführungsbeispiele, Springer Vieweg Wiesbaden (2016), https://doi.org/10.1007/978-3-658-14249-0

11. Klocke, F.: Fertigungsverfahren 1: Zerspanung mit geometrisch bestimmter Schneide, Springer Vieweg Berlin, Heidelberg (2018), https://doi.org/10.1007/978-3-662-54207-1

12. Pimenov, D., Gupta, M., Rosa Ribeiro da Silva, L., Kiran, M., Khanna, N., Krolczyk, G.: Application of measurement systems in tool condtion monitoring of milling: A review of

measurement science approach. In: Measurement 199 (2022), https://doi.org/10.1016/j.mea surement.2022.111503

13. Truong, C., Oudre, L., Vayatis, N.: Selective review of offline change point detection methods. In: Signal Processing 167, 107299 (2020), https://doi.org/10.1016/j.sigpro.2019.107299

14. Wang, M., Zhou, J., Gao, J., Li, E.: Milling tool wear prediction method based on deep learning under variable working conditions. In: IEEE Access 8, 140726–140735 (2020), https://doi. org/10.1109/ACCESS.2020.3010378

15. Boos, W., Arntz, K., Johannsen, L., Prümmer, M., Horstkotte, R., Ganser, P., Venek, T., Gerretz, V.: Erfolgreich Fräsen im Werkzeugbau, Frauhofer-Institut für Produktionstechnologie IPT, Aachen (2018).

16. Xi, T., Benincá, I.M., Kehne, S., Fey, M., Brecher, C.: Tool wear monitoring in roughing and finishing processes based on machine internal data. In: The International Journal of Advanced Manufacturing Technology 113, 3543–3554 (2021), https://doi.org/10.1007/s00170-021-067 48-6

17. Xue, W., Zhou, T., Wen, O., Gao, J., Ding, B., Jin, R.: Make Transformer Great Again for Time Series Forecasting: Channel Aligned Robust Dual Transformer (2023) https://arxiv.org/ abs/2305.12095

Systematic Planning and Early-Stage Development of Industrial AI Systems for Plant Optimization

Jorrit Voigt[1]($\boxtimes$), Marlene Judith Gillmann[2], and Klaus Dröder[3]

[1] Volkswagen Group Components Braunschweig, Gifhorner Str. 180, 38112 Braunschweig, Germany
jorrit.voigt1@volkswagen.de
[2] TU Braunschweig, Universitätsplatz, 38100 Braunschweig, Germany
marlene.gillmann@tu-braunschweig.de
[3] TU Braunschweig, Institute of Machine Tools and Production Technology, Langer Kamp 19b, 38106 Braunschweig, Germany
k.droeder@tu-braunschweig.de

Abstract. Artificial Intelligence (AI) offers promising capabilities for many industrial applications and the use of AI has gained traction within manufacturing processes. However, working procedures lack criteria to decide about the benefit of AI systems for particular use cases and do not describe any uncertainty analysis in the development stage, since they assume readily available significant data. The objective of the present study is to introduce a process model, which includes decision criteria to determine the necessity of AI and evaluate the uncertainty of the system, consisting of suitable sensors and algorithms, in early development phases. The procedure is verified by analyzing a battery production line and identifying the prediction of success for MIG-welding as a suitable point of improvement by an AI system. A qualitative extraction of domain knowledge and data source selection leads to a fuzzy model and a clear distinction between the epistemic and aleatoric uncertainty of the system to estimate the prediction capability. The presented method allows the distinction between such uncertainties in a systematic development process and thus allows for a targeted optimization of AI systems. Extending the introduced systematic to other industrial fields can help to increase the implementation rates of AI in manufacturing.

Keywords: Planning and development · AI-systems · Fuzzy model · Uncertainty quantification

1 Introduction

Predictive analytics in manufacturing has gained high interest in industry and academia over the last couple of years with regard to data processing and predictions. Commonly, machine learning (ML) models are used for tasks such as detecting reoccurring patterns [1], determining machine conditions [2] or performing quality inspections [3] by using

K. Dröder and T. Vietor (Eds.): CD 2024, Proceedings, pp. 331–344, 2025.
https://doi.org/10.1007/978-3-658-45889-8_26

production line data. Conceptual research in this field describes routines for the introduction of these systems. One of the most recognized processes is the 'Cross-Industry Standard Process for Data Mining' (CRISP-DM), which describes six major phases, starting with business (1) and data understanding (2), followed by developing data processing (3) and modelling techniques (4), finalizing the process with evaluation (5) and deployment (6) [4]. However, the industrialization of ML in production remains a challenge since the rate of success continues to be considered low [5]. This led to recent research in this field introducing new phases in the workflow and describing these in more detail [6–8]. The guidelines demonstrate the necessity for an improved data management and modeling, for instance using uncertainty aware models in production. For the development of AI-solutions, the 'smart data' approach tries to minimize the required data by screening large data sets to seek the most relevant features [9]. Andrew Ng emphasizes the need for high quality data instead of big data, naming this approach 'data-centric' AI [10].

As a result, neither the fundamental decision process for applying ML-models in a certain process step nor the data sources itself are questioned in these conceptual models. Researchers describe the lack of domain knowledge (asking the wrong questions), wrong use case selection (forcing AI in applications) and unsuitable data (noisy data, too much data, lack of data) as main reasons for individual project failures [11, 12]. In the present study, an innovative conceptual model is presented, which addresses the described research gap. It includes systematic planning and technology development by concretizing the early stages of a data science project. The planning phase involves a plant data analysis and derives decision criteria to select the use case and right type of model. In the technology development phase, the focus lies on building AI-based systems considering and reducing the uncertainty in each component of the system, i.e. domain knowledge and tailored datasets, the sensors, and ML-models. A distinction between the epistemic and aleatoric uncertainty by careful evaluation guides the way for targeted optimization of the system and estimation of limitations in the early stage. Key aspects of this phase are the extraction of domain knowledge, data source selection, and reflection of process knowledge in the predictive models and data sets. The introduced planning method is designed to enable the detection of beneficial use cases and is demonstrated on a battery frame assembly line. Subsequently, the technology development of an AI system is presented, that aims to predict the rate of success for a metal inert gas welding process (MIG) to reduce defect rates in the plant.

2 Methodology

The integration of data-driven systems often results in failures (see Introduction), which emphasizes the necessity to adapt current process models for the development in manufacturing. The introduced process model is illustrated in Fig. 1 and focuses on systematic planning as well as an all-encompassing view of the development of AI systems. The key concept for the systematic planning is the introduction of defined decision criteria, by avoiding wrong use case selections (compare Fig. 1). The introduced technology development stage focuses on the quantification of uncertainty by the systems components, making this an indicator to determine the lack of knowledge and data quality of the system. The process model has the objective to lay the foundations for a successful AI-project's initiation by concretizing the early stages of other process models, like CRISP-DM or the more specific 'Team Data Science Process' (TDSP), as those assume a correct use case selection and availability of significant data for modelling.

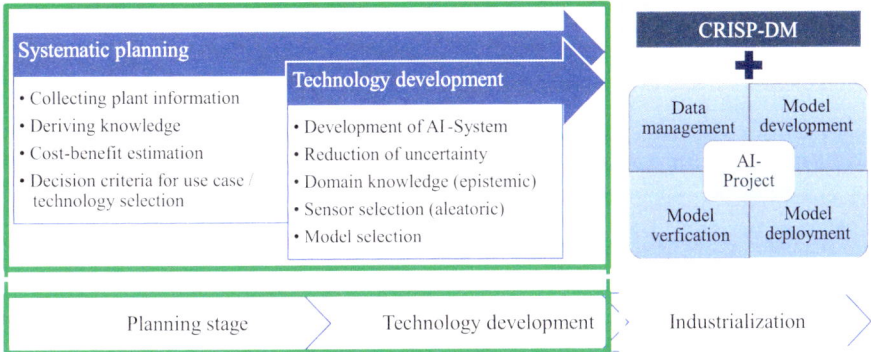

Fig. 1. Method for systematic evolution of AI-based systems in manufacturing laying the early foundation for later process models like CRISP-DM.

2.1 Planning Stage

The implementation of data-driven models in manufacturing aims for overall optimization of the production process, usually by increasing the overall equipment efficiency (OEE) or degree of automation of the plant. For brownfield scenarios, the availability and storage of all OEE relevant information in a maintained database is considered as a prerequisite for the systematic planning and development of AI systems. The whole planning stage with a top-down-analysis is visualized in Fig. 2, summarizing the overall process. The basic concepts for the deductive inference for this planning method are motivated by existing knowledge management and knowledge discovery techniques [13–15].

As shown in Fig. 2, databases usually contain different data types (metric, nominal, ordinal), collected automatically from machinery or manual input. A categorization is used to structure the data, which helps to receive a broader view over the plant performance and issues. In the next step, only categories with a high impact on the OEE are further analyzed and quantified. In this case, the advantage of the top-down-analysis becomes clear, as it provides an overview of the plant and the opportunity to focus on important OEE-decreasing issues. For very few selected issues in the plant, qualitative research (e.g. case studies) can be used for an optimal understanding of the issues with the highest impacts. During this process step, the reason of a particular issue must be clarified and a possible solution is outlined. Subsequently, the required investment for the solution and the estimated savings are evaluated to define the benefit.

To determine the necessity for AI-based models in contrast to rule-based analytics, three questions and their underlying working hypotheses are summarized in Table 1,

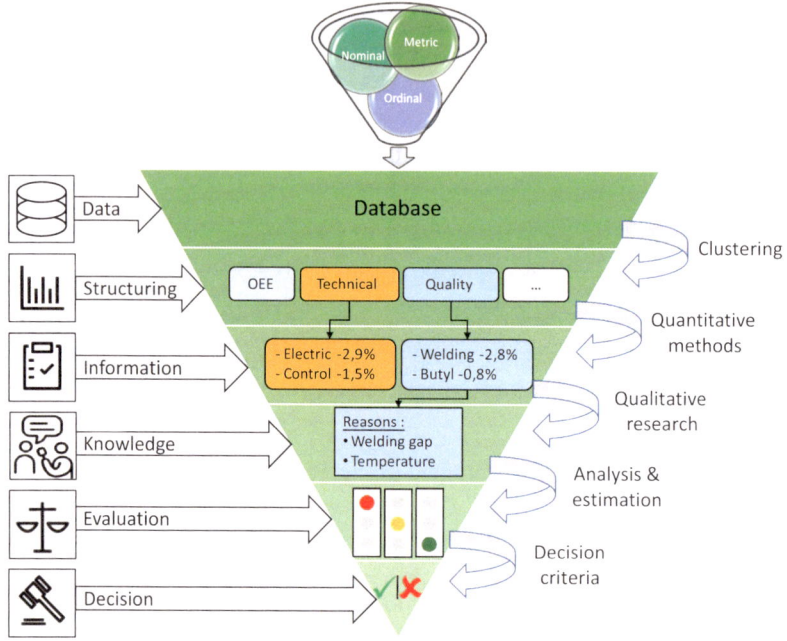

Fig. 2. Systematic planning of AI-based systems to optimize manufacturing plants.

emphasizing the not all-encompassing approach as it is outside the scope of this study. Each question can be attributed to a category in manufacturing and aims to support the choice of a suited technological solution to the identified issue. For simplicity, a binary answer can be used for a first estimation in the planning stage. The cost-benefit evaluation and the answers to the given questions provide the opportunity to make a knowledge based decision about the need and start of an AI-project, which finalizes the systematic planning process.

2.2 Technology Development Stage

In the technology development stage, manufacturing processes are developed and evaluated prior to industrialization in order to guarantee successful integration into manufacturing plants. The aim of this stage in the context of AI optimization in manufacturing is to develop the necessary technological requirements of AI systems for later integration into production. The working hypothesis for the guideline is concluded from fundamental knowledge about AI. In the context of manufacturing ML models are trained on process data to make predictions. The training data can be seen as a sample describing the underlying distribution of a process, which connects ML to statistics and making the model predictions a hypothesis [16, 17]. Consequently, AI systems suffer from the same issues as statistical approaches, as the uncertainty remains a key challenge in data-driven systems. Reasons for project failure can be attributed to uncertainty issues, i.e. insufficient data can be attributed to high aleatoric uncertainty, and modeling issues often derive from lack of domain knowledge and information which is a consequence of

Table 1. Three questions and their corresponding hypotheses are presented to evaluate the necessity of a data-driven model for a particular use case.

Question	Category	Hypothesis
Can the process described only by physical equations or quantities?	Measurement & prediction	If the predictive value is difficult to derive from physical values, then data-driven models may be useful
Is the manufacturing process itself stable and repeatable?	Manufacturing	If the process itself has a large fluctuation, then ML-models can model the underlying process distribution
Is the data dimension small, e.g. 1–3 dimensions?	Data source	If the data has many dimensions, formulating robust rules is a complex task and ML models are a suitable choice

high epistemic uncertainty [18–20]. The working hypothesis for the introduced process model is derived from these inferences: If many issues for successful industrialization of ML-models can be attributed to the effect of uncertainty, then the main effort in the early development stage must be to reduce the epistemic and aleatoric uncertainty by systematically developing each contributing part to develop a complete AI system. The guideline includes four steps shown in Fig. 3, which are mutually dependent and require further steps for industrialization as described by several authors [4, 7].

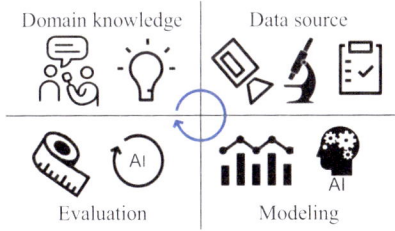

Fig. 3. The four main steps for the development of an AI system.

First, the epistemic uncertainty is reduced by extracting domain knowledge from experts. In particular, the characterization of occasionally occurring situations and fuzzy states plays a key role for robustness of the whole system and needs to be reflected in the models [21, 22]. The deep understanding of the specific task plays an important role for the whole system since the sensor selection, type of model and plant integration must be inferred. Another source of uncertainty in AI systems results from data sources, like sensors, known as aleatoric uncertainty resulting from noise or other disturbances. To reduce this issue, the sensors should be carefully selected and must ensure the clear characterization of rare events. The approach of 'data-centric' AI is followed. For modeling design, the representation of fuzziness, uncertainty-awareness, and the interpretability of the models are addressed. The ML-models are evaluated against defined criteria considering the conditions in the plant environment and quality assurance.

3   Case Study

The introduced process model for systematic planning and development (see Methodology) is demonstrated using a case study on the optimization of a welding process in a manufacturing plant. In this production line, aluminum extrusion profiles are welded airtight using MIG-welding.

3.1   Systematic Planning

Initially, data from a battery assembly line is structured by a top-down analysis to quantify plant capacity losses. The plant capacities and categories contributing to an OEE-loss are evaluated (see Fig. 4 (a)). Almost 20% of the losses are a result of technical defects, followed by quality and logistic problems with 6% loss each. Examining these categories the main issues are quantified, providing a subset of these issues in Fig. 4 (b). As seen in the bar chart, MIG-welding decreases the OEE by almost 3%. Questioning the cause for the welding issues, plant operators agreed that the main reason for unsuccessful welding operations is a varying welding gap. One possible solution to prevent defective welding processes is the inspection of the joining gap with an optical sensor. The investment in such an optical system is comparably low, which results in a positive cost-benefit estimation.

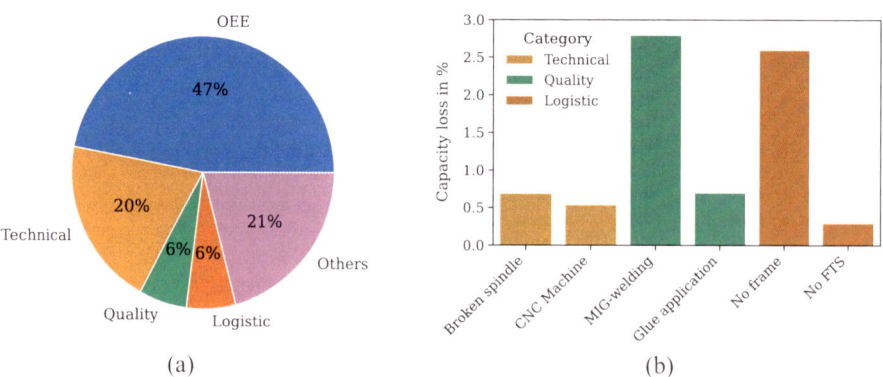

(a)   (b)

Fig. 4. Results of the top-down-analysis with the in (a) evaluated overall capacity losses by category and in (b) the two largest error types for each category.

The introduced decision criteria to choose between a rule-based or AI-based system (see Sect. 2) are answered and explained for the MIG-welding process in Table 2. The answers and the corresponding explanations results in the decision to attempt a plant optimization through an AI project, emphasizing that this conclusion must be verified in the early stages of a technology development phase.

Table 2. The answers to the questions to support the decision in favor or against the implementation of AI-based systems.

Question	Answer	Explanation
Is the prediction based on physical equations or quantities?	No → Data-driven	The outcome of a specific welding process can only be estimated by experience drawn from data
Is the manufacturing process very constant?	No → Data-driven	The MIG-welding process and the corresponding seam shows a large variance
Is the data dimension small, e.g. 1–3 dimensions?	Yes → Rule-based	The main variable for prediction is the gap size, measured at three locations to determine the gap geometry (3-D)

3.2 Domain Knowledge and Experimental Setup

For demonstration purpose of the introduced process model (see 2.2) the knowledge extraction and sensor selection are outlined in this section. A challenge in the welding process is the variation of joining gaps due to component fluctuations as the welding process is designed for a constant gap between the two joining partners. Differences in gap geometries in terms of shape and size are only taken into account to a limited extent, resulting in defective welds and a reduction in system availability due to reworking. The objective of expert interviews is the reduction of epistemic uncertainty early on in the development process, as it introduces a new set of information to the system. Based on the findings of Mieg and Näf [23] as well as Mayring [24] a questionnaire was developed regarding four main topics: Expert's qualifications and background, process design and challenges, varying joining gap, and automated weld seam monitoring. Four interviews were conducted. Diverse statements from experts portray the complexity of the welding process and, as such, they are especially important for the modelling later on.

Regarding the sensor selection, a semantic analysis of the qualitative research concluded four main points: The primary reason for defective weld seams is a varying gap size, not accounted for in the welding robots' programming. While the transition from a weldable gap to a non-weldable gap is fluent, the largest weldable gap for the fillet joint is 0.8–1.0 mm, for the corner joint it is 0.5–0.7 mm. The lower the thickness of the aluminum profile to be welded, the lower the energy that can be applied without risking melting through the base profile. A completely closed joint gap can result in an unpredictable distortion of the aluminum, as the energy input from welding heats up the metal, which can result in warping of the component. The initial temperature of the profiles also influences their distortion and therefore the gap geometry. The profiles demand to acclimatize for at least twelve hours prior the welding process to minimize the risk of porosity in the welding seams. These results provide an answer to the proposed question (cf. Table 1) regarding the manufacturing process: Since it is highly dependent on various unpredictable factors, a data-driven approach is advantageous. To take the joining gap into consideration in the modelling, a sensor for measuring the gap is required. An analysis of the production site sets limitations to the applicable systems: A tactile sensor cannot be used, as it would have to be precisely adapted to the production process and would pose a new risk of failures to the plant. Furthermore, the only suitable location to employ a sensor is in close proximity to the welding process. Thus, an optical sensor insusceptible to ambient light is needed. The object of the experiments was to examine the impact of the reflectiveness of the aluminum extrusions, different gap geometries and ambient light on the performance of all systems and uncertainties resulting from poor performances. The plant setup and the conclusions drawn from the expert interviews are essential to consider the selection of a reliable data source. An experimental setup in correspondence with the production plant is displayed in Fig. 5 (b). It allows the clamping of two aluminum profiles at a 90° angle and their translation in x- and y-direction, as well as the rotation around x to set up the most common gap geometries.

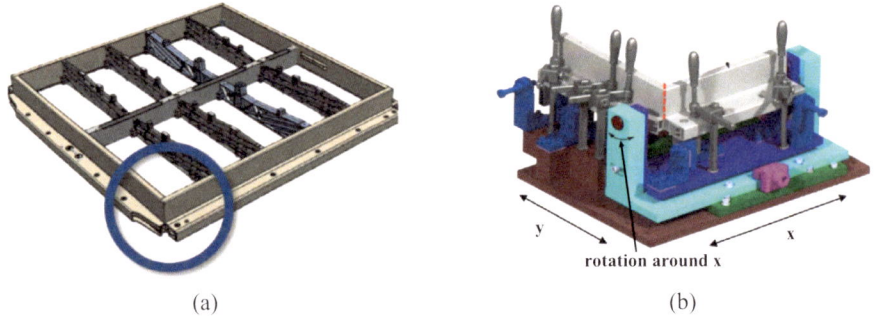

(a)   (b)

Fig. 5. The battery frame with joining partners (a) and the experimental setup for sensor selection (b).

Based on the expert interviews, the independent and dependent variables of the system were identified as in Fig. 6 and using the Reynvaan reduction approach [25], a sub-factorial design of experiments (DoE) was developed. Ambient light has been identified to have a significant impact on the optical measurement of aluminum extrusions [26], as well as the wavelength of LTS used [27] and the machined surface of the aluminum [28]. Based on these findings and the requirements concluded from the experts' knowledge, four optical measuring systems were selected for experiments: Two laser triangulation sensors (LTS) with wavelengths of $\lambda_1 = 405$ nm (Keyence LJ-V7060) and $\lambda_2 = 660$ nm (EngRoTec VisionScanner) as well as two structured light systems (SLS), one using a white ring light projector surrounding one camera (CA-H500MX) and the other consisting of two cameras with a central blue light projector positioned between them ($\mu\varepsilon$ SurfaceControl 3D). A robot was used to move the laser triangulation sensors vertically over the corner joint, while the cameras were positioned in a stationary setup. All systems were set up at the distances given in their respective manuals. Tests were repeated 30 times to enable an assessment of measurement noise.

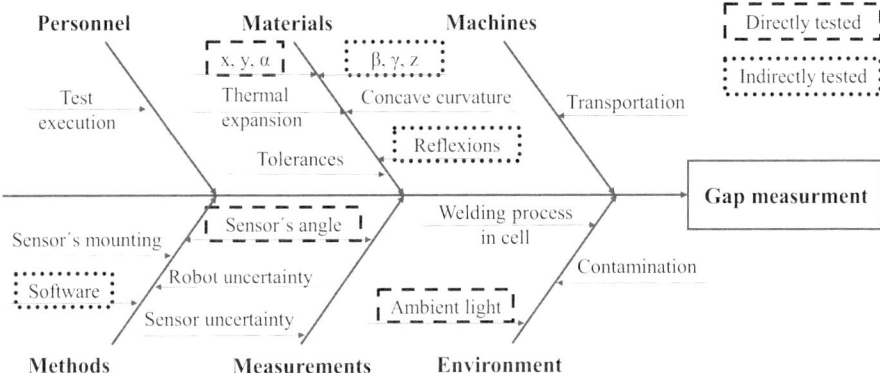

Fig. 6. The Ishikawa diagram shows the influencing variables for the AI system.

3.3 Data Source and Sensor Selection

The poor accessibility of the inner welding gap proved to be an issue for the reliability of both laser triangulation sensors as well as structured light cameras. Especially small to no gaps could not be detected, a crucial case the experts urged to consider. The reason for the difficulties is the geometrical setup of the gap, as it is not located between two edges, but between an edge and a plane, whereby the latter provides no clues for the sensor as to the measurement location. Additionally, the reflection of light between both profiles resulted in obstructive reflections. To eliminate this uncertainty, the inner gap was measured indirectly: the existence of an inner welding gap causes a displacement in z of the profiles relative to each other. By implementing a system to detect the depth displacement of one profile to the other from outside, the inner gap could be measured accurately with no susceptibility to ambient light or reflection issues.

Within these empirical tests, the reflective surface of the aluminum extrusions have a significant impact on the reliability of LTS. As this topic has been intensively discussed by Blanco et al. [26], it is not further outlined. The measurement values of a LTS at different gap sizes are presented in Fig. 7 (a), which demonstrated three incorrect types of measurement due to false reflections. Firstly, at an actual gap size of 0.6 mm a jumping behavior between the correct and a pseudo point of interest (POI) leads to a discrete distribution and large dispersion of the measurement values. Secondly, at an actual gap size of 1.0 mm multiple falsely identified POIs leading to continuously distributed measurement values. Thirdly, at an actual gap size of 2.0 mm one exclusively falsely identified POI leads to a narrow dispersion but wrong determination of the gap size. Such aleatory uncertainties cannot be eliminated or accounted for by software since they stem from a shortcoming inherent to the measuring method, creating systematic yet unpredictable errors. The results of an LTS are shown in Fig. 7 (b), in which a correct measurement of the gap size throughout the range of interest (0.0–2.0 mm) can be seen. The narrow dispersion of the measurement values is an indicator for a low aleatoric uncertainty contributed by the sensor to the whole AI-system.

A third source of uncertainty was identified when applying an external light source to the extrusions. A structured light camera working with white light was highly affected

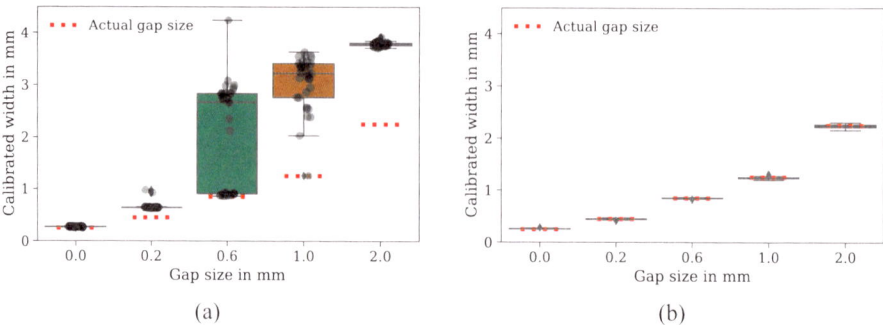

Fig. 7. In (a) the measured values by a LTS shows a large dispersion and significant deviation from the actual gap size. In (b) the SLS correctly determines the welding gap over the complete range.

by the reflections caused by ambient light. Since the sensor will be implemented in close proximity to a welding process, robustness against ambient light is crucial to extract reliable data from the sensor. The results of the DoE showed the most reliable measurements of the joining gap for the SLS working with blue light. While all sensors captured the depth displacement of the extrusion profiles without problems, both LTS were susceptible to false reflections regarding the measuring of the fillet joint gap. The CA-H500MX is affected by ambient light since it works with white structured light. By having identified a reliable data source and its limitations, the key enablers for implementing predictions with a prospective data-driven model on one hand, and judge its results critically on the other hand have been provided.

3.4 Modelling Uncertainty by Fuzzy-Logic

Predicting the outcome of a welding process is a non-binary task, as the inherent process uncertainties influence the outcome. Furthermore, Boolean models would tend to have a large amount of false positive errors in this setup leading to the well-known alert fatigue issue of workers during operations in production lines. Fuzzy models are made to reflect gradually changing probabilities and are used to consider vagueness, which makes them an ideal type of model for the use case. For implementation the Python library SciKit-Fuzzy v0.4.2 is used.

The underlying truth for the weldability with changing welding gaps is drawn from the four expert interviews, which reduces the number of experiments and provides the opportunity to reflect their experience in the model. The experts are questioned regarding the limitations of weldability and transition zones, resulting in input functions shown in Fig. 8 (a) with respect to the welding gap. After defining the fuzzy rules, the model prediction with respect to a varying welding gap is shown in Fig. 8 (b). The figure demonstrates the difference between the fuzzy model and a Boolean model. In contrast to the binary prediction, a fuzzy model is able to capture the gradually changing probability for a successful welding and is much more robust against false alerts. The robustness against accepted production variance is a key performance indicator for predictive models in manufacturing environments.

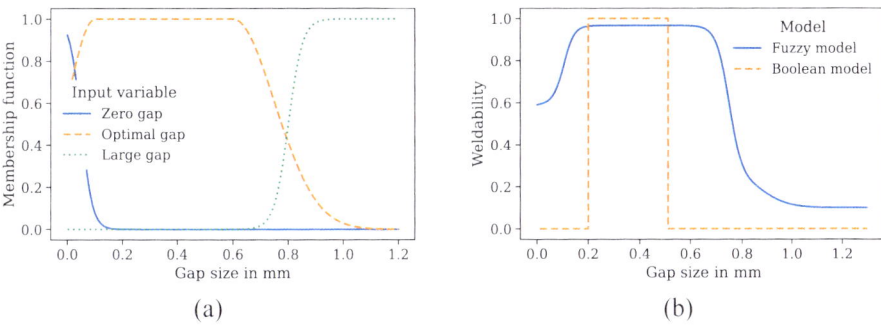

Fig. 8. Development of a fuzzy model with (a) the input variables and in (b) the comparison of the predictions between a Boolean and a fuzzy model.

3.5 Uncertainty Quantification and Experimental Validation

The developed system, which measures the gap size between the frame parts and predicts the weld seam quality is validated with experimental data. For this, 10 frames are welded under constant welding parameters with different gap sizes to verify the outcome of the fuzzy model by new data. The results are presented in Fig. 9 (a) and a discrepancy between the fuzzy model prediction and the experimental data can be seen. The difference is the epistemic uncertainty ($\delta\varepsilon$) of the system caused by lack of information/ knowledge about the welding outcome. The seam welded with a gap size of 0.8 mm shows necking, which lowers the quality quantification for this weld. At a gap size of 1.0 mm the seam is not connected to the joining partners. In Fig. 9 (b) the model predictions with respect to different data sources (see Sect. 3.3) are shown. The measurement noise and the influence on the prediction is visualized by the colored areas providing the 95% interval to quantify the aleatoric uncertainty.

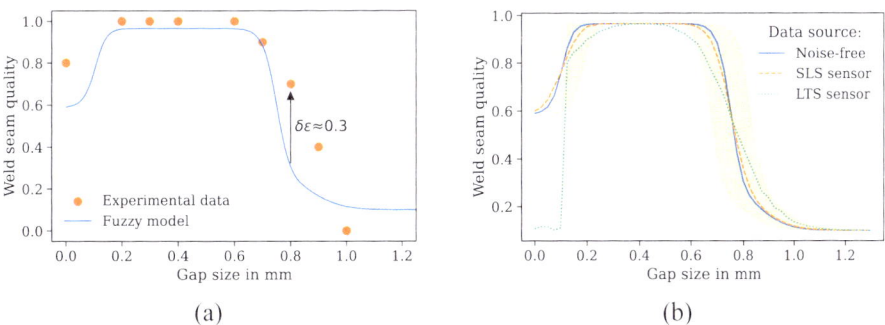

Fig. 9. In (a) the fuzzy model and the experimental data are compared to determine the epistemic uncertainty of the system. In (b) the predictions for different data sources are presented to determine the aleatoric uncertainty.

A hypothetically noise free sensor leads to exclusively epistemic uncertainty in the system. To evaluate the influence of the sensor selection on the aleatoric uncertainty, the prediction intervals for the LTS ($\sigma \approx 0.1$) and the SLS ($\sigma \approx 0.04$) are visualized. Furthermore, the sensor behavior for rarely occurring cases has to be evaluated. The inability of the LTS-sensor to correctly measure close gaps disqualifies the sensor for the system as it leads to a large error in the prediction. The distinction of the two types of uncertainty provides the ability for targeted optimization of the AI system.

In this case, the aleatoric uncertainty of the SLS-sensor and its impact on the prediction is acceptable, and the focus for optimization is the reduction of the epistemic uncertainty. This can be achieved by sampling more data and information.

4 Summary and Outlook

The presented work describes a process model for systematic planning and development of AI-based systems for manufacturing plants. The planning process includes a top-down analysis of plant data to increase the level of knowledge in each analysis step and to estimate the cost-benefit. Criteria and hypotheses to evaluate and answer the necessity of ML-algorithms for the specific use case are formulated to lay the basis for methodological decision to implement an AI project. A process model for the technology development of an AI system is presented. The aim of this model is to lay out fundamental requirements in early development stages. The model is derived from a formulated working hypothesis to focus on the uncertainty quantification and reduction in the system.

The introduced systematic planning and development methods are successfully demonstrated for a use case in a battery frame assembly line to predict the outcome of a MIG-welding process and should result in a reduction of defective parts. The extraction of domain knowledge, the derivation of the DoE for sensor selection, and the development of a fuzzy-logic based prediction model is shown. In a next step, a distinction between the epistemic and aleatoric uncertainty is evaluated and discussed by comparing the model prediction with experimental data for different sensors.

An uncertainty aware development process is a crucial part in any design of industrial AI systems as it provides the possibility to estimate the limiting components of the system (data source, data set). Apart from the case study discussed in this paper, further examples of manufacturing applications in which the components uncertainties evokes challenges are the in situ monitoring of laser welding processes or crack detection in metal forming plants. Both demand for a clear distinction between patterns for defect formation and background noise from process fluctuations, ambient light conditions, or changing material surfaces. An uncertainty analysis provided the ability for a targeted optimization of the system by improving the uncertainty contributing components. Developing reliable data sources are more important then extensive ML modeling in such scenarios to build data sets of high quality. Further scope of research should be the planning of green field applications and the uncertainty quantification for ML algorithms like deep learning models.

Acknowledgements. We gratefully thank the welding experts from Volkswagen Group Components Braunschweig for providing impressive insights into the topic of manufacturing and welding. Further, we thank the competence center for welding at Volkswagen Group Components Braunschweig for experimental support.

References

1. J. Voigt, T. Bock, U. Hilpert, R. Hellmann, and M. Moeckel, "Increased relative density and characteristic melt pool. signals at the edge in PBF-LB/M," *Additive Manufacturing*, vol. 19, p. 102798, 2022, https://doi.org/10.1016/j.addma.2022.102798.
2. F. Bonada, L. Echeverria, X. Domingo, and G. Anzaldi, "AI for Improving the Overall Equipment Efficiency in Manufacturing Industry," in *New Trends in the Use of Artificial Intelligence for the Industry 4.0*, L. Romeral Martínez, R. A. Osornio Rios, and M. Delgado Prieto, Eds.: IntechOpen, 2020.
3. D. Joshi, T. P. Singh, and G. Sharma, "Automatic surface crack detection using segmentation-based deep-learning approach," *Engineering Fracture Mechanics*, vol. 268, no. 3, p. 108467, 2022, https://doi.org/10.1016/j.engfracmech.2022.108467.
4. P. Chapman, "CRISP-DM 1.0: Step-by-step data mining guide," in 2000. [Online]. Available: https://api.semanticscholar.org/CorpusID: 59777418
5. G. Herzberg, R. Panikkar, R. Whiteman, and A. Sahu, "The imperatives for automation success," *McKinsey & Company*, 2020.
6. H. Heymann, A. D. Kies, M. Frye, R. H. Schmitt, and A. Boza, "Guideline for Deployment of Machine Learning Models for Predictive Quality in Production," *Procedia CIRP*, vol. 107, no. 1, pp. 815–820, 2022, https://doi.org/10.1016/j.procir.2022.05.068.
7. R. Ashmore, R. Calinescu, and C. Paterson, "Assuring the Machine Learning Lifecycle," *ACM Comput. Surv.*, vol. 54, no. 5, pp. 1–39, 2022, https://doi.org/10.1145/3453444.
8. A. Paleyes, R.-G. Urma, and N. D. Lawrence, "Challenges in Deploying Machine Learning: A Survey of Case Studies," *ACM Comput. Surv.*, vol. 15, no. 1, p. 22, 2022, https://doi.org/10.1145/3533378.
9. D. García-Gil, J. Luengo, S. García, and F. Herrera, "Enabling Smart Data: Noise filtering in Big Data classification," *Information Sciences*, vol. 479, pp. 135–152, 2019, https://doi.org/10.1016/j.ins.2018.12.002.
10. E. Strickland, "Andrew Ng, AI Minimalist: The Machine-Learning Pioneer Says Small is the New Big," *IEEE Spectr.*, vol. 59, no. 4, pp. 22–50, 2022, https://doi.org/10.1109/MSPEC.2022.9754503.
11. H. Takeuchi and S. Yamamoto, "Business Analysis Method for Constructing Business–AI Alignment Model," *Procedia Computer Science*, vol. 176, pp. 1312–1321, 2020, https://doi.org/10.1016/j.procs.2020.09.140.
12. A. Artasanchez, *9 Reasons why your machine learning project will fail.* [Online]. Available: https://www.kdnuggets.com/2018/07/why-machine-learning-project-fail.html (accessed: Nov. 28 2023).
13. K. North, "Die Wissenstreppe," in *Wissensorientierte Unternehmensführung*, K. North, Ed., Wiesbaden: Springer Fachmedien Wiesbaden, 2016, pp. 33–65.
14. U. M. Fayyad, G. Piatetsky-Shapiro, and P. Smyth, "From Data Mining to Knowledge Discovery in Databases," *AI Mag*, vol. 17, pp. 37–54, 1996. [Online]. Available: https://api.semanticscholar.org/CorpusID:61287995
15. M. Ester and J. Sander, *Knowledge Discovery in Databases: Techniken und Anwendungen.* Berlin, Heidelberg: Springer Berlin Heidelberg, 2000.
16. E. Hüllermeier and W. Waegeman, "Aleatoric and epistemic uncertainty in machine learning: An introduction to concepts and methods," *Mach Learn*, vol. 110, no. 3, pp. 457–506, 2021, https://doi.org/10.1007/s10994-021-05946-3.
17. V.-L. Nguyen, M. H. Shaker, and E. Hüllermeier, "How to measure uncertainty in uncertainty sampling for active learning," *Mach Learn*, vol. 111, no. 1, pp. 89–122, 2022, https://doi.org/10.1007/s10994-021-06003-9.

18. J. U. Hansen and P. Quinon, "The importance of expert knowledge in big data and machine learning," *Synthese*, vol. 201, pp. 1–21, 2023. [Online]. Available: https://api.semanticscho lar.org/CorpusID:256060110

19. R. Gansch and A. Adee, "System Theoretic View on Uncertainties," in *2020 Design, Automation & Test in Europe Conference & Exhibition (DATE)*, Grenoble, France, 2020, pp. 1345–1350.

20. M. Beer, S. Ferson, and V. Kreinovich, "Imprecise probabilities in engineering analyses," *Mechanical Systems and Signal Processing*, vol. 37, 1-2, pp. 4–29, 2013, https://doi.org/10. 1016/j.ymssp.2013.01.024.

21. T. Xiahou, Y. Liu, and T. Jiang, "Extended composite importance measures for multi-state systems with epistemic uncertainty of state assignment," *Mechanical Systems and Signal Processing*, vol. 109, pp. 305–329, 2018, https://doi.org/10.1016/j.ymssp.2018.02.021.

22. M. Wu, T. Xiahou, J. Chen, and Y. Liu, "Differentiating effects of input aleatory and epistemic uncertainties on system output: A separating sensitivity analysis approach," *Mechanical Systems and Signal Processing*, vol. 181, p. 109421, 2022, https://doi.org/10.1016/j.ymssp.2022. 109421.

23. H. A. Mieg and M. Näf, *Experteninterview in den Umwelt- und Planungswissenschaften: Eine Einführung und Anleitung*. Zürich: Pabst Science Publisher, 2006.

24. P. Mayring, "Qualitative Inhaltsanalyse," in *Handbuch Qualitative Forschung in der Psychologie*, G. Mey and K. Mruck, Eds., Wiesbaden: VS Verlag für Sozialwissenschaften, 2010, pp. 601–613.

25. C. H. H. Reynvaan, "Versuchsplanung," in *Wie geht Industrie?*, C. H. H. Reynvaan, Ed., Berlin, Heidelberg: Springer Berlin Heidelberg, 2022, pp. 245–261.

26. D. Blanco, P. Fernandez, E. Cuesta, S. Mateos, and N. Beltran, "Influence of surface material on the quality of laser triangulation digitized point clouds for reverse engineering tasks," in *2009 IEEE Conference on Emerging Technologies & Factory Automation*, Mallorca, 2009, pp. 1–8.

27. L. Song, S. Sun, Y. Yang, X. Zhu, Q. Guo, and H. Yang, "A Multi-View Stereo Measurement System Based on a Laser Scanner for Fine Workpieces," *Sensors (Basel, Switzerland)*, vol. 19, no. 2, 2019, https://doi.org/10.3390/s19020381.

28. P. T. Moe, A. Willa-Hansen, and S. Støren, "Optical Measurement Technology For Aluminium Extrusions," in *AIP Conference Proceedings*, Zaragoza (Spain), 2007, pp. 596–601.

Circular Product Design

Evaluation of Component Suitability for Hybrid Remanufacturing Process Chains

Veronika Marquart[1]([✉]), Julian Meißner[1], Andreas Hofer[2], and Michael Milde[2]

[1] roeren GmbH, Consulting for production and management, Ludwig-Erhard-Street 13a, 84034 Landshut, Germany
{veronika.marquart,julian.meissner}@roeren.eu
[2] Aperion Analytics GmbH, Leopoldstr. 32, 80802 Munich, Germany
{andreas.hofer,michael.milde}@aperion-analytics.de

ABSTRACT. The topic of sustainability is currently not only omnipresent in research and politics but is also gaining significant importance in the industry. Different customer groups are increasingly attaching importance to sustainable products and adapting their purchasing behaviour accordingly.

One important field of action for achieving ecological sustainability is the circular economy. Circular economy describes a cycle-based model of production and consumption, whereby materials and products are to be recycled after use in such a way that further use of the product, or the materials used is possible. This can be done in different degrees of recovery, for example by reuse, repair, recycling, remanufacturing etc. Remanufacturing offers an industrially attractive recovery option with high value retention. However, it also places the highest demands on the product to be recycled as the quality is described "as new".

The Landshut Level Model of Remanufacturing (LEMOR) is a comprehensive model for the step-by-step assessment of the suitability of a component for remanufacturing from a technical, economic and strategic perspective. This ranges from a review of the basic technical suitability to the influence of the individual damage pattern on the remanufacturing process. The assessment can be divided into a generic and a component-specific spectrum. The generic evaluation is specified in more detail by defined component properties which were collected in the component-suitability-assessment. This initial part of the evaluation shows the possibility of integrating different production processes (in particular additive manufacturing processes in combination with subtractive processes) to expand and realize high-quality and complex remanufacturing processes. A software tool was developed for the automated and user-friendly collection of component properties and their automated comparison with the requirements catalogue for testing remanufacturing suitability.

Keywords: Circular Economy · Remanufacturing · Additive Reprocessing · Sustainability · Production

K. Dröder and T. Vietor (Eds.): CD 2024, Proceedings, pp. 347–358, 2025.
https://doi.org/10.1007/978-3-658-45889-8_27

1   Introduction

The political and social endeavours to achieve greater sustainability are supported, among other things, by the UN Sustainable Development Goals [1] and the EU's Green Deal [2] in particular. The aim is always to take a holistic view at an economic, ecological and social level [3]. A central means of achieving a more sustainable economy is the so-called circular economy, i.e. cycle-centred value creation [4]. The importance of the circular economy has increased noticeably in recent years due to numerous external factors [5]. Therefore it is not only omnipresent in research and politics but is also gaining significant importance in the industry [6]. Increasingly, customers are attaching importance to sustainable products and are orienting their consumer behaviour correspondingly. [7].In addition to the political decisions mentioned above, more difficult and more expensive access to resources is also increasing interest in circular economic processes. Access to resources is becoming increasingly difficult due to geopolitical tensions (e.g. the Ukraine war) and the associated sanctions regimes as well as disrupted supply chains (e.g. Covid and the Suez Canal blockade).

In addition, more sustainable product design is increasingly becoming the focus of politics and business. One example of this is the stricter draft of the End-of-Life Vehicles Directive published in July 2023 [8]. This not only represents a legal requirement for the environmentally friendly recovery of end-of-life vehicles (e.g. by dismantling components containing valuable and raw materials), but also for the recovery and recycling-friendly design of future vehicles and the mandatory use of secondary raw materials.

This paper presents a model that enables the evaluation of components with regard to their suitability for complex remanufacturing. The focus here is on a so-called hybrid process chain, which combines additive and subtractive production processes. This initially enables the build-up of material at a damaged component location, which is then subtractively manufactured into its original or newly defined shape.

2   Foundations and State of the Art

Within the circular economy, there are various recycling strategies to contribute to circular value creation models, which are often summarised as so-called "R-strategies" [9]. These include for example the reuse, repair, remanufacture and recycle approaches [10]. The reuse approach is based on the further use of the product in a different range of applications. Repair describes the repair of a defective product. Remanufacturing is the mechanical reworking of a defective product to restore it to as-new condition. Recycling is characterised by the dismantling and material recovery of the product in order to reuse the raw materials [9]. Due to the endeavour to maximise the value and quality of the product, the other R-strategies are preferable to recycling [11]. A product that fulfils the requirements for remanufacturing can generally also be recycled using the other strategies, as remanufacturing is subject to the "as good as new" requirement [12].

Classic remanufacturing process chains and sequences can be used to successfully implement remanufacturing. These are essentially centred on five main process steps: Disassembly, cleaning, diagnosis, refurbishment and reassembly (Fig. 1) [13] [14].

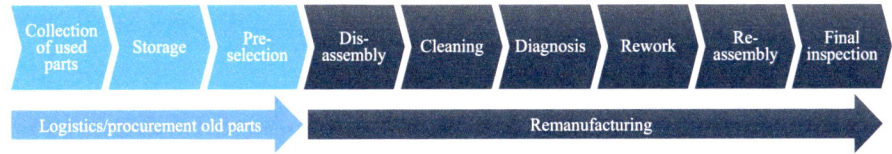

Fig. 1. Process chain of remanufacturing, own illustration based on [13]

The rework process step mainly refers to material-removing processes such as drilling, milling, honing or grinding. However, these classic methods are limited, for example if the target state differs geometrically from the initial state. However, this may be necessary due to the desire to customise a newer component variant/version. The repair of certain types of damage is also limited by conventional approaches. Therefore, there is the possibility of hybridization existing remanufacturing approaches by adding additive manufacturing processes [14]. However, this hybrid process chain for remanufacturing, consisting of a combination of additive and subtractive processes, places increased demands on the properties of the product to be remanufactured. For this reason, the focus below is on hybrid remanufacturing process chains.

To ensure that a product can be remanufactured, attention must already be paid to recyclable design and construction during product development. If it is unclear whether a product intended for remanufacturing complies with the relevant guidelines, a check must still be carried out to determine whether the product is suitable for remanufacturing before a decision on remanufacturing is made (analysis of the current state). There are already established methods for assessing remanufacturing suitability.

DIN EN 4553:2020–11, for example, describes a general procedure for assessing the recyclability of energy-related products and defines seven general process steps that are important for remanufacturing. The process steps for testing the actual condition are associated with product properties [15]. Lahrour & Brissaud [14] performed a fundamental analysis to examine the suitability of additive manufacturing technologies for remanufacturing purposes. Additionally, Goodall et al. [16] present an overview of the state of the art on tools and techniques used to evaluate remanufacturing feasibility in general. However, there is currently no comprehensive methodology that can be used to assess the remanufacturing suitability of products for process chains which combine additive and subtractive technological elements. The aforementioned assessment methodology in accordance with DIN EN 4553:2020–11 also only includes technical properties. It does not include an examination of the economic efficiency and strategic suitability of a product. A methodology is therefore needed that can also be applied to products that have not yet been designed with circularity in mind. The actual state of the component should be assessed from a technical, economic and strategic perspective.

In the following, a model is presented with which the suitability for remanufacturing can be assessed on the basis of a step-by-step evaluation (LEMOR model). In the first two stages, general component properties are analysed. The component-suitability-assessment (CSA) offers a general procedure for querying these generally valid properties. The remanufacturing suitability is standardised as a score using a point-based evaluation. The associated software tool enables quick and uncomplicated access to the component-suitability-assessment in the form of a user interface and automatically derives the score presented from the properties recorded.

3 Landshut Level Model of Remanufacturing (LEMOR)

The Landshut Level Model of Remanufacturing (abbreviation LEMOR from German: **L**andshuter **E**benen **M**odell des **R**emanufacturing) is a comprehensive model for the step-by-step assessment of a component's suitability for remanufacturing from a technical, economic and strategic perspective. This ranges from checking the basic technical and physical suitability to the influence of the individual damage pattern on the remanufacturing process. The assessment can be divided into a generic, a component-specific and a wear-specific area (see Fig. 2). As the individual levels are passed through, the suitability profile of a component for remanufacturing becomes increasingly clear.

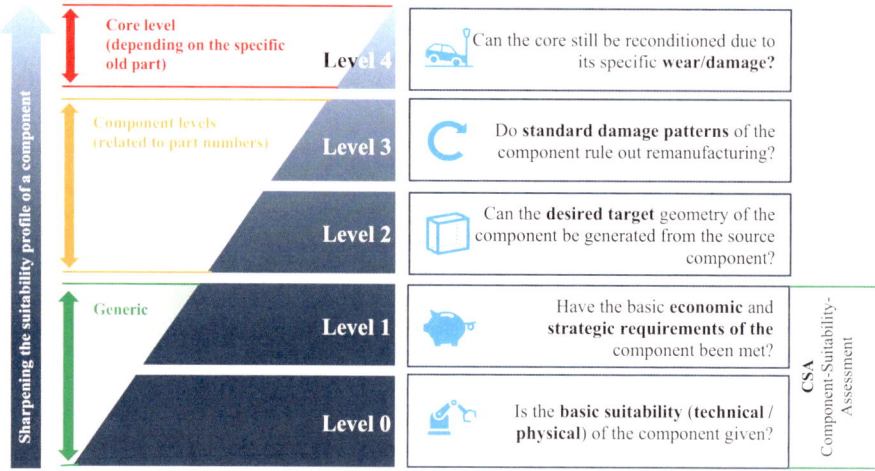

Fig. 2. The Landshut Level Model of Remanufacturing—LEMOR for short (own illustration)

The starting point of the model is **level 0**, which evaluates the remanufacturing suitability of a component in terms of technical feasibility in the generic area. In this context, generic means that any component can be considered, regardless of whether it is a classic automotive component or a product from the consumer sector. If, for example, the engine component crankshaft is to be analysed, it is not yet necessary to narrow it down to a part number or variant within this level, as this is initially a superordinate analysis of the crankshaft product family. Within this level, the basic physical component properties are assessed. These include, for example, geometry, weight and material properties.

Following on from level 0, economic and strategic component aspects are assessed in **level 1**, which is still generic. The economic aspects include the number of units required, the process costs and the transport costs. For example, economies of scale can be evaluated based on the required quantity of a component to be remanufactured, which influence the economic efficiency of remanufacturing. In addition to the economic dimension, strategic aspects are also evaluated in level 1. Among other things, the availability of raw materials, sustainability and the origin of the old parts (referred to as cores) influence the strategic suitability of remanufacturing. The so-called component-suitability-assessment (CSA) is a tool to evaluate the suitability of a component for remanufacturing regarding level 0 and level 1 of the LEMOR model. The examined perspectives on remanufacturing feasibility match the findings of Goodall et al. [16] and thus generate a comprising assessment of components. The CSA is further described in the following chapter.

If the technical and economic feasibility from levels 0 and 1 is not given, the component is not suitable for remanufacturing and a more in-depth analysis based on subsequent levels is not necessary. If it turns out that remanufacturing of the evaluated component is technically feasible in principle and makes economic and strategic sense, the component-specific geometry is then assessed in **level 2.** Component-specific means that within a superordinate product family, e.g. crankshafts, the selection must be limited to a specific variant/part number. If several variants are to be reprocessed, the following levels must be checked separately for each variant under consideration. In level 2, the original geometry (initial geometry) of the component is compared with the target geometry (target state). The aim of the comparison is to find out whether it is fundamentally possible to produce the desired target state from the initial geometry. It should be emphasised that the initial geometry can deviate from the target geometry in terms of dimensions and volume. This is permissible because in individual cases the component is to be reworked to a different component variant through reprocessing.

After successful geometry evaluation, component-specific standard damage patterns are analysed within the subsequent **level 3.** The aim of the analysis is to define the most common damage patterns that could categorically rule out reconditioning of the component.

The **fourth** and final **level** evaluates an individual worn and/or defective component (synonym: core) with regard to its suitability for remanufacturing, taking into account the individual damage pattern independent of the standard damage pattern from level 3. If a core is too worn or irreparably damaged, it is excluded from remanufacturing. If, on the other hand, it turns out that the core is suitable for remanufacturing, the necessary remanufacturing steps are shown, for example as part of a hybrid process chain. Yet there is no tool existing to automatically calculate the suitability of a component regarding the levels 2, 3 and 4. Within the research project EREP, it has been taken into account to make research if such tools could be developed and implemented.

4   Component-Suitability-Assessment (CSA)

As soon as hybrid reprocessing is considered, a suitability test must be carried out in advance due to the increased requirements for the recyclability of a product. The so-called component-suitability-assessment can be used for an initial analysis (for the generic levels of the LEMOR, level 0 and level 1). This is an extension of existing assessment methods in remanufacturing (Fig. 3), such as the general product attributes for reprocessability according to DIN EN 4553:2020–11, and evaluates the technical, economic and strategic suitability of products for hybrid remanufacturing (process sequence of WAAM processes (wire arc additive manufacturing—wire-based laser deposition welding) and machining).

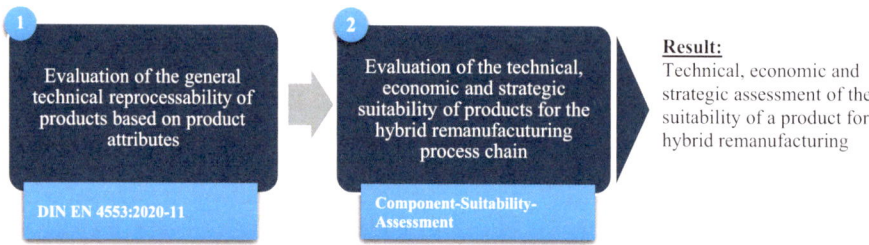

Fig. 3. Classification of the Component-Suitability-Assessment (own illustration)

Only individual components can be assessed, i.e. if an entire system or assembly is present, the assessment must be carried out separately for all individual components. The dismantling capability of the systems or assemblies must be assessed in advance.

The component-suitability-assessment is used to record and assess the actual condition of components. For each component property, there is a score that evaluates the property in terms of suitability for WAAM processes and subtractive processes. Finally, an overall result per category (technology, economy, strategy) can be derived from the individual assessments of the properties. A scoring system from at least 1 to maximum 5 points is used for the evaluation (whereby 1 means KO criterion—not suitable; 2 means less suitable; 3 means suitable, 4 means well suitable and 5 means very suitable).

This implies that as soon as a product property, e.g. the material, has been assessed as "1" for the WAAM process or for machining processes according to the component-suitability-assessment, the component is not suitable for reprocessing in the hybrid process chain.

The component-suitability-assessment is therefore a generic assessment that can be used to make an initial scoring-based assessment of the remanufacturing suitability for products of all types. It therefore forms the basis of the LEMOR model (level 1 and level 2) and should be carried out in advance of more detailed analyses of geometry, damage patterns and processes. Level 0 describes the basic technical and physical suitability of components and is thus described by the technical score of the component-suitability-assessment. This contains relevant properties for WAAM processes and machining processes (Table 1).

There are two scores for each characterisation of the technical requirement properties (one for WAAM and one for machining processes). Each individual score was based on literature research and expert knowledge. As a detailed list and justification for each score would go beyond the scope of this paper, it will not be discussed in detail.

If the technical suitability has been established by appropriate testing of the characteristics listed in Table 1 the economic and strategic suitability of a component can be checked via level 1 of the LEMOR model. For this purpose, the corresponding score is

Table 1. Technical properties for evaluating the suitability of components for the hybrid remanufacturing process chain

Technical properties		
Requirement property	Category	Characterisation of the requirement property—examples
Material	Metallic materials	Various steels, cast and wrought aluminium alloys, titanium alloys, magnesium alloys, copper alloys, nickel alloys, etc
	Non-metallic materials	Plastics, glass, ceramics, semiconductors, etc
Geometry	Feature Size	–
	Bounding Box	–
	Symmetry	Symmetrical, non-symmetrical
Weight		–
Processing status	Coating	Painting, coating, etc
	Hardening	Hardened and tempered, case-hardened etc
Special features	Clampability/measurability	–
	Possibility for measurement	Allowance of at least 1 mm possible
	Technical complexity	Sharp edges, radii, overhangs, undercuts, etc

again determined by means of a component-suitability-assessment. The following criteria are essentially relevant for analysing the economic viability of remanufacturing: Number of units to be remanufactured, degree of automation in production, product complexity of the product to be remanufactured, manufacturing costs of the new part, production location and investment requirements. The number of units is a key factor for economic efficiency, as (fixed) costs are distributed over the number of units. The degree of automation plays a role to the extent that labour costs and, in some cases, process costs fall as the degree of automation increases. However, the economic efficiency of automation depends on the number of units and product complexity [17]. Product complexity, in turn, is decisive for the design of the remanufacturing process chain. The more complex the product, the more extensive the remanufacturing processes usually are. The higher the manufacturing costs of the new part, the more lucrative remanufacturing is in most cases, as it is often associated with lower manufacturing costs. The location of production has an influence on costs, especially labour costs. However, a location decision should not only consider labour costs, but also other factors such as product complexity and the degree of automation. The smaller the investment requirement and consequently the larger the existing remanufacturing machinery, the more economical a remanufacturing scenario will be. Uncertainties regarding the quality of used parts and the quantity available make investment planning for remanufacturing even more difficult [18].

A one-dimensional score evaluation of the individual properties is not sufficient for calculating economic efficiency, as there are direct and indirect dependencies between the evaluation criteria for economic efficiency. An example of this is the number of units and the degree of automation. A low number of units is not per se a criterion for inefficiency, as the production of small quantities can prove to be quite economical in the context of manual production. Higher quantities, on the other hand, are generally only economically viable with a high degree of automation. The Fraunhofer IPT, for example, describes this dependency on the number of units and degree of automation in its conference proceedings "Empower Green Production" using the example of the remanufacturing of washing machines as a challenge to ensure economic efficiency [17]. The component-suitability-assessment therefore utilises a multidimensional assessment system that correlates the dependencies between the aforementioned properties.

In addition, strategic requirements for the product and remanufacturing also play an important role. Remanufacturing should also create added value for the company from a strategic perspective. With regard to the strategic characteristics, aspects of ecological sustainability are predominantly considered, as it is of central importance that remanufacturing can be mapped in an ecologically meaningful way in comparison to new production [19]. Sustainability is also a central key factor in corporate strategies these days. In addition to the legal requirements and demands of society, the circular economy is a suitable strategy for companies to position themselves favourably against the competition. An image and brand advantage can also be generated, as targeted presentation of sustainability activities, for example through marketing, can fulfil the demands of society and politics [20].

The strategic remanufacturing suitability score therefore indicates the extent to which remanufacturing can contribute to the achievement of companies' environmental sustainability goals compared to new production. For calculating the score for strategic suitability of a component, the key figures are the availability of used parts and the potential for saving energy and resources. The availability is considered regarding logistics, quantity and quality of the product. The logistics describes how far away the cores are from the remanufacturing plant as this is an essential factor regarding the CO_2 footprint. Quantity of cores means if enough old parts are available to meet the demands. In general, you need more cores than there is demand for remanufactured parts, as there will be some bad cores that can't be remanufactured. This leads to the next point of quality which states the damage patters of the cores: the worse the damage, the more efforts need to be made in remanufacturing. Regarding the potential for saving energy, it can be said the more extensive the remanufacturing process is (because of the product complexity), the more energy is used in remanufacturing and the less ecological the process is. The potential to save resources mainly depends on the weight and the materials of the product.

5   Software Tool for CSA

A software tool was developed to increase the industrial applicability of the developed methodology and the underlying components-suitability-assessment. In addition to pragmatic accessibility of the evaluation methodology, this should also ensure simple interpretation of the results. The software tool currently only maps level 0 of the Landshut level model to provide potential users with a quick and simple assessment of suitability for hybrid remanufacturing.

To this end, users are guided through the three evaluation categories of material information, geometric information, and other features as part of a simple interface structure. Component characteristics are queried for each category, which are processed into an overall assessment in the final step. The final assessment is made on a scale of 1 to 5, with a high score indicating good suitability for remanufacturing using a combination of additive and subtractive technologies. The basis for calculating the score is the previously presented components-suitability-assessment for hybrid remanufacturing. This logic is derived from expert knowledge and state of the art literature and transferred in a survey based logic. As the software tool aims to represent the developed methodology, the different influencing factors can be weighted by their technical importance and are ultimately merged to an overall technical score consisting of the described categories. From a practical point of view, some of the parts characteristics can disqualify it from hybrid remanufacturing like certain materials (e.g. wood) or shapes (e.g. ultra-thin walls). Future releases of the software tool strive to cover economic as well as strategic evaluation perspectives as well. Figure 4 shows an exemplary evaluation report of the technical suitability of an example part which represents a realistic component from the rail industry.

6   Evaluation and Industrial Application

In order to validate the developed methodology and to check the industrial suitability of the developed software tool, expert interviews were conducted with test users in various partner companies. The range of companies was chosen to be as representative as possible of the German economy. To this end, the methodology and the software tool based on it were presented to a large German industrial corporation, a medium-sized metalworking company and a small custom manufacturer and independently tested and evaluated by the experts. In addition, a validation was carried out by experts from a consulting company that specialises in the manufacturing industry.

The evaluation was conducted in different dimensions, with the methodology itself being assessed on the one hand and the industrial feasibility of the methodology in the form of the software tool on the other. In particular, the pragmatic approach to pre-categorization within the framework of evaluation level 0 of the Landshut level model was emphasized. According to the experts, the quick and intuitive assessment of a component's suitability for hybrid remanufacturing offers relevant added value. The intuitive user guidance, the manageable extent of the assessment and the visualisation of the software tool were also mentioned positively.

According to the experts, there is particular potential in further detailing the component specification, for example in the material configuration or geometry description.

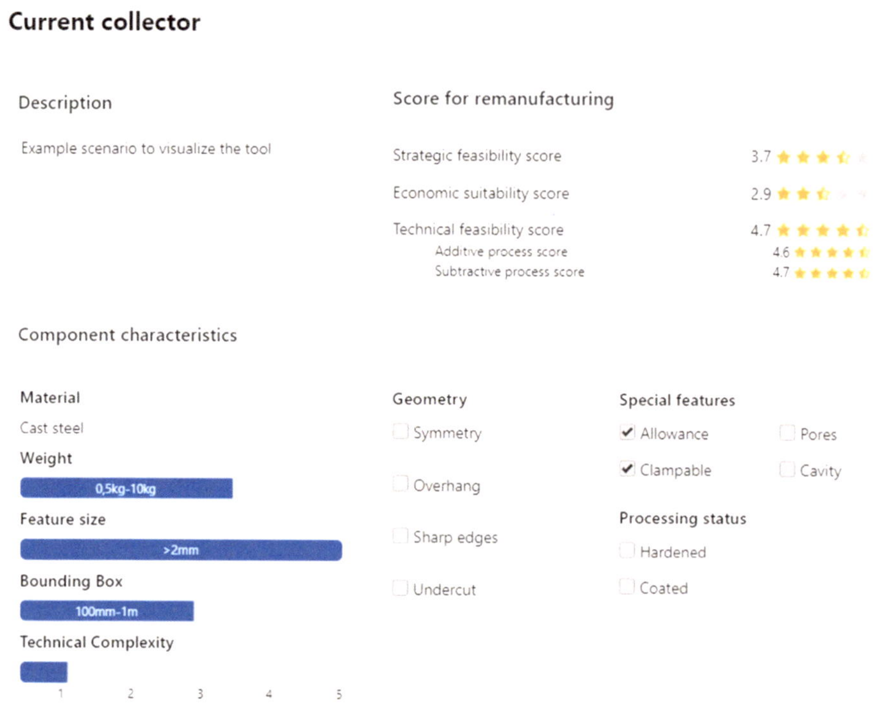

Current collector

Description

Example scenario to visualize the tool

Score for remanufacturing

Strategic feasibility score   3.7 ★ ★ ★ ☆
Economic suitability score   2.9 ★ ★ ☆
Technical feasibility score   4.7 ★ ★ ★ ★ ★
    Additive process score   4.6 ★ ★ ★ ★ ★
    Subtractive process score   4.7 ★ ★ ★ ★ ★

Component characteristics

Material	Geometry	Special features	
Cast steel	☐ Symmetry	☑ Allowance	☐ Pores
Weight		☑ Clampable	☐ Cavity
0,5kg-10kg	☐ Overhang		
Feature size		Processing status	
>2mm	☐ Sharp edges	☐ Hardened	
Bounding Box			
100mm-1m	☐ Undercut	☐ Coated	
Technical Complexity			

1   2   3   4   5

Fig. 4. Exemplary evaluation scenario within the software tool

Particularly in the case of more complex components that have been partially treated or coated, for example, the differentiation of component properties is currently only mapped to a limited extent.

In summary, both the methodology and the software tool based on it were positively evaluated by the experts and rated as very relevant and helpful for industrial applications.

7   Conclusion and Outlook

The subject of sustainability is becoming increasingly important socially, politically and economically. Remanufacturing is a relevant option for realizing more sustainable production. If conventional remanufacturing, which is based on subtractive manufacturing technologies, is supplemented by preceding additive technologies, this is referred to as hybrid remanufacturing. In the context of the present work, this extended approach to remanufacturing is considered and offered as a basic option for industry and research. The aim is to define a starting point for further research into the development and technological design of these hybrid remanufacturing processes. The Landshut level model forms the methodological framework for the evaluation of component suitability along an operationally logical sequence. With the help of a software tool based on the Landshut level model, the industrial accessibility of the methodology has been improved and validations and tests in industry have been made possible.

The work presented here and the results already obtained can be seen as a starting point for further research activities. These can focus in particular on the further detailing of the assessment levels. In addition, the approach must be analyzed in greater depth as part of comprehensive expert studies in industry and science in order to gain an in-depth understanding of the corresponding needs and development potential. At an operational level, technological concepts and solutions for mapping the different levels in the industry must be developed in the further course. In particular, the cost-effective implementation of technically sophisticated hybrid remanufacturing processes still poses a number of challenges.

Acknowledgements. The methods and models presented were developed as part of the EREP research project. This research and development project is funded by the German Federal Ministry of Education and Research (BMBF) within the "The Future of Value Creation—Research on Production, Services and Work" program (02J21E110) and managed by the Project Management Agency Karlsruhe (PTKA). The authors are responsible for the content of this publication. The authors thank the funders for the generous support.

References

1. G. Gratzer and V. Winiwarter, "Chancen und Herausforderungen bei der Umsetzung der UN-Nachhaltigkeitsziele aus österreichischer Sicht," *KIOES Opinions,* vol. 8, pp. 13–26, 2018.
2. C. Fetting, "The European Green Deal," ESDN Office, Vienna, 2020.
3. A. Kleine, Operationalisierung einer Nachhaltigkeitsstrategie—Ökologie, Ökonomie und Soziales integrieren, Gabler Edition Wissenschaft, 2009.
4. J.-P. Schöggl, L. Stumpf and R. J. Baumgartner, "The narrative of sustainability and circular economy - A longitudinal review," *Resources, Conservation & Recycling,* vol. 163, 15 August 2020.
5. M. Oliveira, M. Miguel, S. K. van-Langen, A. Ncube, A. Zucaro, G. Fiorentino, R. Passaro, R. Santagata, N. Coleman, B. H. Lowe, S. Ulgiati and A. Genovese, "Circular Economy and the Transition to a Sustainable Society: Integrated Assessment Methods for a New Paradigm," *Circular Economy and Sustainability,* vol. 1, pp. 99–113, 2021.
6. F. Tonelli, P. Taticchi and S. Evans, "Industrial Sustainability challenges, perspectives, actions," *International Journal of Business Innovation and Research,* vol. 7 (2), pp. 1751–0252, 2013.
7. D. Feber, A. Granskog , O. Lingqvist and D. Nordigarden , "Sustainability in packaging: Inside the minds of US consumers," McKinsey, 2020.
8. S. Proff , A. Scholz, C. Welter and S. Marmulla, "IN4climate.RR 2023: Die Verwertung von Altfahrzeugen: Status Quo, Herausforderungen und Potentiale im Hinblick auf eine effizientere Kreislaufwirtschaft in Deutschland und dem Rheinischen Revier," IN4climate.RR.
9. P. Morseletto, "Targets for a circular economy," *Resources, Conservation & Recycling—ELSEVIER B.V.,* no. 153, 2020.
10. Ellen-MacArthur-Foundation, "Finding a common language — the circular economy glossary," 2021. [Online]. Available: https://ellenmacarthurfoundation.org/topics/circular-economy-introduction/glossary. [Accessed 02 12 2023].
11. A. Münger, Kreislaufwirtschaft als Strategie der Zukunft—Nachhaltige Geschäftsmodelle entwicken und umsetzen, Freiburg—München—Stuttgart: HAUFE, 2021.

12. S. Butzer, S. Schötz and R. Steinhilper, "Remanufacturing Process Assessment—A Holistic Approach," *Procedia CIRP—Changeable, Agile, Reconfigurable & Virtual Production,* pp. 234–238, 2016.

13. D.-I. U. Lange, "Ressourceneffizienz durch Remanufacturing—Industrielle Aufarbeitung von Altteilen," VDI Zentrum Ressourceneffizienz GmbH, 2017.

14. Y. Lahrour and D. Brissaud, "A Technical Assessment of Product/Component Remanufacturability for Additive Remanufacturing," *Procedia CIRP,* vol. 69, pp. 142-147, 2018.

15. *DIN Deutsches Institut für Normung e. V.—DIN EN 4553:2020–11,* Beuth Verlag GmbH, 2020.

16. P. Goodall, E. Rosamond and J. Harding, "A review of the state of the art in tools and techniques used to evaluate remanufacturing feasibility," *Journal of Cleaner Production,* no. 81, pp. 1–15, 2014.

17. T. Bergs, C. Brecher, R. Schmitt and S. Günther, "Empower Green Production," 2023.

18. J. M. Kafuku, M. Z. M. Saman, S. M. Yusof, S. Sharif and N. Zakuan, "Investment Decision Issues from Remanufacturing System Perspective: Literature Review and Further Research," in *12th Global Conference on Sustainable Manufacturing,* 2015.

19. Recycling_Magazin, "Recycling Magazin: Trens, Analysen, Meinungen und Fakten zur Kreislaufwirtschaft," 19 05 2023. [Online]. Available: https://www.recyclingmagazin.de/2023/05/19/remanufacturing-besser-als-neu/.

20. F.-M. Belz and M. Bilharz, Nachhaltigkeit-Marketing in Theorie und Praxis, Deutscher Universitäts-Verlag, 2005.

Strategic Requirements-Based Product Design: A Tri-Tool Methodology For Advanced Circular Economy

Phillip Wallat[1]([⊠]), Jiwei Low[2], and Eldwin Darryl Tanzil[3]

[1] DfACE Strategy Systems, Gerhard-Rauschenbach-Straße 14, 38678 Clausthal-Zellerfeld, Germany
`wallat@dface.eu`

[2] RWTH Aachen, Templergraben 55, 52062 Aachen, Germany
`jiwei.low@rwth-aachen.de`

[3] TU Darmstadt, Karolinenplatz 5, 64289 Darmstadt, Germany
`eldwin.tanzil@stud.tu-darmstadt.de`

Abstract. Redesigning and updating products represent time-consuming processes, even if the same methodology as the original design is applied. New political influences, economic factors and newly developed techniques must be taken into account. Reliable tools and robust methods to streamline the process are therefore necessary to support engineering designers in their decision making and potentially decrease the time to market for the revised product. Three tools with different focuses will be used to structure an efficient and comprehensive methodology either for new or redesign of a product in a circular economy. The first tool aligns the class of good of the product with four different circular approaches, establishes the basic framework of the methodology and provides the designer with a suitable circular approach. The next tool is a method-variants analyzer that serves as a guidance tool for the designer to understand the dynamic relationship of product type and design methodology in order to determine the most efficient method in each design phase The third tool will assist designers with specific tasks regarding the material of a certain part. Utilizing numerical values of the inherent chemical and physical properties of the material, it is possible to assess and align the parts' performance in fulfilling assigned functions and the parts' behavior throughout the recycling process. This subsequently improves the assessment of the product in correlations of recycling and functions fulfillments, and its evaluations for potentially better alternatives. Hence, with the combined approach of those three tools into one methodology, the design engineer will be assisted throughout the whole process with: swift feedback regarding the circular approach, most suitable for the goods type of the product; proposals for the necessary depth of the design process for each assembly as well as balancing functionality and recyclability of single parts.

Keywords: Engineering Design · Recycling · Circular Economy · Materials · Design Efficiency · Circular Design

© The Author(s) 2025
K. Dröder and T. Vietor (Eds.): CD 2024, Proceedings, pp. 359–373, 2025.
https://doi.org/10.1007/978-3-658-45889-8_28

1 Introduction and Motivation

Besides the current VDI 2243 and VDI 2343 especially targeting recycling [1, 2], there are no newer or enhanced versions to represent the current recognized rule of technology for the circular engineering design. Most scientific design handbook include chapters about 'design for recycling' which establish a well-founded base of sources for the design for recycling and represent the state of the art of the engineering design [3, 4]. The state of the scientific knowledge addresses different specific areas regarding *Circular Economy* (CE), with case studies in a practical manner [5] and [6] or theoretical investigations on a strategic level [7, 8]. Although numerous general design strategies for CE exist [9, 10], there is a lack of in-depth discussions on the methodologies to achieve them. Furthermore, in recent years there are upcoming legislative [11, 12] and socioeconomic [13, 14] requirements for the whole economy which are not yet addresses properly with the existing tools and methods. To mitigate possible innovation delays and economic losses, it is necessary to utilize digitized methods to address general requirements systematically to start a development process for CE; which subsequentially will be addressed with familiar tools in the specific field.

In the topic of developing products within the CE, designers frequently struggle with intricate and conflicting technical challenges, especially when CE-based requirements are factored in. Unable to achieve a balance between informative, circular-oriented, and well-structured in a product design or redesign process, while also nurturing creativity, is a sign of an inefficient design methodology. In light of this, beyond the surface level design strategy, the market is in need of a coherent, universal framework for product design within CE—from major decision making to detailed engineering development. This paper is the result of collaborative research of three independent works from the authors of this paper: the first tool (T1), a data-based tool to formulate strategy and direction in the scope of *Circular Approach* (R-proach) and joint technology. The second tool (T2) to guide adaptive methodology in design process. And lastly, the third tool (T3) to assist in material selection and evaluation. With the aforementioned challenges in mind, each of the tools provides the necessary assistance within the *product development process* (PDP) and focuses on achieving the following goals as a combined system.

Goals for the Tri-Tool Design System (TTDS):

1) Develop a universal system that effectively assigns the appropriate R-proach to products across different classifications of goods and varying levels of complexity, including highly complex assemblies within a product.
2) Streamlines the PDP by optimizing the gathering of solution ideas and evaluating results.
3) Evaluate and balance possible circular-sustainable material combinations depending on the functions of the product, concurrently finalizing decision on joint technology.

To accomplish the first goal in terms of including both complex products and complex assemblies, the TTDS is designed to accommodate both cases, as indicated in Fig. 1. Consequently, this paper addresses the product's structure level with the terms "top hierarchy" and "detailed hierarchy", which can denote Product and Assembly, or

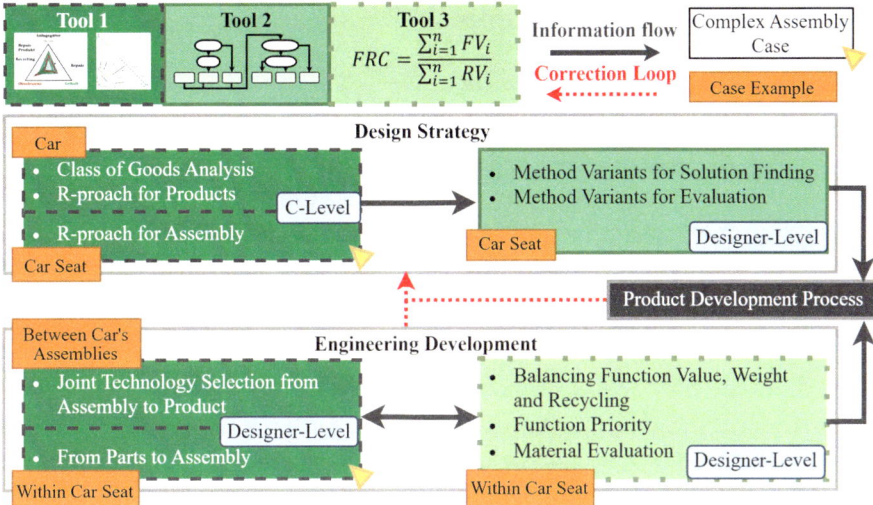

Fig. 1. Overview of Tri-Tool Design System for assisting in Product Development Process

Assembly and Parts respectively. For a clearer understanding of the flow and interdependencies between the tools, Fig. 1 presents an overview of the TTDS, divided into two segments: the overall *"Design Strategy"* segment and the more detailed *"Engineering Development"* segment. Within the *"Design Strategy"* segment, T1 aids strategic decisions sufficiently to guide the design direction toward a specific circular approach in the first iteration, T2 then delves into the operative level of design methodology based on the direction. Therefore, this segment is responsible for providing information to actualize a modified PDP with the chosen R-proach from T1 and subsequently fulfil the second goal. Following the design solution from the modified PDP, incorporating details like function priority and structure hierarchy, T3 collaborates with T1 iteratively in *"Engineering Development"* segment to achieve the third goal mentioned earlier.

During the design process there are different company levels involved. Strategic-Level decisions regarding the development of a product are mostly made on the C-Level. Later during the design process, choices will be made by the head of the development department or managers of certain special areas. Later the engineers will design the product in detail and render accurate decisions regarding single parts and assemblies. Depending on the size of the developing company the decision level may change. While additional details of each tool will be further elaborated in their respective sections, the interdependencies, information flow (black bold arrows) between the tools and correction loops (red dotted arrows) will be discussed with the help of a complex assembly case example of an electrically adjustable car seat throughout the paper, and concluded in Sect. 6, to address contradictions and showcase the flexibility and coherency of TTDS.

2 Circular Economy and Design Strategies

Within the *Circular Economy*, engineering design can follow different approaches. Targeting a more circular product while redesigning, there are potentially more than one of these approaches embedded in the new product design. At least, depending on the part and business model of the product, the approach will differ. Complex industrial goods may depend on Reuse and Repair. Whereas fast consumer goods focus more on Recycling. Additionally, the economical feasible time and effort investment in the corresponding design process depends on the market value of the product.

To structure the circular approaches, Fig. 2 is given. Within the figure, ten R-proaches are aligned for product (Ordinate) and material (Abscissa) recycling. The dashed green line between 100% product and material recycling represents optimal properties for the CE. Marked as a threshold for "Rethink" a dotted red line between the two variable cut-off factors Y_{PR} and X_{MR} is drawn to symbolize products with insufficient circular characteristics. Furthermore Reduce, Refuse, and Recover are not explicitly within the 1st Quadrant of the coordinate system; those will not be discussed further in this paper. All other R-proach are structured and ordered between product and material recycling, but not exclusive for the optimal case within the CE.

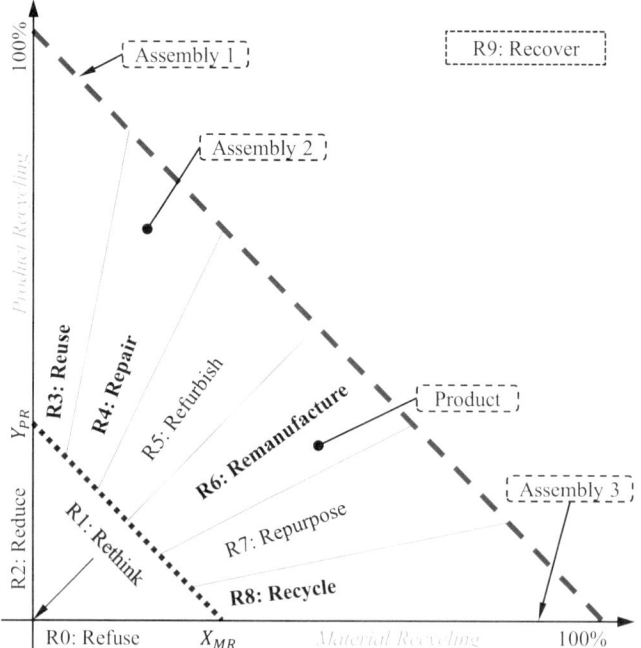

Fig. 2. Overview for Circular Approaches regarding Product and Material Recycling. Included are three âssemblies for approaches. "RX-Terms" based on work from KIRCHHERR [15].

For better understanding of the graph's purpose, three Assemblies and a possible resulting product are drawn marked. Assembly 1 is marked directly on the dashed green line, therefore representing a assembly which is 90% suitable for *product recycling* (PR) and 10% for *material recycling* (MR) and located in the area of *Reuse*—ideal parameters for a normative CE, but not reachable in a descriptive reality. With ~ 60% PR and ~ 30% MR the second assembly is located within *Repair*, missing ~ 10%, which represents the loss of material. The third assembly is located on the Abscissa with 0% PR and ~ 90% MR, again with a loss for the CE but suitable for *Recycle*. All three assemblies may be combined to a product which is designed for *Remanufacturing*. The values for the assemblies and the product are ambiguous based on the combined materials and joint technology. This represents a short insight in the complexity of the engineering design for CE.

As hinted in Fig. 2 there are multiple possible R-proach combinations for products, assemblies, and the parts within them. During the design process a certain structure of plausibility should be taken into account.

Example: A product designed for reuse shall not be composed out of assemblies designed for repair. In contrast a complex product designed for repair, like a car, could not only be made with reuse parts, but with a combination out of reuse and recycling. Furthermore, all products, assemblies and parts must be recycled at some point, Recycling has therefore considered for every single case / part, which might cause some contradicting demands within the design process.

3 Tool 1: Adepted Engineering Design for Advanced Circular Economy

The engineering *product design process* (PDP) referred to in this paper is described in PAHL / BEITZ.[4] To adapt the PDP for the CE further information is necessary. Different decisions along the process will influence operations later in the product life cycle. As overview, there are four simplified steps in the process: Defining the functions, search for suitable solutions for each function, evaluation, and selection of the found solutions, specific design for the solution including material choice. Each step will have influences on the CE properties of the product. This paper focuses on the impacts of the used materials, the joint technology, and the architecture.

There are R-proaches which are more suitable for certain classes of goods. Most convenience goods should be designed for recycling, due to their short lifespan and simplicity. In contrast, complex industrial goods should focus on remanufacturing for the product itself, but also consider other R-proaches in the overall product structure.

Materials are the key reasons for recycling, different procedures were developed to extract certain materials. Those include different mechanical, chemical and thermal processes which will be combined in various layouts, depending on the input and desired output, into a whole procedure [5]. Joint technologies are therefore a driving factor for the necessary equipment, not only for the recycling process at the end life, but also for repair or remanufacturing operations during the product lifetime. Those planned lifecycle prolonging operations will also lead to inevitable consideration regarding the products architecture to make those parts accessible with justifiable expenditure. Hence the applicable joint technology is different depending on the R-proach. *Example: Welding the rim onto the wheel hub, instead of utilisation of a screw set, will have substantial negative impact on the repairability of the braking system.*

There are additional influences which will further affect the implementable design process of a certain company, such as: the class of goods and suitable R-proach; capabilities and accessibility to work with certain materials as well as the expertise and equipment to apply different joint technologies. In general, longer lasting products will possess a higher circular complexity based on the utilisation of different R-proaches depending on the part, assembly, and the whole product. Those are important factors when redesigning a product for the CE. To use Tool 1 in a more practical and efficient manner a software is in development. The software provides easy access to evaluate the product during the iterative steps of the design process. Depending on the development state of the product, varying software inputs will generate data which will be transformed into information by the system and provide valuable knowledge to different actors throughout the company to aid the following design steps afterwards. For further information regarding the workflow and evaluation, details are described in WALLAT [16].

This data-based tool was designed to take all the former factors into account while *designing* a product *for advance circular economy* (DfACE). Joint technologies were set in context to the goods type, the R-proach and the materials they will combine, with regards to their inherent and specific properties for the later recycling process. Furthermore, data exchange will be necessary to provide affected internal and external stakeholders with information to assist them with their decisions to enhance the CE operations in their economic environment, therefore the approach carries the addition of *advanced* circular economy.

Along the PDP Tool 1 can be utilized to design the product more *cyclophane*[1] with each iteration. Depending on the allocated resources to the design process, the utilization of the tool can vary, which offers the possibility to increase the products suitability for the circular economy in balance with the available resources. Hence it is necessary for the C-Level, the engineers, and all involved levels in between, to know the different targets of each other's decision (see Fig. 1). Thus, the tool also provides harmonized information flow along the levels. This is also valid for later revised decisions depending on the technical developments of engineers in the detailed phases of the PDP.

This could involve an alternation of a certain R-proach (see Fig. 2), which might lead to the exchange of joint technologies for assemblies or parts to fulfil the requirements unique to each R-proach. Joint technologies can be ordered in material-/, form-/ and force-fitting, those have different properties for the later separability of the linked materials, ordered in non-/, partially-/ and destructive disassembly. Those criteria have interdependencies with each other as well as for the R-proach; also valid for the combined application of two or more joint technologies—e.g. a screw (non-destructive; force-fitting) glued with 'Loctite' (partially-destructive; material-fitting) to secure a generic latch (non-destructive, form-fitting). The overall lifespan and the operational environment of the product will also have an influence on the time-related properties of the joints later in the product lifetime.

---

[1] "More suitable for the circular economy"; opposite of *cyclophobe.*

*Case-Example: A car as a more complex product shall be designed for Remanu-
facturing, stipulated by the C-level. Therefore, the product contains assemblies for
Reuse (e.g. car body, transmission, electrically adjustable car seat), Repair (e.g.
braking system, seats) and Remanufacturing (e.g. turbo charger, alternator). The
engineers later in the process realize, some parts (adjustment drives) of a Reuse-
assembly (electrical adjustable car seat) will have insufficient life expectancy due
to the applied forces, leading to a mandatory re-evaluation of joint technology.
If a modification it is not adequately viable, a change of the assembly R-proach
is also possible—e.g. back to Reuse with the use of more or a different material
or purely Recycling with the use of lesser material, or Repair as a combination
of both (see example in previous section). Again, demanding adjustments of the
overall applied joint technology for the whole product. Furthermore, there might
be other restrictions which have to be taken into account like lightweight design
in the automotive sector.*

4   Tool 2: Method-Variant Analyzer

While equipped with insights provided by T1, where the element of CE is embedded
into the overall strategy of the PDP, numerous decisions within individual phases of
the process hold substantial sway over the resulting outcomes. T2 is thus essential for
contributing to the development of a systemic and efficient design process, that extends
beyond the strategic level, into the fundamental process of the operative level. By leveraging the methodology established in one of our author's prior study [17], each PDP-
Phase encompasses its distinctive "core" method. These core methods further integrate
elements of the circular economy into the foundational model of the PDP. With the goal
of establishing a robust and comprehensive guidance framework, *method-variants* (MV)
are incorporated in addition to the designated core method during the solution finding
phase and evaluation phase. These added variants adequately address multiple factors,
including the class of goods, applications, and development direction of the company,
which are crucial information discussed on the C-Level during the application of T1.
Through the utilization of T2, designers gain the ability to comprehend the intricate
dynamic relationships among these factors and the underlying design methodology. To
showcase this feature, Fig. 3 is a representation of T2 in the form of flow diagram to guide
designer in selecting a suitable MV according to the class of goods and desired design
model, such as adaptation design or explorative design. This version of the guidance
flowchart is a modified version of LOW [17], where in TTDS, the suitable R-proach is
already determined before PDP using T1 as discussed in Sect. 3, whereas some methods
proposed in LOW [17] selects the suitable R-proach during the evaluation phase.

In the solution finding phase, the core method revolves around identifying pertinent
design aspects of products that influence the *Main Elements of the Circular Approach*,
namely MECA in the work of LOW [17]. In encapsulating the MECA concept, each
R-proach can be characterized at different levels of **MECA 1**: Lifespan, **MECA 2**:
Accessibility, and **MECA 3**: Circular Measures against Failures. Depending on the R-
proach determined by T1, the solutions gathered should strongly emphasize on one
MECA aspect and to a lesser extent on the others. Simultaneously, the MV in this

phase, namely SF 1–3 in Fig. 3, revolve around the techniques and design mentalities employed in the collection of solution ideas, all while preserving the core method. Naturally, a judicious combination of different MV can prove advantageous, given that some products may comprise a simple set of assemblies alongside a complex set of assemblies. When dealing with product that consist of assemblies with considerable complexity, such as the electrical adjustable car seat, the design process for the car seat necessitates traversing the entire TTDS independently, as highlighted in Fig. 1. Finally, in the last phase of the PDP, the evaluation phase plays a pivotal role in shaping the essence of the results. The core method of this phase is inherently flexible, requiring solely a point-based system and a criteria list. Consequently, the criteria list is the major factor that influences evaluation. Depending on the product analysis, diverse depths of evaluation—whether on the top or detailed hierarchy level—can lead to a range of outcomes through the suggested MV, as shown in Fig. 3 as E 1–3.

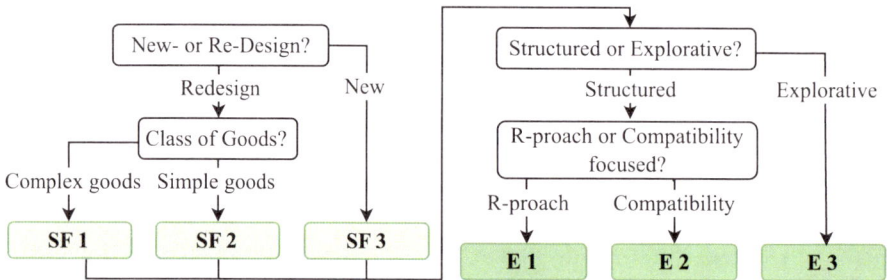

Fig. 3. Method-Variants Decision Flowchart for Solution Finding and Evaluation Phase [17]

SF 1—Opportunistic Technique by PAHL / BEITZ [16]: Solution ideas are grounded in the state of the art and existing proven designs, making it more suitable for adaptation design or redesign of complex products. Furthermore, designers can prioritize found solutions that significantly influence MECA or those that enhance the current design to integrate MECA.

SF 2—Problem-Targeting Technique/R-proach Design: Solutions generated through this method span from technical design to R-proach-focused-Design. This variant is highly effective for designing simple goods, where the entire product can be completely redesigned to address a specific problem, to aligned with the R-proach by focusing on integrating MECA, or to balance both types of design.

SF 3—Abstraction Technique: The product is deconstructed from its main function to the smallest subfunction, and solution ideas are formulated to address these subfunctions, emphasizing the avoidance of preconceptions during the solution finding process, which efficiently creates more innovative ideas.

E 1—R-proach based Evaluation: Solution ideas for each detailed hierarchy undergo evaluation using a criteria list derived from MECA. This results in the best design corresponding to the R-proach for each detailed hierarchy. However, it may lead to potential incompatibility between the detailed hierarchies.

E 2—Compatibility based Evaluation: Solution ideas for each detailed hierarchy are individually evaluated based on their compatibility with the solution ideas from other lower hierarchies. This approach necessitates a compatibility-based criteria list to identify the most suitable combination of lower hierarchies as the final solution.

E 3—Intuitive Product Evaluation: Solution ideas for each detailed hierarchy are intuitively combined by the designer to create different variations of solutions in the top hierarchy. This method relies heavily on the expertise and experience of the designer, requiring a comprehensive criteria list that holistically evaluates the solutions.

These proposed variations offer guidance for adapting the foundational framework of PDP, concluding the "*Design Strategy*" segment shown in Fig. 1. Each variant entails trade-offs among creativity, degree of freedom, difficulty, repeatability, and time investment. These recommendations—tailored for various application scenarios and different results in product analysis—are aimed at achieving an optimal balance between the design iteration efficiency and quality of solution ideas in the context of circularity optimization, while minimizing correction loops within the design process. The following electrical adjustable car seat case example applies the guidance flow chart, exploring the dynamic interplay between tools, and demonstrates how designers can factor in various elements in the decision-making process, creating a nuanced and comprehensive approach towards a particular design:

> *Case-Example: Suppose that Repair R-proach is considered from T1 for the complex assembly electrical adjustable car seat. The core method for gathering solutions in this scenario therefore entails a strong emphasis on accessibility (MECA 2), with a lesser focus on lifespan (MECA 1). Simultaneously, the solutions should possess the capability for maintenance to introduce circular measures against failures (MECA 3)* [17]. *Moving forward, the chosen method variant for solution finding is the "Opportunistic Technique" (SF_1), which allows designer to gather solutions that significantly influence the accessibility of the assembly. Regarding evaluation, the responsibility for enhancing compatibility with other car assemblies lies in the joint technology decision with T1 in the subsequent development stage, as seen in Fig. 1. Moreover, considering that the lifespan aspect is largely dependent on material, which falls under the purview of T3, the designer can therefore strategically focus on optimizing the accessibility of parts (detailed hierarchy) within the car seat (top hierarchy) for maintenance using "R-proach based Evaluation" (E_1).*

5   Tool 3: Material Evaluation

The recyclability of the part's material has a large impact on the circularity of the product. The core problem occurs in combination with different materials used when formulating a design. Several materials are intrusive to the recycling process of other materials, as a result, an optimal extraction of the desired secondary materials could not be achieved. This problem can be mitigated by arranging materials combinations that will not interfere with each other during the recycling process [5]. After the first iteration of product design and material selection is produced in the PDP assisted by T2, proceeding to this tool for an evaluation will determine how optimal is the materials selection, regarding its function and recyclability. Characteristics and the property's value of the materials are used as a reference to quantify the relations between functions and the recycling capability of certain material combinations. Firstly, suitable materials are selected in the design process to fulfill certain functions and are often considered according to their ability to perform their task. This performance is evaluated based on their individual mechanical, physical, and chemical properties of the materials. The numerical values of the material properties will be used to determine the function's value. By calculating the quotient between all the materials and the material with the highest value, the *Function's Value* (FV) of each material can be quantified. When multiple material properties are involved, they each create one FV and the overall value will be calculated for example, for "power transfer" the ultimate tensile strength and the yield strength could be considered as each individual FV and the overall can be calculated. Another case is when the FV is derived from one or more materials properties. For example, when selecting materials for a lightweight design the specific strength of a material is evaluated, which is the ratio between the tensile strength and the density of the materials. The specific strength of each material will be divided by the highest value, in this case Titanium [Table 1]. Therefore, titanium will represent the best material available with the FV of 1, followed by aluminum with the second highest, having the closest specific strength to titanium. The more accurate the representing material properties, the more reliable the functions value will be. Furthermore, an adequate and large database will increase the effectiveness of the functions value. This value will later then be evaluated in conjunction with the recycling behavior of each material.

The following will be an approach to determine the compatibility between materials based on their behavior in a recycling process. Depending on the specific recycling plant, the parts will undergo a series of automated processes starting from dismantling, sorting, until thermal and chemical processing [5].

Table 1. Generating Function Value for "light weight" based on the specific strength [18]

Materials	Tensile Strength (MPa)	Density (g/cm^3)	Specific strength (Nm/g)	Functions value
Ti	**334**	**4.5**	**74.22**	**1.00**
Al	140	2.7	51.85	0.70
Fe	400	7.874	50.80	0.68
Cu	210	8.92	23.54	0.32

The Recycling process utilizes the materials properties, such as density sorting and separation through the magnetism. This will represent a crucial link between the fulfillment of the function and the suitability for recycling. Problems occur when the assembly / input consist of materials with high similarities in the numerical value utilized by the process, e.g. Copper (8.92 g/cm^3) and Nickle (8.908 g/cm^3) having similar density. Moreover, materials such as copper and zinc from an alloy, could lead to a deterioration of steel's cold-drawn property after steel recycling [5]. However, the reverse is not true when the main goal is copper recycling. Consequently, the interaction between the materials within the recycling process will differ depending on the objective of the recycling, hence the unsymmetrical matrix shown in Fig. 4. With this knowledge, the compatibility of each material to one another could be summarized and given a qualitative value based on their interaction. The compatibility will be narrowed according to the material with the highest weight percentage, this is calculated under the assumption that the whole product will undergo the same recycling process for its respective material, and the rest will be the accompanying material. The recycling compatibility are broken down into its behavior and have value assigned to each quality. The combinations that are labeled "possible" require high effort to be recycled and "poor combinations" should be avoided. *KO-criteria* are given as a value to the poor combination to indicate that the material choice has a poor combination for the circularity, the value will be the maximum number of functions and materials combinations.

		Main Materials						
		Fe	Al	Cu	Ti		0.1	No Affect / same Material
Accompanying materials	Fe	0.1	0.5	0.25	n[a]		0.25	Good
	Al	0.25	0.1	0.25	0.5		0.5	Possible
	Cu	n[a]	0.5	0.1	n[a]		0.75	Inadequate Data
	Ti	0.25	0.75	0.25	0.5		n[a]	Poor

Increasing effort (left of legend)

[a] Maximal combination of materials and functions

Fig. 4. Estimated Recycling Value for different material combinations according to the main material [18]

Once the values for the functions and the recycling have been determined, the relations between the two values are depicted as a quotient that can be compared with other available combinations. During the product development process, the material will be chosen according to the desired function which then be evaluated for its recyclability. The *Recycling's Value* (RV) of the different materials will be evaluated within its column according to its main material, which corresponds to the intended recycling process. The quotient of both FV and RV resulted as the *Function-to-Recycling Coefficient* (FRC) as shown in Eq. (1). The higher the FRC, the more optimal the circularity in accordance with its function's fulfillment. In the case of unsuitable material selections for a circular design, the value will be less than 1.

$$Function\ Recycling\ Coefficient = \frac{\sum_{i=1}^{n} Functions\ Value_i}{\sum_{i=1}^{n} Recycling\ Value_i}; FRC = \frac{\sum_{i=1}^{n} FV_i}{\sum_{i=1}^{n} RV_i} \quad (1)$$

Fixed predetermined material that is critical for the desired main function often exists in PDP. In this context, it is advisable to revisit Table 2 and re-adjust the materials on the parts for the detailed hierarchy. Altering the material with the highest weight percentage will also shift the overall RV. In instances where material options, to further improve the RV, are limited, re-evaluating the joint technology on Tool 1 (see Fig. 1) will mitigate the use of less suitable material combinations and their impact for the recycling approach.

This method provides the designer an evaluation whether the materials selected are suitable for the circular economy, as well as how optimal the materials selections for

Table 2. Materials selection table according to *Function Value* and their respective *Recycling Value* in regards on the main material compatibility [18]

Function's value			Recycling's value according to the main material by weight percentage							
Legend	Fe / Cu	Al / Ti	Fe		Al		Cu		Ti	
Power Transfer	0.55	0.24	0.1	0.25	0.5	0.1	0.25	0.25	3	0.5
	0.32	0.84	(3)	0.25	(0.5)	0.75	(0.1)	0.25	(3)	0.5
Cable	0.21 (b)	0.59	0.1 (b)	0.25	0.5	0.1	0.25	0.25	3	0.5
	0.75	0.22	3	0.25	0.5	0.75	0.1	0.25	3	0.5
Light weight	0.68	0.70	(0.1) ← 0.25 (a) (0.5)			0.1	(0.25)	0.25	(3)	0.5
	0.32	1.00	(3)	0.25	(0.5)	0.75	(0.1)	0.25	(3)	0.5

the desired functions in relations to its recyclability and to assess for potentially better material alternatives and also acquiring information that will support the decision on the joint technology (Tool 1) based on the recycling quality of the materials combination, regarding the separation percentage during the recycling's process.

> *Example for Table 2 (shown in green cells):in a car seat, aluminium will be used as "light weight" frame and due to the design, a high material volume will be used, thus resulting in highest weight percentage. Furthermore, Copper is chosen for "Cable" for the seat heater, and Titanium for "power transfer" for electric seat adjustment. This results in an overall ΣFV of 2.29 and ΣRV of 1.35 Table 2 and thus an overall FRC of 1,70 can be determined (see Eq. 1). In an adaptation scenario where the seat frame ("light weight") is restricted to iron[2] (shown by a red arrow "a"), the RV of other materials will shift according to its compatibility with iron as new main material (shown in yellow highlight). However, the previous combinations of copper for 'Cable' will prove to be detrimental to the recycling quality of the product. In this case (shown by green arrow "b"), aluminium could be chosen as an alternative for copper due to its better RV whilst still providing the second highest FV out of all the alternatives. Overall, by using iron as the main material and substituting copper with aluminium, the FRC increases to 3,52, which indicates a more optimal combination despite the change in materials-functions combination.*

6   Interdependencies within Tri-Tool design system

The TTDS provides the necessary assistance and flexibility for the designer without imposing limitation as seen in the case example in Sect. 4, this stays true when it comes to the "*Engineering Development*" segment with T1 and T3, involving complex scenarios of joint technology and materials selection, particularly in the context of CE. As depicted by double ended arrows in Fig. 1, T1 and T3 are used back and forth and iteratively to

---

[2] For in this example the material is simplified as pure iron, it's likely to be light weight steel alloy (iron-based material) in practice.

increase product's circularity in context of material and joint technology in top and detailed hierarchy by finding the most optimal selections.

For material combinations which are not suitable for a certain R-proach but crucial to sufficiently fulfil the high-priority functions, this conflict can be resolved through selection of joint technology, and vice versa. Otherwise, the efforts for a prolonging R-proach can be undermined by complications within the final recycling approach. Such design contradiction can be seen in the last case example in Sect. 5, where the seat frame is required to withstand multiple maintenance since Repair R-proach is assigned to it, which leads to the material restriction in iron in this scenario. In rare cases, where none of the available decision for a product design, material and joint technology is viable as a whole, designers are bound to either opt for another design solution in PDP, or iterate the PDP through changing the methodology in operative level using T2, or finally when all else fails, fully reset by adjusting R-proach in the strategy level as the last resort. These cases are illustrated as the correction loops flowing back to the "*Design Strategy*" segment, the red dotted arrows in Fig. 1.

7   Conclusion and Outlook

By combining these three tools into one coherent methodical guidance aligned with the goals outlined in the introduction, it becomes feasible to integrate the circular economy within the design process of a company. Furthermore, the methodology's minimal barriers enable it to be implemented at any point in the design stage, combined with an efficient workflow, served as a strong incentive for engineers. This encourages them to utilize the system, transforming easily attainable results for the products as a first step towards CE. Using the TTDS, companies have the opportunity to integrate their products gradually into the circular economy. This is even valid for designing them in one country, manufacturing them in another and selling over a whole region, continent or worldwide, although regions did not adapt to the CE yet. The beneficial products are in the economic system, thus can be treated accordingly and the benefits for societies and the environment will have a positive effect towards the worldwide climate goals.

Further Research.
Tool 1: To further extend the software research data is needed. Most important is the experimental investigation of joint technologies. Therefore parts, assemblies and products will be shredded, and the outcome will be investigated depending on the grain size and separation of the processed materials.

Tool 2: Naturally, more MVs can be added into the account in each phase of PDP since the quality of the solutions heavily depends on the methodology. This also further increases the flexibility of TTDS to integrate elements of CE. Consequently, a more in-depth guide will then be necessary to navigate the interdependencies among the MVs and with other tools, ensuring the adaptability of TTDS to all kind of product design.

Tool 3: The use of a larger database that includes more materials (such as polymer and alloy) and determining more precise material properties will enable the designer to create a personalized table according to needs and assist them in selecting the right materials when designing a circular product. The whole process could be digitalized. It also enables the possibility to further specify the recycling's value by collaborating with a recycling plant, by knowing the recycling series, the recycling's value can be tailored and quantified to match the plant, thus giving a more reliable tools for its needs. To validate the whole TTDS it is necessary to transfer the system into practical use and verify the potential impacts to economics, societies, and the environment. Therefore, it is essential to track a product and its materials which will require the participation of involved stakeholders.

References

1. *Recycling-oriented product development: VDI 2243,* Verein Deutscher Ingenieure, Beuth Verlag GmbH, 10772 Berlin, Jul. 2002.
2. *Recycling of electrical and electronic equipment: VDI 2343,* 1–7, Verein Deutscher Ingenieure, Beuth Verlag GmbH, 10772 Berlin.
3. B. Alber-Laukant, *Handbuch Konstruktion,* 2nd ed. München: Hanser, 2018.
4. B. Bender and K. Gericke, Eds., *PAHL/BEITZ KONSTRUKTIONSLEHRE: Methoden und anwendung erfolgreicher Produktentwicklung,* 9th ed. Springer Vieweg Berlin, Heidelberg: MORGAN KAUFMANN, 2019.
5. H. Martens, *Recyclingtechnik: Fachbuch fur lehre und praxis,* 2nd ed. Springer Vieweg Wiesbaden: MORGAN KAUFMANN, 2016.
6. B. Bookhagen, "Metallic resources in smartphones," none, 2020.
7. A. Kampker, S. Wessel, F. Fiedler, and F. Maltoni, "Battery pack remanufacturing process up to cell level with sorting and repurposing of battery cells," *Jnl Remanufactur,* vol. 11, no. 1, pp. 1–23, 2021, https://doi.org/10.1007/s13243-020-00088-6.
8. P. Beerten, F. Ostuzzi, and L. Brosens, *Barriers and Enablers for the Implementation of Design for Sustainability in Flemish Design Agencies,* 2023.
9. N. M. P. Bocken, I. de Pauw, C. Bakker, and B. van der Grinten, "Product design and business model strategies for a circular economy," *Journal of Industrial and Production Engineering,* vol. 33, no. 5, pp. 308–320, 2016, https://doi.org/10.1080/21681015.2016.1172124.
10. M. C. den Hollander, C. A. Bakker, and E. J. Hultink, "Product Design in a Circular Economy: Development of a Typology of Key Concepts and Terms," *J of Industrial Ecology,* vol. 21, no. 3, pp. 517–525, 2017, https://doi.org/10.1111/jiec.12610.
11. European Parliament, *MEPs adopt new law banning greenwashing and misleading product information,* 2024. Accessed: Mar. 13 2024. [Online]. Available: https://www.europarl.eur opa.eu/news/en/press-room/20240112IPR16772/meps-adopt-new-law-banning-greenwash ing-and-misleading-product-information
12. *Regulation (EU) 2023/956 of the European Parliament and of the Council of 10 May 2023 establishing a carbon border adjustment mechanism (Text with EEA relevance): CBAM,* 2023. Accessed: Mar. 13 2024. [Online]. Available: https://eur-lex.europa.eu/legal-content/ EN/TXT/PDF/?uri=CELEX:32023R0956
13. European Commission. Directorate General for Internal Market, Industry, Entrepreneurship and SMEs., *Critical raw materials for strategic technologies and sectors in the EU: a foresight study*: Publications Office, 2020.
14. European Commission. Joint Research Centre., *Supply chain analysis and material demand forecast in strategic technologies and sectors in the EU: a foresight study*: Publications Office, 2023.
15. J. Kirchherr, D. Reike, and M. Hekkert, "Conceptualizing the Circular Economy: An Analysis of 114 Definitions," *SSRN Journal,* 2017, https://doi.org/10.2139/ssrn.3037579.

16. P. Wallat, "Entwicklung einer digital gestützten Konstruktionsmethodik unter Berücksichtigung von recyclingtechnischen und wirtschaftlichen Einflussfaktoren für die Produktentwicklung innerhalb einer Kreislaufwirtschaft," Universitätsbibliothek der TU Clausthal, 2023.
17. J. Low, "Mehrfache Adaptionskonstruktion von Planetengetrieben zum Einsatz innerhalb der Kreislaufwirtschaft," Bachelorarbeit, Technische Universität Clausthal, Clausthal-Zellerfeld, 2022.
18. E. D. Tanzil, "Zusammenhänge von Funktionen, Werkstoffen und Recyclingeignung in der Produktentwicklung für die Kreislaufwirtschaft," Bachelorarbeit, Technische Universität Clausthal, Clausthal-Zellerfeld, 2022.

Life-Cycle Assessment

Sustainable Plastic Packaging from Sorted Municipal Waste and Its Life Cycle Assessment Using Digital Product Passports

Elena Berg[1]([✉]), Pia Wagner[1], Gonsalves Grünert[2], Johannes Mayer[2], Lisa Leuchtenberger[1], Philipp Niemietz[2], Thomas Bergs[2], and Rainer Dahlmann[1]

[1] Institute for Plastics Processing (IKV) in Industry and Craft at RWTH Aachen University, Seffenter Weg 201, 52074 Aachen, Germany
Elena.Berg@ikv.rwth-aachen.de
[2] Manufacturing Technology Institute (MTI), RWTH Aachen University, Campus-Boulevard 30, 52074 Aachen, Germany

Abstract. Plastic waste from packaging is being repurposed to create new products with similar processing and usage properties as virgin plastics. Understanding its composition, degradation processes, and recycling steps is essential to open up new fields of applications. To shed light on sustainable uses of recycled plastics in packaging, this paper presents interdisciplinary research which investigates and connects three areas to a holistic product view from the material to the end of production. The first area focusses on the exploration of recyclate applications by analysing and processing different recyclates and design-recyclates to identify their potential in plastic packaging products. In the second research area an evaluation model to assess the ecological sustainability of plastic packaging is developed, considering factors like material composition and processing methods. The final area focuses on developing a platform-supported information system for sharing relevant sustainability data, considering mechanisms to enhance data sharing.

The challenges and opportunities of using recyclates in packaging are examined through a demonstrator product—a recyclable stand-up pouch which also contains recyclates. The analysis involves production, life cycle assessment, and utilising platforms for data sharing throughout its life cycle. The paper delves into methods to combine the practical analysis of material behaviour, the mechanisms to use and share the resulting material, machine and process data and the sustainability assessment of this data.

Keywords: Digital product passport · Life cycle assessment · Plastic packaging · Recyclate

1 Introduction

Today, less than 10% of plastics from municipal waste in Europe are mechanically recycled [1]. Reasons for the low mechanical recycling rates are non-recyclability due to improper product designs, an unknown or complex material-composition of such plastic

K. Dröder and T. Vietor (Eds.): CD 2024, Proceedings, pp. 377–392, 2025.
https://doi.org/10.1007/978-3-658-45889-8_29

products and a general lack of a clear and commonly binding definition of "recyclability". Especially packaging plastics are considered difficult to recycle due to their complex multimaterial product composition [2, 3]. Those products are produced almost exclusively from fossil fuels and not from recycled plastics. However, the production of packaging plastics made from municipal waste recyclates faces different challenges along the value chain. Due to the different origins of municipal waste and the numerous grades of plastics, recyclates are subject to material variations and their properties are difficult to predict. This is further complicated by the numerous effects that contaminants can have on recycling and manufacturing as well as product properties. Various methods exist for examining the properties of recycled plastics, but there are no established standard specifications that can be used to conclude the processing and usage properties. Existing standards (e. g. EN 15344 and EN 15345 for PE and PP) for the characterisation of recycled materials are often not sufficient for a comprehensive examination of the material quality, as they contain few general-use test methods [4]. In order to increase the proportion of PCR in plastic packaging and to quantify the impact on the environment compared to packaging made from virgin material, a holistic product analysis is required.

2   State of the Art for an Interdisciplinary Approach to the Ecological Assessment of Plastic Packaging

Regardless of the type of recycling, recyclates must meet the general legal framework conditions, e. g. producer liability and achieving the same required product specifications. The multitude of stresses that recyclates experience during their repeated processing and usage results in an overall higher level of degradation. During processing, the polymers are exposed to strong forces (mostly shear with some extension) and high temperatures, which affect the polymers and lead to thermo-oxidative and shear-induced chain scission, chain branching, or crosslinking of the material [5–8]. The amount of degradation depends on the chemical properties of the polymer and the selected processing conditions. With the specific choice of processing parameters, degradation and thus direct effects on process stability and product quality can be controlled to a certain extent [9–13]. While degradation mechanisms can significantly affect the material properties, this is further influenced by the material composition and other impurities [14, 15]. Such impurities contain e.g. printing ink systems, adhesives, barrier materials, additives (e. g. processing aids, compatibilisers) and fillers (e. g. minerals). Overall, the various components influence the material properties to an unknown extent, depending on their type and concentration. In addition to the direct influence of the impurities on the processing behaviour, they also influence the degradation of polyolefins. For example, metallic impurities or pigments used to colour plastics can accelerate thermo-oxidative degradation [16, 17].

In contrast to the decreasing material and product quality of PCR-products, the sustainability and ecological environmental impact improves compared to products from virgin material. So-called life cycle assessments (LCA) are therefore suitable to evaluate the sustainability of products and processes. In this way, the sustainable performance of a product made from recyclates can be compared with products from virgin material.

Moreover, different manufacturing and recycling processes as well as different usage scenarios of recyclates are considered using defined impact indicators to describe the environmental impact. In a circular economy, an ecologically and economically use of recyclates consequently places requirements on each step of the product life cycle. The European Union is currently investigating the extent to which digitisation in form of product passports and the associated transparency of information on product properties can support the transition to a more resource-efficient internal market [18, 19]. In such a product passport, necessary data regarding the product history of packaging (e. g. net CO_2 emissions) is stored and exchanged between different actors. Provided that such data is collected online, product passports offer enormous potential with regard to the calculation of environmental impacts in life cycle assessments. It is thus possible to make greater use of primary data than of secondary data, i. e. resource consumption and processing conditions measured on production plant machinery. This enables improved transparency for consumers and stakeholders to make more informed decisions by means of LCAs. In order to quantify the environmental impact of product systems, LCAs been carried out since the early 1970s. [20] With the introduction of ISO standard 14040, the first internationally recognised standards for conducting LCAs were implemented [21] whose phases and interaction (cf. Figure 1) are described in detail in ISO 14044 (2006) [22].

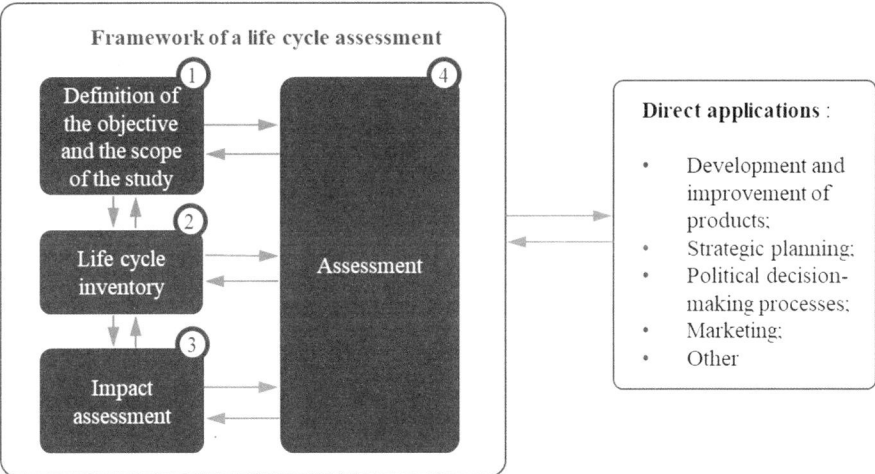

Fig. 1. Phases of a life cycle assessment according to ISO 14040 and ISO 14044 [21, 22]

The LCA method is designed to be applicable to any product or process. Therefore, the method offers the advantage allowing for many degrees of freedom to analyse the desired object, but also makes it difficult to compare different LCAs. The approaches and methods chosen for each LCA lead to different results for the same product system [23]. Only broad assumptions regarding the environmental effects of comparable items can be made.

Digital platforms offer the opportunity to fundamentally improve the comparability of LCAs in the future. They enable pre-competitive exchange along the value chain and they are an important tool for developing and coordinating sustainable products and services. [24]. One challenge for the comprehensive utilisation of LCAs is the lack of data exchange between individual companies to evaluate a product or functional unit beyond the company boundaries. One solution for the corresponding cross-company data exchange are digital platforms on which data can be stored or made accessible, shared and, depending on the application, downloaded. The provision of production, quality, and utilisation data can be used for impact and sustainability analysis. Platforms are characterised by properties like real-time capability, flexibility, geographical independence, and maximum scalability. This is advantageous in terms of network effects, as the benefit of individual participants increases with the number of participants. However, the use of platforms by companies is barely established [25]. Concerns about trust between collaborative partners and data security, sovereignty, and integrity are high. Those challenges are met by GAIA-X as a suitable standard/initiative. GAIA-X is a concept for data-driven ecosystems where the six principles, openness, transparency, interoperability, federation, authenticity and trust, are enforced through technical guidelines. In GAIA-X compliance creates digital sovereignty, independence, and security according to the Data Protection Regulation (GDPR) [26]. The initiative introduces concepts that deal with data storage and cloud connected elements with a focus on sovereign cloud services and cloud infrastructure. Regarding the willingness to share data, the definition of governance structures is important. Governance here refers to a system of decision rights and responsibilities for information-related processes that are executed according to agreed models. [27]. The link between the Gaia-X initiative and the product passport is based on another initiative—Catena-X. This initiative is collaborative, open data ecosystem of trust for the automotive industry in Europe with the goal of increasing resilience, innovation, and revenue opportunities and creating equal access to information flows for all actors along supply chains (e. g. digital carbon footprint or traceability in accordance with the Supply Chain Act) [28]. Its system architecture is based on GAIA-X with e.g. the data sovereignty and security aspects. The value proposition is the first time offer of a sovereign, multi-tier data sharing and use case collaboration across the entire value chain. In Catena-X multiple use cases are examined. In the use case of circular economy the community is working on a product passport. Due to relevancy of a product passport as a use case for data sharing, this example is part of the overall research goal.

As the challenges for ecological plastic packaging encompass various industries and aspects along the value chain, an interdisciplinary approach is necessary. This research provides a possibility to connect the different necessary areas and to generate knowledge about material behaviour as much as data generation, data usage and interpretation in context of product sustainability.

The research objective is the ecological optimisation of plastic packaging using digital technologies. For this purpose, a general description model for evaluating plastic packaging is being created on the basis of a stand-up pouch as a demonstrator product.

3 Methodology and Preliminary Results Using the Example of a Mono-Pe-Pouch

The first strand of research is dealing with the possibilities and limits of using mechanically recycled polyolefins in packaging with the aim of increasing the use of recyclates in the packaging sector. The influence of different material compositions or impurities in combination with the degradation of plastics has not yet been sufficiently investigated and studies with pure virgin material cannot be directly transferred to recyclates. Many analytical methods, as for example differential scanning calorimetry (DSC), infrared spectroscopy (IR), capillary rheometry or gel permeation chromatography (GPC), already try to assess the quality of recycled plastics [29]. Nevertheless, there are still uncertainties about necessary quality parameters, to ensure the recyclate's performance in new applications regarding for example resulting rheological, fatigue and breakage behaviour or processability. According to the researchers, the current non-binding and partially incomplete standards for recyclates and virgin materials lead to plastic packaging manufacturers agreeing individual and bilateral on different quality parameters and delivery conditions. Plastics analysis and testing is a central component in generating new standards. At the same time, it is important to determine the possibilities and limits of various processes such as cast packaging films, blown films and injection moulding and related applications to increase the use of recyclates and thus meet current EU requirements (e. g. Packaging and Packaging Waste Directive, PPWD) [30].

With the aid of a product demonstrator the theoretical product life and the real material and product properties can be correlated. Design for recyclability plays an important role. The more recyclable a packaging is designed, the easier it is to separate the individual plastics by type and feed the waste into the highest possible value stream. As the use of flexible plastics can save packaging material compared to rigid packaging solutions a stand-up pouch for a cleaning agent with a spout is chosen as a demonstrator product [31]. The stand-up pouches currently available on the market are fully recyclable, but contain little or no post-consumer recyclate (PCR). Therefore, a complex mono-material structure with a high proportion of recyclate was developed (100% recyclate in the spout and > 60% recyclate in the film laminate). As no fractions or recyclates from flexible PP waste are currently available on the market, the bag is made exclusively from PE [32]. Figure 2 shows a schematic representation of the pouch and the corresponding laminate structure of the film consisting of a blown PE film, a laminating adhesive, printing ink and a biaxial oriented PE film. Furthermore, the challenges in the production of the packaging components are presented.

3.1 Challenges During Processing of Recyclates

In a first step, commercially available recyclates were processed and analysed to assess the extent to which the various materials differ and are suitable for the respective production processes. For this purpose, PCRs from different providers (e. g. Vogt-Plastic GmbH, Der Grüne Punkt Holding GmbH & Co. KG, Ecoplast Kunststoffrecycling GmbH, Interzero Holding GmbH & Co. KG, Morssinkhof-Rymoplast) were analysed. The purity of such recyclates compared to virgin materials is relatively low, making them the greatest challenge for processing. Figure 3 shows an example of a blown film made

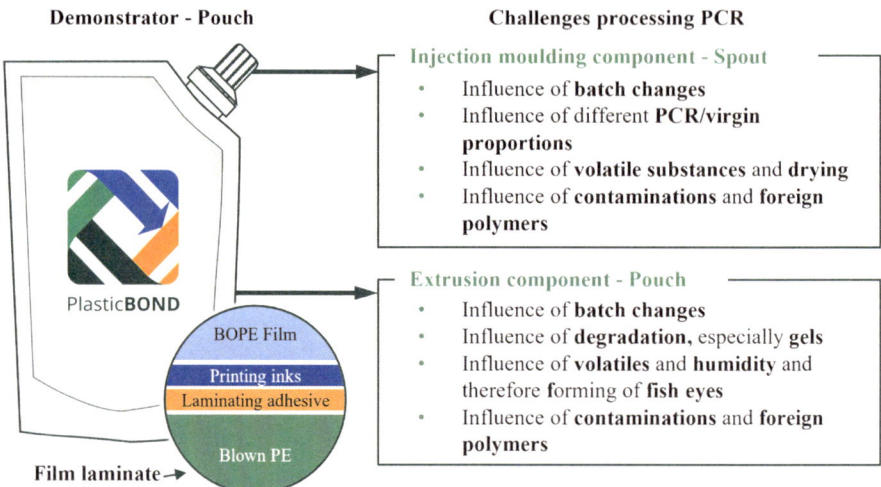

Fig. 2. Design of the demonstrator pouch including the challenges of processing PCR.

from mechanically recycled PCR with a possible composition based on the material analyses conducted.

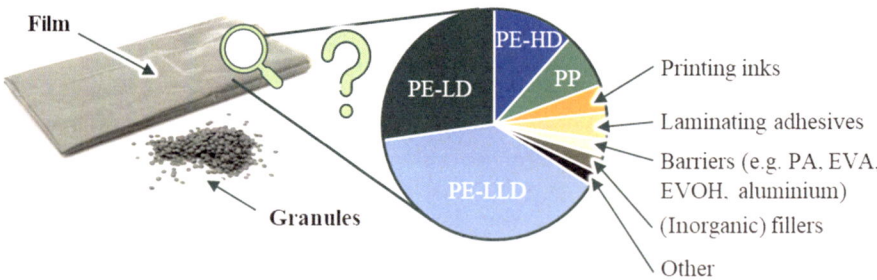

Fig. 3. Exemplary composition of a film recyclate

To investigate the influence of impurities and material degradation on the processability, recyclates and so-called design recyclates (contaminated virgin materials [33]) were repeatedly processed in blown film and regranulated afterwards. Various PEs from SABIC, Riyadh, Saudi Arabia, were used for this purpose. The impurities used included an impact-resistant polypropylene (PP) copolymer for film applications, an ethylene-vinyl alcohol copolymer (EVOH) with an ethylene content of 32 mol%, a PE-based compatibiliser grafted with maleic anhydride (MAH) for physical bonding to the PE phase and chemical bonding to polar phases (PA or EVOH), binders and a calcium carbonate as a filler. The gained knowledge enables to select and develop simple methods for determining and improving mechanically recycled PCR quality (e.g. adapting existing design guidelines). Furthermore, the prediction of processing options depending on the PCR quality is targeted. In the long term, this should enable the creation of a material and process specification to assess the quality of recycled materials, e.g. in the form of digital product passports (Research strand 3).

The first pouch component, the film laminate, consists of a biaxially stretched top film and a blown film for sealing. Both the top film and the blown film consist of three layers: a relatively thick layer of recycled material ($> 70\%$) surrounded by two thin outer layers of virgin material. The outer layers of virgin material improve the processability, sealing properties and odour formation during and after processing. One of the most critical challenges during film extrusion of mechanically recycled PCR is the amount of high volatile components and moisture in the material. While virgin PE generally does not absorb moisture, this does not appear to be the case with PE recyclates due to impurities (e. g. ethylene vinyl alcohol (EVOH)). During processing, this leads to fish eyes, i. e. holes in the film, and thus to rejects. Pre-drying can provide a solution, but only to a limited extent. Furthermore, lower processing temperatures e. g. 180 °C in PE film extrusion lead to fewer or no fish eyes. At higher temperatures (e. g. 230 °C), on the other hand, the proportion of fish eyes increases and film breaks occur. In addition, large differences were found between PCRs from different recyclers (e. g. Vogt-Plastic GmbH, Der Grüne Punkt Holding GmbH & Co. KG, Ecoplast Kunststoffrecycling GmbH). It is therefore assumed that a higher processing temperature during recycling leads to increased degassing and therefore fewer volatile components escape in subsequent processing steps. Another major challenge in film extrusion and the further processing of the films into a laminate are so-called gels. Gels are any small defects that change a film product (e.g. cross-linked material, highly oxidised material, filler agglomerates, fibres, remelted polymer) [34, 35]. Microscopic investigations have shown that a large proportion of the gels in mechanically recycled materials consist of cross-linked or high-molecular structures [36]. Investigations on design recyclates also showed that the recycling process and the amount of recycling steps influences the size and number of gels depending on the material composition of the recyclates [33]. In addition to impurities such as polypropylene (PP) and EVOH, the use of a MAH-based compatibiliser in particular catalyse gel formation. High fluctuations were also observed with regard to the PP content in commercially available recyclates. For example, the PP content in film recyclates ranged from $< 1\%$ to greater than 15%, which in turn influences gel formation. The mechanical properties, on the other hand, are less influenced by the mechanical recycling process. Only a slight decrease in the tensile strength of the mixtures as well as the maximum elongation after recycling was detected. No influences on the impact strength were observed [33].

In addition to the properties of the extruded packaging film, the production of the spout by injection moulding is considered and analysed. The spout needs to fulfil various quality requirements. On the one hand, the part must have sufficient rigidity, strength and precision concerning the rotating mechanism and, on the other hand, it must have a high level of geometric accuracy for a reproducible sealing process between the spout and the film pouch. During the injection moulding trials different aspects of the material-process interaction are analysed. The influence of the set parameters (e.g. injection speed, holding pressure, mould temperature etc.) is just one example, but due to the differences in the material composition also other effects like the influence of volatile substances, batch changes and material conditioning (e. g. drying) are considered. First trials of two materials show that the PCRs show similar reactions to changes in process settings as virgin material. For example an increase in melt temperature (+ 10%) and an increase in injection speed (+ 60%) tend to decrease the strength (- 3%). Furthermore, the conditioning of the PCR influences the mechanical behaviour significantly. Drying, even though recommended by the recycler, can lead to further ageing of the material which is detectable during processing (slight increase in injection pressure) and results in brittle behaviour during mechanical testing. The results of the injection moulding processing of the recyclates show a strong influence of the material composition and the resulting varying processability on the process stability and part quality. Indicators such as the viscosity and the part weight can be used to monitor reproducibility, but analysis of the correlation between machine data, viscosity, part weight and part strength suggest that similar processes can still result in considerably different part qualities. A product passport in which the material history and composition is documented is therefore useful for estimating the resulting component quality depending on the material and process control.

3.2   Assessment Model for the Ecological Sustainability of Packaging

The second research strand focuses on LCA. In addition to the identification and characterisation of all relevant material and energy flows a general description model for the evaluation of plastic packaging is developed. The focus of LCA is always on comparability between potential alternatives, such as the comparison of sustainability between linear recycling and closed-loop recycling of packaging. The models and methods established are then validated using the demonstrator product. An increase in the recycling rate does not necessarily mean an increase in resource efficiency. With the determination of the product design, the processes for manufacturing and further processing of the products as well as their recycling, the energy and material requirements of a production are predetermined. Additionally, transport routes also influence the environmental impact. To estimate the environmental impact, the multitude of energy and material flows of the individual process steps of a product system must be combined into an overall balance. These balances are analysed for their environmental impacts such as greenhouse gas emissions or human toxicity. This is done in LCA [21]. Emissions from greenhouse gases are described by CO_2 equivalents as an indicator of the impact on global warming [22]. The calculation of environmental impacts such as eutrophication requires the use of characterisation models based on emitted energy and material flows [21].

With the preliminary studies from [37], a model was developed with which the main environmental impacts of a packaging pouch along its life cycle can be assessed with the help of an LCA. The model can be used to evaluate different end-of-life scenarios depending on the scope of the study (cradle-to-gate, cradle-to-grave, cradle-to-cradle). The product system and the specified process modules for a pouch's life cycle are shown in Fig. 4.

The four stages of the life cycle: raw materials, production, usage, and end of life, are represented by the product system. Some manufacturing procedures that are part of the pouch's value chain are also applicable to the majority of other plastic products. All pouch components are made using different production processes from mechanically recycled plastic and partly from virgin plastic. When evaluating the packaging only the

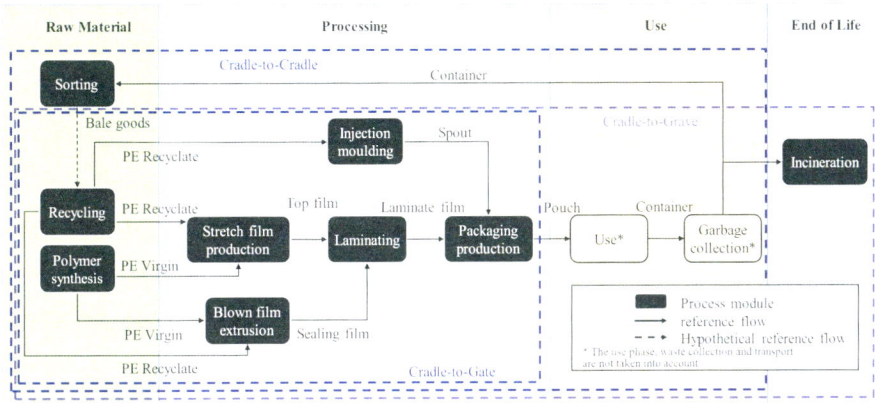

Fig. 4. Product system of pouch manufacturing

manufacturing itself is considered, parts of the use phase for example the content of the packaging are discarded.

In a preliminary study, an LCA was conducted for a pouch to identify the relative influence of the individual production steps [37]. For this purpose, additional modules such as transportation or pre-drying of the granulate were considered. The reference flow considered for the LCA corresponds to the production of a stand-up pouch. For the LCA the open Source Software OpenLCA was used. [38] The environmental impacts were calculated with the ReCiPe Method. [39] The used LCI are from ecoinvent 3.9. [40] This preliminary study endeavors to delve into the intricate landscape of individual process modules within the framework of Life Cycle Assessment (LCA). Life Cycle Assessment, a widely utilised methodology for evaluating the environmental impacts of products and processes, involves various interconnected components. However, a detailed understanding of the distinct process modules is imperative for a comprehensive analysis. The primary aim of this study is to acquire nuanced insights into these individual modules, setting the stage for a more profound comprehension of the broader LCA methodology. The results of the different environmental impacts can be seen in Table 1.

The largest share of greenhouse gas emissions is attributable to the extraction and provision of raw materials. Nevertheless, the production of the pouch has a non-negligible share of greenhouse gas emissions. Based on this study, the next step is to specify the modelling of raw material production and manufacturing to derive the potential for the circular economy [37]. The next step is to enrich the developed model with primary data of the pouch production.

3.3 Design of a Platform for Sharing Material Data

The third research strand focuses on the information and data flow-related aspects of LCA and the design of a platform for sharing data while maintaining security, sovereignty, and integrity aspects according to GAIA-X. In this case, a platform should primarily enable data-driven LCA as well as information-based cooperation between the players in the plastics processing value chain. The first task is to identify all data flows and characterise

Table 1. Relative results of a preliminary LCA study for a stand-up pouch [36].

Impact Category	Unit	Process modules [%]									
		Raw material	Pre drying	Film-extrusion	Spout	Post process	Refine-ment	Filling	Transport	Incinera-tion	Landfill
Global Warming	kg CO_2-eq	70.8	1.0	7.3	3.1	4.5	4.5	4.5	0.9	2.9	0.4
Acid[1] Soil	kg SO_2-eq	69.9	0.7	8.6	3.5	3.7	3.7	3.7	2.0	3.1	0.9
Eutro[2] Freshwater	kg P-eq	83.6	0.1	0.9	0.4	0.6	0.6	0.6	0.3	1.0	11.9
Eutro[2] sea	kg N-eq	69.7	0.8	7.5	2.7	3.5	3.5	3.5	3.7	4.2	1.0
Particulate Matter (PM2.5)	Illness-rate	62.5	0.9	11.9	4.5	4.0	4.0	4.0	1.4	5.8	1.2
Water use	m^3-eq	45.7	1.2	9.1	4.0	5.0	5.0	5.0	0.3	24.3	0.3

Legend: [1] Acidification, [2] Eutrophication

them in terms of collection type, storage location, and format. This is important for defining process interfaces. The challenge in the context of the PlasticBond research project lies in dealing with new technologies for storing, managing and orchestrating manufacturing process data for which there is still no sound empirical evidence. While maintaining compatibility with established standards, solutions and norms, secure data exchange is to be ensured. This requires the definition of all desired functionalities of such a marketplace and their continuous completion. When determining the requirements for a platform, the first step is to define feature classes to make the individual requirements clearer and more transparent. Requirements for a data marketplace were derived with the help of analyses of scientific literature, expert discussions and methods of error analysis of technical systems such as FMEA. For structuring reasons, these were divided into 14 feature classes (cf. Figure 5). Each of these feature classes contains at least one feature that can be fulfilled in different solution specifications.

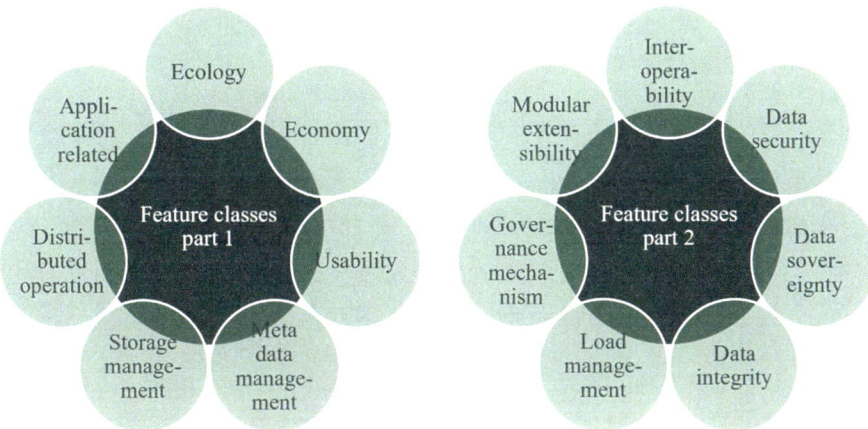

Fig. 5. Categories of the specification sheet for requirements of data marketplaces

All the requirements of potential stakeholders for the platform to be used are defined in a specification sheet. This ensures a certain scalability. Based on the determined requirements of the stakeholders, such as data providers and data users, the requirement specification follows on the part of the platform developer including technology recommendations for the implementation of specific customer requirements, for example, security and compliance demands. Based on these specifications, the platform developer offers a functional specification in which all (safety) technical and infrastructural features are prepared in relation to the customer's requirements. Without harmonisation of the requirements with the solution approaches, customers will lack the necessary acceptance and (intuitive) understanding to join the platform and share their data.

Following, each actor uploads the material and energy flow data required for LCA into the platform while maintaining data sovereignty. A mathematical model makes use of the LCA data identified and calculates the environmental footprint of each process step and/or the entire value chain. Results can be viewed by others upon request and, if necessary, against payment. The element of potential payment for access to data leads to the final aspect of the third research strand: New business models are identified and developed for the plastics processing value chain and their potential is demonstrated.

The main objective of the ecological optimisation of plastic packaging is to provide a wide range of options for exchanging information as a basis for calculating a life cycle assessment. The focus is on the provision and exchange of primary data for maximum accuracy of the results as well as increased flexibility and transparency regarding the development of the calculation logic. Aggregated data or theoretical values reduce these factors and may involve greater effort. In addition to the calculation logic to be developed, the configuration of the platform poses a major challenge. It is subject to three central requirement categories to counteract the current scepticism of companies in data sharing. Data sovereignty, integrity and security must be guaranteed for network participants. This means that data is recorded transparently and stored in a trustworthy manner, whereby data sovereignty always remains with the owner and ownership is not transferred to centralised bodies or third parties (data governance) [41]. In addition, data formats and their content must be homogenised, and correspondingly flexible interfaces defined. This enables the integration of many physical assets and existing systems without major additional effort in data acquisition [42]. Furthermore, governance and incentivisation mechanisms must be developed and established to motivate participants in the platform's collaboration network to provide authentic data within the network [43].

The prototype of the data sharing platform was implemented using an existing product from senseering GmbH, which had already integrated the identified core requirements and was therefore able to focus on the development of missing functionalities. The prototype is a distributed platform consisting of various nodes. Each node is owned by a user of the platform. They act as data storage and enable the management and monitoring of connected data sources. A NoSQL database is used to store the data for greater flexibility. An SQL database is used to manage the corresponding metadata. To ensure data integrity, hash values are generated for each incoming data record and stored in a distributed ledger. Due to the security against manipulation, these hash values can be used to maintain the integrity of the data. Scripts between the data source and node consume a data stream and forward it to a node via HTTPS. All nodes are linked via a central marketplace, which manages routing information, controls access control and can be used as an authentication instance, for client management and data exchange. The prototype developed enables companies to store data in dedicated rooms and share it with other parties as required. Individual interfaces to the data storage locations were designed to connect the project partners.

The LCA analyses are to be automated via a separate application on the digital platform. This necessitates the establishment of a standardised data format. The process involves the comprehensive integration of data collection across various stages of value creation. It further requires the development of a sophisticated data model that delineates distinct value creation steps. This model should facilitate individualised data mapping for each company, carefully considering data availability. Crucially, it avoids the granularity of breaking down data per machine or company. In the project, an LCA-orientated ontology was developed for this purpose, which acts as a source map for obtaining information for each process step.

4   Conclusions

In conclusion, the challenges surrounding the ecological optimisation of plastic packaging are multifaceted and demand a comprehensive interdisciplinary approach. Current recycling rates remain low in Europe, primarily due to factors such as improper product designs hindering recyclability and a lack of clear definitions for "recyclability." Packaging plastics, particularly, pose challenges due to their complex composition and reliance on fossil fuels. Mechanically recycled plastics face various challenges along the value chain, including material variations, degradation during processing, and the influence of impurities. Standards for assessing recycled materials are insufficient, leading to individual agreements between manufacturers and a lack of comprehensive quality parameters. The presented research shows opportunities for advanced analysis and evaluation of the material itself and the resulting processing data.

Despite the shown challenges, life cycle assessments (LCAs) are crucial for evaluating the sustainability of packaging solutions. LCAs enable comparisons between products made from virgin materials and recyclates, considering factors such as production, usage, and end-of-life scenarios. The evaluation of the demonstrator shows the positive effects of recyclates and recyclability on the sustainability of a product.

To further allow and improve LCAs, enhanced transparency and data exchange is necessary. Therefore initiatives like GAIA-X, which are presented in this paper, need to be understood, used and enforced to enhance transparency and data sovereignty.

Acknowledgements. The investigations set out in this report received financial support from the German Bundesministerium für Bildung und Forschung (No. 02J20E540), to whom we extend our thanks. We would like to thank our consortium partners Brückner Maschinenbau GmbH & Co. KG, Henkel AG & Co. KGaA, Reifenhäuser Blown Film GmbH, Interzero Circular Solutions Germany GmbH, Arburg GmbH + Co. KG, Pöppelmann GmbH & Co. KG Kunststoff-Werkzeugbau, Carbon Minds GmbH and Reifenhäuser GmbH & Co. KG Maschinenfabrik for the excellent cooperation.

References

1. J. Geier, M. Bredács, A. Witschnigg, D. Vollprecht, G. Oreski, "Analysis of different polypropylene waste bales: Evaluation of the source material for PP recycling". Waste Management & Research. 2024; 0 (0)
2. C. T. de Mello Soares, M. Ek, E. Östmark, M. Gällstedt, S. Karlsson, " Recycling of multi-material multilayer plastic packaging: Current trends and future scenarios", Resources, Conservation and Recycling, vol. 176, pp. 1–10, 2022
3. F. Vilaplana, S. Karlsson, "Quality Concepts for the Improved Use of Recycled Polymeric Materials: A Review", Macromolecular Materials and Engineering, vol. 293: pp. 274-297, 2008
4. H. Endres, M. Shamsuyeva, "Warum die Kreislaufwirtschaft bessere Standards braucht", Plastverarbeiter, vol. 71, pp. 46-53, Mar 2020

5. J. Brandrup, M. Bittner, W. Michaeli, G. Menges, "Die Wiederverwertung von Kunststoffen", München Berlin: Carl Hanser Verlag, 1995
6. Z.O.G. Schyns, M.P. ShaverMec, "Mechanical Recycling of Packaging Plastics: A review", Macromolecular Rapid Communications, vol. 42, pp. 1-27, Sep 2021
7. S. Pongratz, "Alterung von Kunststoffen während der Verarbeitung und im Gebrauch, Ph.D. dissertation, Universität Erlangen-Nürnberg, Erlangen, 2000
8. N. Rudolph, R. Kiesel, C. Aumnate, "Einführung Kunststoffrecycling". München: Carl Hanser Verlag, 2020
9. H. Bayazian, V. Schoeppner V, "Investigation of molecular weight distributions during extrusion process of polypropylene by rheometry experiment", in *AIP Conference Proceedings*, 2019
10. S. Canevarolo, A. Babetto, "Effect of Screw Element Type in Degradation of Polypropylene upon Multiple Extrusions", Advances in Polymer Technology, vol. 21, pp. 243-249, Apr 2002
11. W.O. Drake, J.R. Pauquet, R.V. Todesco, H. Zweifel H, "Processing Stabilization of Polyolefins", Advances in Polymer Technology, vol. 176/177, pp. 215–230, Jan 1990
12. G.N. Foster, S.H. Wasserman, D.J. Yacka, "Oxidation Behavior and Stabilization of Metallocene and Other Polyolefins", Die Angewandte Makromolekulare Chemie, vol. 252, pp. 11-32, Jan 1997
13. G. Xuemei, L. Zenan, W., Yingjun, H. Zhangping, W. Mengmeng, J. Gang, "In-Line Monitoring the Degradation of Polypropylene under Multiple Extrusions Based on Raman Spectroscopy", Polymers, vol. 10, pp. 1–11, Oct 2019
14. R. Marsh, A.J. Griffiths, K.P., Williams, S.L. Evans, "Degradation of recycled polyethylene film materials due to contamination encountered in the products' life cycle", Journal of Mechanical Engineering Science, vol. 210, pp. 593–602, May 2006
15. G. Menges, W. Michaeli, M. Bittner, "Recycling von Kunststoffen", München Wien: Carl Hanser Verlag, 1992
16. G. Ehrenstein, S. Pongratz, "Beständigkeit von Kunststoffen", München: Carl Hanser Verlag, 2007
17. L.M. Gorghiu, S. Jipa, T. Zaharescu, R. Setnescu, I. Mihalcea, "The effect of metals on thermal degradation of polyethylenes" Polymer Degradation and Stability, vol. 84, pp. 7-11, Jan 2003
18. European Union, "Circular Economy Action Plan—For a cleaner and more competitive Europe", [Online] 2020. Available: https://environment.ec.europa.eu/strategy/circular-economy-action-plan_en
19. European Union, "Digital product passports: enhancing transparency & consumer information", [Online] May 16 2022. Available: https://www.europarl.europa.eu/committees/de/digital-product-passports-enhancing-tran/product-details/20220510CHE10181
20. J. Guinée, H. Reinout, G. Huppes, A. Zamagni, P. Masoni, R. Buonamici, T. Ekvall, T. Rydberg, "Life cycle Assesment: past, present and future", Environmental Science & Technology, vol. 45, pp. 90-96, Jan 2010
21. Environmental management—Life cycle assessment—Principles and framework, ISO 14040, 2006
22. Environmental management—Life cycle assessment—Requirements and guidelines, ISO 14044, 2006
23. M. Pizzol, P. Christensen, J. Schmidt, M. Thomsen, "Impacts of "metals" on human health: a comparison between nine different methodologies for Life Cycle Impact Assessment (LCIA)", Journal of Cleaner Production, vol. 19, pp. 646–656, May 2011
24. M. Brudermüller, R. Hoffmann R, H. Kagermann, R. Neugebauer, G. Schuh, "Innovationen für einen europäischen Green Deal", acatech—Deutsche Akademie der Technikwissenschaften e.V., München, 2020
25. S. Ransbotham, P. Gerbert, M. Reeves, D. Kiron, M. Spira, "Artificial intelligence in business gets real", MIT Sloan Management Review and The Boston Consulting Group, 2018

26. International Data Spaces Association: Reference Architecture Model—Version 3.00; 2019 https://internationaldataspaces.org/wp-content/uploads/IDS-Reference-Architecture-Model-3.0-2019.pdf

27. A. Tiwana, B. Konsynski, A.A. Bush, "Research Commentary—Platform Evolution: Coevolution of Platform Architecture", Governance, and Environmental Dynamics. Information Systems Research, vol. 21, pp. 675-687, Nov 2010

28. N.N., "Catena-X—Your Automotive Network", [Online] March 7 2024, Available: https://catena-x.net/de/

29. E.G. Thoden van Velzen, S. Chu, F. Alvarado Chacon, M.T. Brouwer, K. Molenveld, "The impact of impurities on the mechanical properties of recycled polyethylene. Packaging Technology and Science, 34, S. 219–228, 2021

30. N.N., "Proposal for a Regulation of the European Parliament and of the council on packaging and packaging waste, amending Regulation (EU) 2019/1020 and Directive (EU) 2019/904, and repealing Directive 94/62/E", European Parliament, Brussels, 30.11.2022

31. N. Caye, S. Marasus, K. Schuler, "Aufkommen und Verwertung von Verpackungsabfällen in Deutschland im Jahr 2021", GVM Gesellschaft für Verpackungsmarktforschung mbH, Mainz, Germany, 2023

32. N.N., "2023 Flexible films market in Europe—State of play", Plastics Recyclers Europe, Brussels, Belgium, Tech. Report, 2023

33. E. Berg, L. Leuchtenberger, R. Dahlmann, "Influence of varying material components in representative polyethylene recyclates on the mechanical properties of blown films", in *32st International Colloquium Plastics Technology*, vol. 32, (in press), Feb 2024

34. R. Dahlmann, E. Berg, M. Schön, "Usage potentials for different qualities of recyclates in plastics packaging", in *31st International Colloquium Plastics Technology*, vol. 31, pp. 190–204, Sep 2022

35. E. Berg, "Recyclingfähige Polyolefin-Compounds für flexible Verpackungen", in Symposium Kunststoffkreisläufe schließen—Neue Ansätze für Verpackungen, vol. 1, pp. 675–687, Aachen, Germany, 2022

36. E. Berg, R. Dahlmann, "Investigation of gel formation and degradation of representative polyolefin recyclates", ANTEC, St. Louis, March 4–7, 2024

37. G. Grünert, C. Lürken, S. Barth, "Ein Nachweis, der bald von jedem verlangt wird—Ökologische Lebenszyklusbetrachtung von Kunststofferzeugnissen", Kunststoffe, vol. 113, pp. 90-93, Mar 2023

38. A. Ciroth: "ICT for environment in life cycle applications openLCA — A new open source software for life cycle assessment". In: *The International Journal of Life Cycle Assessment*. 12. Jg., 2007, 4, S. 209–210

39. M. A. J. Huijbregts, Z. J. N. Steinmann, P.M.F. Elshout, G. Stam, F. Verones, M. Vieira, M. Zijp, A. Hollander, R. van Zelm, "ReCiPe2016: a harmonised life cycle impact assessment method at midpoint and endpoint level". In: *The International Journal of Life Cycle Assessment*. 22. Jg., 2017, 2, S. 138–147

40. G. Wernet, C. Bauer, B. Steubing, J. Reinhard, E. Moreno-Ruiz, B. Weidema, "The ecoinvent database version 3 (part I): overview and methodology". In: *The International Journal of Life Cycle Assessment*. 21. Jg., 2016, 9, S. 1218–1230

41. J. Pennekamp, M. Henze, S. Schmidt, P. Niemietz, M. Fey, D. Trauth, T. Bergs, C. Brecher, K. Wehrle, "Dataflow Challenges in an Internet of Production: A Security & Privacy Perspective", in *ACM Workshop on Cyber-Physical Systems Security & Privacy*, London, United Kingdom, 2019

42. F. Betti, F. Bezamat, M. Fendri, B. Fernandez, D. Küpper, A. Okur, "Share to gain: Unlocking Data Value in Manufacturing". White Paper, World Economic Forum in collaboration with Boston Consulting Group, Jan 2020

43. D. Trauth, T. Bergs, C. Gülpen, W. Maaß, J. Mayer, H. Musa, P. Niemietz, A. Rohnfelder, M. Schaltegger, S. Seutter, J, Starke, E. Szych, M. Unterberg, "Internet of Production Turning data into value—Monetizing manufacturing data", In: *Internet of Production—Turning Data into Value*, Status Reports from Production Technology, Stuttgart, 2020

A Systematic Approach for the Reliable and Automated Selection of Life Cycle Assessment Data Sets Exemplified by the Automotive Industry

Johannes H. L. Sturm[1]([✉]), Sebastian Gehrke[1], and Christoph Herrmann[2]

[1] Volkswagen AG, Berliner Ring 2, 38440 Wolfsburg, Germany
{johannes.sturm,sebastian.gehrke}@volkswagen.de
[2] Institute of Machine Tools and Production Technology (IWF), Sustainable Manufacturing and
Life Cycle Engineering, Technische Universität Braunschweig, 38106 Braunschweig, Germany
c.herrmann@tu-braunschweig.de

Abstract. For the automotive industry, national and EU-wide regulations have been announced for the coming years, which require evidence from the manufacturer on the carbon footprint of a product or sub-product. Due to the resulting urgency, the state of the art of life cycle assessments (LCA), the "measuring instrument" for such characteristic values, is subject to accelerated further development. The decisive core element of an LCA is the transparent and consistent life cycle modelling of materials and manufacturing processes of a product, which is a major challenge, especially for complex product systems, such as vehicles. This challenge arises from the variety of different product information flows, e.g. material and structural information flows, which must be completed, homogenized, and validated. In order to conduct LCAs for such systems in appropriate time, the outlined information flows are usually mapped to secondary data sets. Their selection is generally made with extensive dictionaries, which are often simply structured, have been evolved over a longer period by different editors, and are only valid for explicit individual assignments. In the case of new materials, modified identifiers and standards, or subjectively made decisions, incorrect LCA mappings can occur. Time-consuming rework or manual corrections are then necessary. Therefore, a generally applicable, transparent, and consistent as well as extendable assignment methodology is presented and applied to the process of automotive LCAs. The central approach of this methodology is to secure data assignments by means of different prioritized information sources. Among other things, data on standards, similarity to known assignments as well as textual and material information are used. Assignments are made cascaded so that a lack of information does not lead to the termination of the modelling process and an assignment is returned for each valid input. The developed assignment mechanism also offers various options for fine-tuning the system, such as the creation of exception rules, the implementation of function-based rules, or the optional execution of supervised learning in the integration of new materials.

Keywords: Life cycle assessment · Automation · Standard-based data mapping

© The Author(s) 2025
K. Dröder and T. Vietor (Eds.): CD 2024, Proceedings, pp. 393–406, 2025.
https://doi.org/10.1007/978-3-658-45889-8_30

1   Introduction

In view of new national and EU-wide legislation and subsidy strategies to reduce *green-house gas* (GHG) emissions, the automotive industry is facing major challenges. As the electrification of the automotive sector progresses, legislators are increasingly focusing on the production stage of a vehicle and individual main components, such as the battery, which is now becoming increasingly dominant in terms of GHG emissions. This includes, in particular, the recently announced funding strategy for electric vehicles in France for the year 2024 and the recently adopted EU Regulation 2023/1542 on batteries and waste batteries, which will come into force in 2025 and requires, among other things, proof of their carbon footprint [1, 2]. Thereby, it can be assumed that the number of similar requirements from both legislators and customers will increase in the future.

The ability of vehicle manufacturers to prepare *life cycle assessments* (LCA) for the entire product portfolio within a reasonable period of time is therefore an essential prerequisite for successfully meeting the above-mentioned requirements. Due to the large number of individual subsystems, the large amount of data to be taken into account and its completeness and completion, special importance is also given to the modeling of the production stage when preparing an LCA [3]. In addition to modularization, parameterization, and the simplification of the data space required for the calculations, automation methods in particular are being discussed to overcome the aforementioned challenges [4]. The core issue for many automation solutions is the reliable and automated compilation of the so-called *life cycle inventory* (LCI), i.e. the detailed listing of incoming and outgoing flows of the product system [4, 5]. In this context, many automation solutions rely on the linking of LCA software tools and the company's own product information platforms [4]. The aim of this linkage is to model the product information relevant to the LCA using the data sets provided in the LCA software or LCA database. As the granularity of the product data exceeds that of the LCA databases, the product data, also referred to as foreground data, must be projected into the master data of the available LCA data sets, also referred to as background data. This projection process is usually realized via reference works, see [6] or [7]. Whereby, the quality of the projection from foreground to background data is decisive for the quality of the final LCA, see also [3].

Derived therefrom, a key factor for successfully fulfilling the legal requirements and customer requests regarding the carbon footprint is in the transparent, reproducible, and technically correct linkage of foreground and background data. Currently often used reference works show weaknesses in the projection process. In particular, the management of equivalence lists, the integration into an increasingly dynamic environment, and the traceability of assignments have some drawbacks. To overcome these limitations, this paper aims to analyze the state of the art and further approaches to develop a new concept adapted to the automotive industry.

The paper is structured as follows: Sect. 2 presents the results of a literature review reflecting the state of the art and discussed further developments. Based on the findings from Sect. 2, Sect. 3 presents a new concept for the projection of foreground data. Sect. 4 discusses the resulting concept, reveals limitations, and provides an outlook. Sect. 5 presents the conclusion.

2   Methods and Materials

The following Sect. 2.1 provides an overview of automated LCA in the automotive industry. Building on this, advanced studies will be presented and analyzed in Sect. 2.2. The results of the following sub-sections are based on a systematic literature review in which over 2,300 sources from three different databases (IEEE, Scopus, Web of Science) were examined. The core question of the review focused on the integration of LCA into a dynamic product development process. As a result of this review, it should be noted that only a few publications can be specifically associated to the automotive sector. Other summarizing publications such as Haun *et al.* [8] reinforce this impression, as they present a similar picture of the source materials. Haun *et al.* [8] have, for example, also identified the work of Koffler *et al.* [6], Yu and Kim [7] and Oliveira *et al.* [3] as relevant sources for the state of the art [8]. Therefore, for the state of the art (Sect. 2.1) and for advanced approaches (Sect. 2.2), the following sources were used primarily, see Table 1:

Table 1. : Primarily assessed literature of Sect. 2

Topic (Section)	Source	Summary
State of the art (Sect. 2.1)	Koffler *et al.* [6]	Well-known approach (cited in [3, 4, 8]) for the semi-automated creation of LCA; reference work-based projection of foreground data into background data master; Volkswagen
	Yu and Kim [7]	Approach for semi-automated creation of LCA; reference work-based projection of foreground data into background data master; GM Korea Company
	Oliveira *et al.* [3]	Study on the influence of the level of abstraction of the foreground data on the accuracy of the LCA results and workload
Advanced studies (Sect. 2.2)	Meron *et al.* [9]	Method for selecting a proxy data set with different regionalization, in the case of a missing region-specific background data set; classification-based
	Zhao *et al.* [10]	Method for completing information flows of unit processes in LCI modeling using classification and regression
	Kuczenski *et al.* [11]	Presentation of various prototypes resulting from a software development project; automated creation of an LCA product model from a parts list

2.1   Automated LCA in the Automotive Industry

As already presented in Table 1, the sources from Koffler *et al.* [6], Yu and Kim [7] and Oliveira *et al.* [3] are particularly relevant for the state of the art of automated LCA in the automotive industry. Analyzing these methods, it can be summarized that they follow the procedure outlined in Fig. 1:

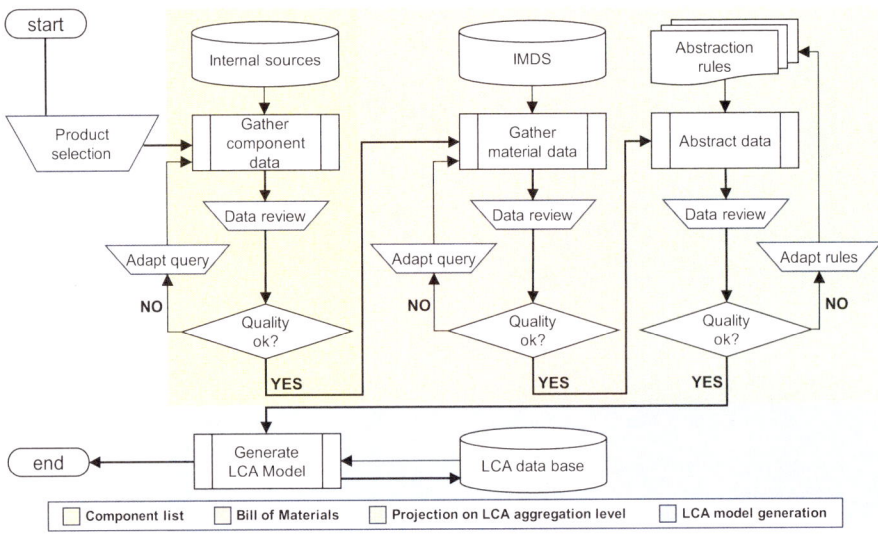

Fig. 1. : Simplified flow chart for an automated generated LCI in the automotive industry (derived from [6, 7])

Figure 1 shows how an LCI model of a vehicle is typically generated in the production stage, starting with the selection of a specific product via the consolidation of component lists and data of the *International Material Data System* (IMDS), and the application of abstraction rules. Besides the consolidation of foreground information, i.e. vehicle structure (components) and material information (IMDS), the projection process of the primary data to the aggregation level of the LCA database (background data) represents a major module of the concept in Fig. 1. Whereas in the preceding modules primarily different vehicle-specific information is merged, for the first time the projection process represents a specific change to the input data with a correspondingly significant influence on the LCA. With regard to Yu and Kim [7] and Koffler *et al.* [6], this step is mainly defined by one or more control files storing abstraction rules in tabular form [6, 7]. The quality of these files is the responsibility of the expert team in charge [6], see also the similar work by Oliveira *et al.* [3] or the summary by Haun *et al.* [8].

Assuming a quasi-static environment, e.g. constant parameter naming, this approach generates a manageable amount of manual work. Especially if, as suggested by Koffler *et al.* [6], the aforementioned abstraction rules respectively the look-up tables are only created for the first vehicle project to be analyzed and expanded accordingly in the course of subsequent projects [6]. However, if the boundary conditions change (identifiers, calculation assumptions, etc.), the maintenance of these look-up tables can represent a bottleneck for the (partially) automated LCA creation process. In practice, the following aspects have proven to be a hindrance to the process when using a similar approach:

- **Modeling of similar materials and processes:** If the look-up tables are maintained over a longer period or by several people in charge, it is not guaranteed that the most representative data set is selected for each material. Different editors or newly added data records in the LCA database can lead to inconsistencies between related materials and process descriptions.
- **Changing assignments at material master level:** As there is a dedicated assignment of an LCA data record for each material, it is complicated to change the assignment for an entire material data cluster. Possible semantic relationships between different materials and their attached material information can only be mapped to a limited extent using a look-up table.
- **Orientation guidance for new materials:** In particular, changing designations using freely selectable designators, such as the trade name, the standard name, or own internal names, can result in an increased research workload.
- **Integration of multi-level dynamic fallback options for missing input data:** If information is missing for the adequate projection of a material, usually only a few fallback options, such as the material grouping of the *German Association of the Automotive Industry* (VDA), are used, see [3, 6, 7]. A situation- and material-dependent, arbitrary fallback strategy for individual material categories can generally only be achieved with a high workload when using the existing approach with control files.

The aspects mentioned above clearly show that a field of research is emerging in this regard. In the following, these aspects will serve as points of reference for further investigations. First, possible known solutions should be discussed which address the set of questions specified here.

2.2 Known Approaches to Improve Projection Quality

Many of the advanced approaches discussed in the literature deal with the completion of the LCI with the use of computer-aided methods, such as machine learning or other complex statistical models, and attempt to categorize or regress the data on the basis of specific accompanying parameters [9, 10, 12, 13]. The exemplary work of Meron *et al.* [9] or Zhao *et al.* [10] should be mentioned in this context, which show how specific data gaps can be closed with the help of classification and regression. Meron *et al.* [9], for example, present a methodology for the selection of proxy data sets to describe a region-specific water supply system using classification algorithms [9]. Zhao *et al.* on the other hand show how data gaps of unit processes in the LCI modelling can be closed using regression [10].

For very large, heterogeneous, and complex product systems, however, it cannot be guaranteed that the boundary conditions, i.e. the availability of all accompanying information, necessary for such approaches can be established for each component. This fact can be underpinned by the initial filling levels of the IMDS by Koffler *et al.* [6]. Despite the promising study results of scientist like Meron *et al.* [9] and Zhao *et al.* [10], an automated LCA in the automotive sector would therefore often be preceded by the expert-dependent selection of a sufficiently representative entry data set. Other more general approaches with a focus on the basics of process modeling, i.e. the assignment of foreground and background data, must first be identified, see also [11].

With the help of the review conducted by Köck *et al.* [5] suitable relevant publications could be found. In this context, particular attention should be paid to the work by Kuczenski *et al.* [11]. Kuczenski *et al.* [11] present the results of an open-source software development experiment in which three developers were independently asked to automatically generate a so-called *product system model* (PSM), an LCA model of a product, based on a parts list [11]. In accordance with the aim of automated model creation, four decisive steps were to be addressed in the course of the project [11]:

1. Import and interpretation of the parts list
2. Enriching the parts list with essential data such as material composition and weight
3. Linking the product information (foreground) with the information from a selected LCA database (background)
4. Transfer of the model into an interoperable format

The focus of the three prototypes *pslink*, *antelope* and *perdu* was set differently in dealing with this task, so that the complexity of the approaches varies from prototype to prototype, especially in the third aspect that is decisive for the presented work [11]. Table 2 summarizes the main differences and characteristics of the data projection approaches based on [11].

Table 2. : Data linking of the prototypes

Prototype	Linking of foreground and background data
pslink	1. Manual creation of a semantic network for materials (relationships in the network (score descending): equivalent, broader, narrower, derived) 2. Manual listing of all available background data records 3. Linking the background data set list with the semantic network using the similarity of the network node names and the process names in the background list (binding), setting a similarity score 4. Linking a material with an input node for the semantic network, setting a similarity score 5. Creation of an absolute score value for all nodes with a binding starting from the respective input node (multiplication of the individual scores along the path) 6. Selection of the background data set with the highest score in each case
antelope	1. Specific linking rules between foreground and background data
perdu	1. Classification of processes (background data) using a user-selectable product classification database (Global Product Classification, North American Industry Classification, etc.) 2. (Manual) classification of product information at part level (foreground data) 3. Picking the background data set based on the set group

With regard to the limitations of approaches analogous to those of Koffler *et al.* [6] or Yu and Kim [7], the prototypes presented by Kuczenski *et al.* [11] outline initial approaches for addressing the current limitations of automated model creation with a focus on the selection of data sets. The *pslink* and *perdu* prototypes deserve special mention here, as these two systems have provided deeper insights into the respective new type of linking process used. Prototype *perdu*, for example, responds directly to the question of an orientation aid for novel products by using a semantic network for materials. The *pslink* system, on the other hand, uses lexical analysis to show a way of consistently mapping related and similar materials. However, in view of the legal requirements in the automotive industry, there are still some limitations to the operational use of these systems that still need to be addressed:

- **Handling variable naming of materials background data records:** Background data sets are selected by the prototypes based on their name. Particularly in the case of an inhomogeneous data master, with changing identifiers from different sources, it cannot be guaranteed that the most suitable data set can always be found using the identifier. Data records with generic or short names can sometimes be given less consideration than data records with more descriptive identifiers. Furthermore, when using the *perdu* approach, an already set classification does not seem to be considered further, as the model is built on the basis of individual flows [11]. This can result in different assignments for a process master. In addition, a fallback option does not appear to be provided for missing assignments in *perdu*, see [11].
- **Handling the use of variable material naming:** Like the selection of a background data set, inconsistent assignments of background data sets and foreground data can occur because of an inhomogeneous master of material identifiers for a material group. For example, the trade name of a material does not have to correspond to the standard identifier of the same material, although both material entries should ideally be modeled in the same way.
- **Modeling of complex subjects:** All approaches shown by Kuczenski *et al.* [11] focus on the assignment of one data set to describe one attribute. However, analogous to Koffler *et al.* [6], if different processing methods for the same material are to be considered, it should also be possible to model simple process chains. The approaches presented by Kuczenski *et al.* [11], though, do not address this aspect.
- **Audit-compliant archiving of assignments:** To ensure the reproducibility of the results of a model, changes in the assignment mechanism may only be actively triggered. If the database is changed (added processes), e.g., it must still be possible to reproduce a result that has already been created. It must therefore be possible to record the corresponding data statuses and settings.

In summary, the findings from Sect. 2 indicate a comprehensive need for research regarding the projection of foreground data in the context of more restrictive requirements and an increasing need for LCA product evaluations.

3   Model Development

Based on the limitations from Sects. 2.1 and 2.2, a *projection interface* was conceptualized and newly developed as a bridge module between the foreground and background data for the framework in Fig. 2.

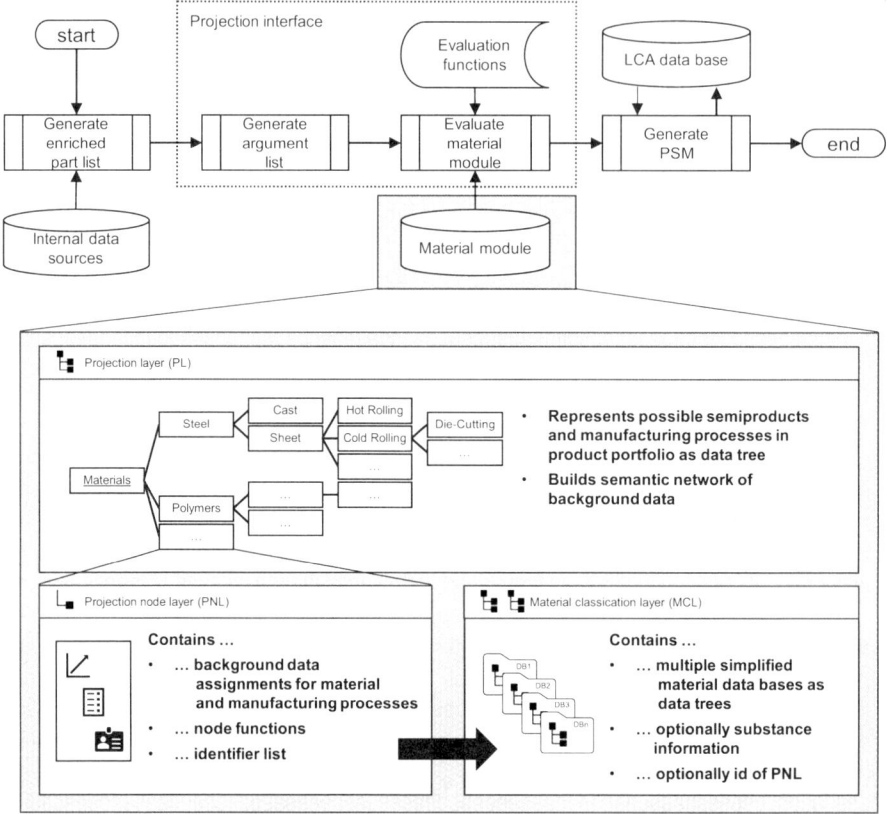

Fig. 2. : Framework for a new approach

The provision of a part list enriched with material information and corresponding interfaces for PSM creation are considered as given. The core element of the developed approach is the so-called *material module*, which provides basic functions for linking foreground and background data and is made up of two interlinked core components: the *projection layer* (PL) and the *material classification layer* (MCL). These determine how a relevant characteristic in the enriched parts list or *bill of materials* (BOM), i.e. an object to be assessed, is linked to the background data.

3.1 Projection Layer

The key component of the material module is the so-called *projection layer*, an ordering tree that primarily contains all expected process descriptions of the product spectrum from an LCA perspective and creates a semantic framework for background data similar to *pslink*. Advantageously compared to *pslink*, the background data is not added directly through a lexical analysis but is assigned to the individual nodes independently of the name. One material data set and any number of processing data sets can be assigned to each node of the PL, so that each relevant characteristic can be described via a simple modular process container (material → processing 1 → processing N), analogous to Koffler *et al.* [6]. In addition, data set assignments are hereditary, so that all children receive the assignment of the last parent node in the case of missing their own setting. Each setting is inherited individually so that, for example, deviating settings of a single processing method at child node level are possible without affecting the other inherited assignments, see also Fig. 3. The PL thus also enables different processing methods for the same input material or semi-finished input product. In addition to the links to the background data, evaluable functions and term lists can also be added to each node of the PL and, together with the background data links and a distinct node identifier (ID), form the *projection node layer* (PNL).

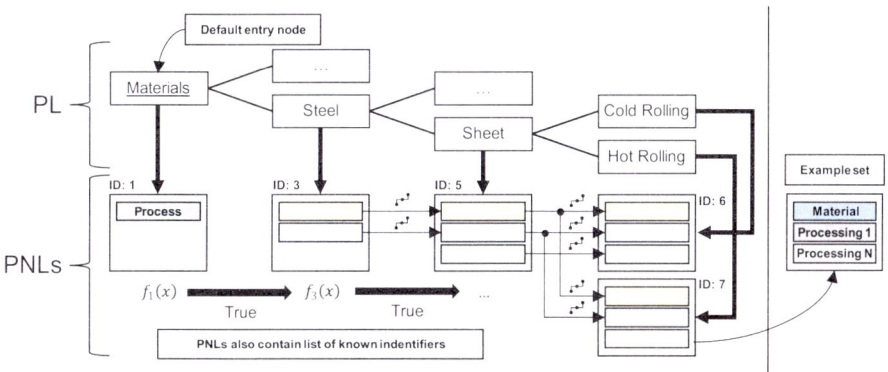

Fig. 3: Inheritance logic of the projection layer

The described PL is the basis of the mapping mechanism and can already be used for initial linking of foreground and background data independently of the MCL, as shown in Fig. 3. For this purpose, the PL must be evaluated using an associated recursive evaluation function and two additional arguments: an entry node for the PL and an argument list formed from the BOM for evaluating the functions of the PNL. Starting from the set entry node, the evaluation function then evaluates the function stored in the respective PNL with the argument list. If the entry node is not specified, the root node of the PL is used by default. If the evaluated node function returns the logical value "True", the evaluation function for the children of the active node is called recursively until the value "False" is returned for the first time. The last node of a branch whose node function has returned the value "True" or is equal to the entry node is then adopted as the setting for the respective characteristic value of the examined BOM. In contrast to the other approaches discussed earlier, the node function can be designed to be as complex as desired, if its return value can be mapped as a logical value. For example, string similarities, RegEx commands, or statistical functions can be used. All attributes of the BOM can also be used as arguments for the node functions if the layout is designed accordingly. Individual nodes of the PL are prioritized according to their position in the tree to ensure unambiguity. For example, if the node functions of all child nodes of a parent node return "True", the first child of this node is weighted higher than its sibling nodes.

Alongside the evaluation of the node functions, assignments can also be made using the term lists of the PNL. These term lists act as exception rules to skip the evaluation of the node functions partially. In the currently implemented prototype, for example, material identifiers can be stored for each node of the PL. If one of these stored material identifiers occurs in a characteristic of the BOM, the corresponding node of the PL is set as the entry node for the previously mentioned evaluation function. The evaluation of the parent nodes can then be skipped, for instance.

3.2   Material Classification Layer

To be able to identify material groups in the described concept whose naming does not follow any general lexical rules, the *material classification layer* can be connected upstream of the PL, see Fig. 4. The MCL enables existing classification systems to be considered so that an exception rule does not have to be enforced for every special lexical case of a material. The MCL represents a collection of different official and internal material databases that can be expanded as required and are interpreted as classification trees similar to the PL. Each category of a material database corresponds to a node of a classification tree with a distinct ID. Furthermore, the materials stored in the database are attached to each node with associated information, such as identifiers, synonyms, or substance information. Each node of the MCL can also be given an ID of the PL, a so-called foreign key, in order to create a direct link to the PL. The material information, the node ID, and the foreign key then form a *material classification node layer* (MCNL) in line with the PNL. All child nodes of an MCNL inherit the same PNL node ID, so that when the decisive parent node is selected, the entire branch of the MCL is automatically linked to the selected PNL, see also Fig. 4. When using the MCL, the argument list mentioned at the beginning is therefore used in the first step of the overall evaluation of the material module to assess the MCL with an associated further evaluation function. For example, a material identifier in the BOM can be checked for identical names with an entry in the MCL. The MCL then returns the ID of the matching node and consequently a node of the PL. This node then serves as the entry node for the evaluation of the PL, as shown in Fig. 4. Any exception rules of the PL remain unaffected by this.

Materials that cannot be classified directly are checked for similarity to the existing classifications (Sørensen-Dice coefficient, Jaro-Winkler distance, etc.), similar to the procedure in *perdu*, and are presented to the user for confirmation. Furthermore, if substance information is available, the developed concept also allows a classification based on this accompanying information alongside the lexical comparison. The master of the known settings of a material database tree can be sustainably expanded once the classification has been completed. If no user input is made for an unknown material, the MCL does not return an ID, so that the routines set in the PL are used, starting from the root node of the PL.

In summary, the presentation of the interlinking of the individual elements, such as PL, PNL, MCL and MCNL, fully describes the concept required in Fig. 2. Based on this, a prototype was realized in the programming language R with the help of the packages {shiny}, {shinyTree}, {shinyAce} and custom-written libraries, so that a practical evaluation was possible.

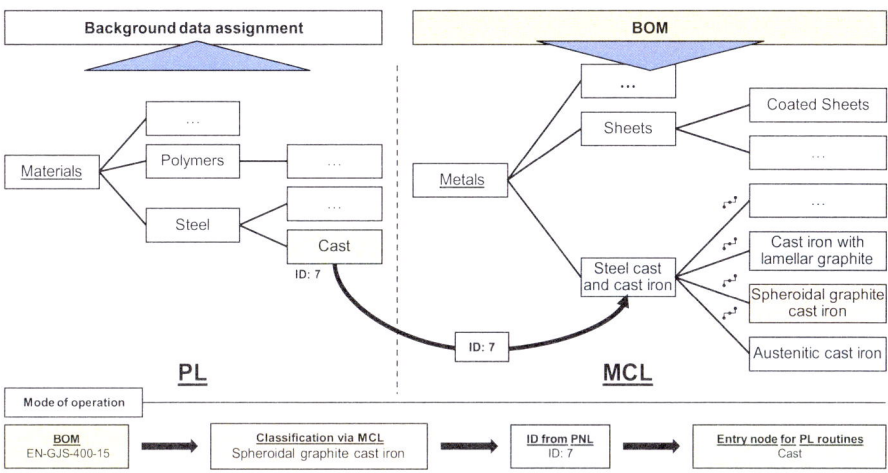

Fig. 4: Integration of the material classification layer

4 Discussion and Outlook

The limitations of the known approaches discussed in Sects. 2.1 and 2.2 directly set requirements for the developed concept, so that the resulting prototype can also be discussed and evaluated on the basis of these specifications, see Table 3.

Aspects not yet addressed, but to be considered in the further development of the concept as well as of the prototype, are the processing of scenarios with different region-alization, optimized materials, and components with targeted values for certain impact categories, and corresponding data sets. In addition, the direct linking of LCA results, such as carbon footprints of individual materials or even entire components, is also conceivable within the framework of the concept presented.

Table 3. : Addressing of fulfillments of the presented approach

Requirement	Addressing
Modeling of similar materials and processes	Related materials can be categorized by name using the MCL and transparently linked to the PL. Various manufacturing processes and further classification using the PL are also possible when using the MCL. Unknown materials are provided to the user with a corresponding suggestion for classification in the MCL. Various material databases, even describing the same material master, can be stored in the MCL for adequate classification
Changing assignments at material master level	If the PL is ideally filled, background data is only set for the decisive nodes so that any child nodes inherit their assignment. Changing the assignment of a central node therefore has a direct influence on all child nodes, so that settings can also be made for the entire root of a material, such as changing a manufacturing process
Orientation guidance for new materials	When using the MCL, materials can be classified by technical aspects. In addition, the MCL is designed as a growing database so that existing classifications with systemic assistance provide orientation for new materials
Integration of multi-level dynamic fallback options for missing input data	The layers stored in the material module serve as dynamic fallback options for each other. As the PL has a hierarchical structure, at least the root node of the PL and therefore a background data set is returned for each valid entry of a characteristic of a BOM
Modeling of complex subjects	The PL is designed so that any number of consecutive manufacturing data records can be added in addition to a material background data record
Audit-compliant archiving of assignments	As all background data to be set must be set once when the PL is created, a changed background database status does not lead to a changed assignment of background processes. The setting can only be actively adjusted when the PL is updated

5 Conclusion

Based on the underlying challenge of linking foreground and background data in a clear, flexible, and transparent way, a new concept was developed, implemented, and tested using a prototype. The core element of the concept is the so-called material module. This contains the so-called projection layer and an additional material classification layer. In combination, these allow both multi-layered rule-based evaluations of material information and the processing of product data using external material classification systems when linking data. The main advantages of the shown concept compared to existing solutions are the multi-level, unambiguous and seamless fallback strategy, the linking of background data with the aid of semantic correlations from different sources, and the demand-oriented modeling of relatively complex product subsystems.

References

1. *REGULATION (EU) 2023/1542 OF THE EUROPEAN PARLIAMENT AND OF THE COUN-CIL of 12 July 2023 concerning batteries and waste batteries, amending Directive 2008/98/EC and Regulation (EU) 2019/1020 and repealing Directive 2006/66/EC*, 2023.
2. a. R. Julien Bonnet, "Bonus écologique 2024: les constructeurs ont jusqu'au 15 décembre pour déposer leur dossier," *BFM Auto*, 10 Oct., 2023. https://www.bfmtv.com/auto/bonus-eco logique-2024-les-constructeurs-ont-jusqu-au-15-decembre-pour-deposer-leur-dossier_AV-202310100632.html (accessed: Nov. 25 2023).
3. F. B. de Oliveira, A. Nordelöf, B. A. Sandén, A. Widerberg, and A.-M. Tillman, "Exploring automotive supplier data in life cycle assessment – Precision versus workload," *Transportation Research Part D: Transport and Environment*, vol. 105, p. 103247, 2022, https://doi.org/10.1016/j.trd.2022.103247.
4. S. Kiemel, C. Rietdorf, M. Schutzbach, and R. Miehe, "How to Simplify Life Cycle Assessment for Industrial Applications—A Comprehensive Review," *Sustainability*, vol. 14, no. 23, p. 15704, 2022, https://doi.org/10.3390/su142315704.
5. B. Köck, A. Friedl, S. Serna Loaiza, W. Wukovits, and B. Mihalyi-Schneider, "Automation of Life Cycle Assessment—A Critical Review of Developments in the Field of Life Cycle Inventory Analysis," *Sustainability*, vol. 15, no. 6, p. 5531, 2023, https://doi.org/10.3390/su1 5065531.
6. C. Koffler, S. Krinke, L. Schebek, and J. Buchgeister, "Volkswagen slimLCI: a procedure for streamlined inventory modelling within life cycle assessment of vehicles," *IJVD*, vol. 46, no. 2, p. 172, 2008, https://doi.org/10.1504/IJVD.2008.017181.
7. M. Yu and Y. Kim, "Development of Environmental Assessment System of Vehicle," in *Lecture Notes in Electrical Engineering, Proceedings of the FISITA 2012 World Automotive Congress*, Berlin, Heidelberg: Springer Berlin Heidelberg, 2013, pp. 1151–1160.
8. P. Haun, P. Müller, and M. Traverso, "Improving automated Life Cycle Assessment with Life Cycle Inventory model constructs," *Journal of Cleaner Production*, vol. 370, p. 133452, 2022, https://doi.org/10.1016/j.jclepro.2022.133452.
9. N. Meron, V. Blass, and G. Thoma, "Selection of the most appropriate life cycle inventory dataset: new selection proxy methodology and case study application," *Int J Life Cycle Assess*, vol. 25, no. 4, pp. 771–783, 2020, https://doi.org/10.1007/s11367-019-01721-8.
10. B. Zhao, C. Shuai, P. Hou, S. Qu, and M. Xu, "Estimation of Unit Process Data for Life Cycle Assessment Using a Decision Tree-Based Approach," *Environmental science & technology*, vol. 55, no. 12, pp. 8439–8446, 2021, https://doi.org/10.1021/acs.est.0c07484.

11. B. Kuczenski, C. Mutel, M. Srocka, K. Scanlon, and W. Ingwersen, "Prototypes for automating product system model assembly," *The international journal of life cycle assessment*, vol. 26, no. 3, pp. 483–496, 2021, https://doi.org/10.1007/s11367-021-01870-9.

12. S. Zargar, Y. Yao, and Q. Tu, "A review of inventory modeling methods for missing data in life cycle assessment," *J of Industrial Ecology*, vol. 26, no. 5, pp. 1676–1689, 2022, https://doi.org/10.1111/jiec.13305.

13. M. Akhshik, A. Bilton, J. Tjong, C. V. Singh, O. Faruk, and M. Sain, "Prediction of greenhouse gas emissions reductions via machine learning algorithms: Toward an artificial intelligence-based life cycle assessment for automotive lightweighting," *Sustainable Materials and Technologies*, vol. 31, e00370, 2022, https://doi.org/10.1016/j.susmat.2021.e00370.

Life Cycle Assessment of End-Of-Life Scenarios for an Automotive Hybrid Component

Felix Wanielik$^{(\boxtimes)}$, Sebastian Weise, and Christoph Herrmann

Institute of Machine Tools and Production Technology, Technische Universität Braunschweig,
Langer Kamp 19B, 38106 Braunschweig, Deutschland
f.wanielik@tu-braunschweig.de

Abstract. Hybrid lightweight components, e.g., made of glass fibre reinforced plastics and steel, have the potential to reduce the environmental impacts during the use-phase of automotive vehicles. However, the end-of-life (EoL) of lightweight materials such as glass fibre reinforced plastics still poses a challenge, since material separation is challenging and no consequent recycling system is established, as it is e.g., for steel. Furthermore, the application of secondary materials can reduce the environmental impacts in the raw material stage. These aspects lead to uncertainty about the EoL impacts of such hybrid components. In order to assess the environmental impact of automotive components, the EoL is a relevant life cycle stage and should be included in Life Cycle Assessment (LCA) studies. To do so, it is required to define possible recycling scenarios and applications of the recycled material to enable a calculation of different EoL routes. This approach is demonstrated within this paper, by defining multiple recycling routes of a hybrid lightweight component, depending on the depth of disassembly and calculating the environmental impacts of these different EoL routes through LCA.

Keywords: Hybrid component · Composites · End-of-life · Recycling · Life Cycle Assessment

1 Introduction.

One method to reduce the environmental impacts of automotive components is by reducing the components mass through light weighting in order to decrease the energy consumption in the use stage. This can be done through the material lightweight approach, where a component's material is replaced by another material with improved mass-specific properties [1]. Especially the combination of different materials, e.g., combining steel with glass fibre reinforced plastic (GFRP) has a high lightweight potential whilst keeping valuable properties of the metallic part, such as weldability or the ductile failure mode. These components with a combination of different materials are also referred to as hybrid components. One big disadvantage of hybrid components is related with the end-of-life (EoL) treatment and especially the recycling, since the separation of the different materials is often difficult [2]. In addition, legal requirements such as the revised directive on end-of-life vehicles [3] aim to increase the share of materials

K. Dröder and T. Vietor (Eds.): CD 2024, Proceedings, pp. 407–419, 2025.
https://doi.org/10.1007/978-3-658-45889-8_31

that will be recycled. Furthermore, it is not easy to predict, which EoL treatment the materials will go through (i.e., recycling, incineration or landfill), since this strongly depends on the future material separation and recycling technology, the grade of disassembly during EoL treatment and the materials themselves. The different EoL routes are also connected with different environmental impacts. Even though a high share of recycling without loss of material properties is striven for, it might not always lead to the lowest environmental impacts, e.g. due to high energy demand during recycling [4]. To avoid a problem shifting of environmental impacts from the use stage to the EoL stage it is thus required to analyse the environmental impacts through Life Cycle Assessment (LCA). For such LCAs it is essential, that a viable methodological approach is applied, that includes the EoL-stage as well as environmental benefits from recycled materials. Furthermore, different EoL routes should be analysed to identify the EoL routes that lead to the lowest environmental impacts and to reduce uncertainty connected to future EoL scenarios. Therefore, the environmental impacts of a hybrid automotive lightweight component will be analysed with respect to the above-mentioned methodological challenges. To do this, a viable LCA methodology will be chosen and different EoL routes for the hybrid component will be defined and analysed (Sects. 2 and 3), in order to then recommend an EoL route with the lowest environmental impact (Sect. 4). Finally, a conclusion and an outlook are provided (Sect. 5).

2   Life Cycle Assessment Methodology.

The environmental impacts of the hybrid components and the different EoL routes are determined by conducting an attributional LCA. The LCA is carried out with the conventional steps of goal and scope definition, life cycle inventory analysis, life cycle impact assessment and interpretation of the results.

The intended application of the performed LCA is the quantification of the environmental impacts associated with the EoL processing of an automotive hybrid component. Therefore, the functional unit of this LCA is the EoL processing of a GFRP reinforced steel bumper crossbeam in an automotive application from the front crash management system.

The focus lies on the comparison of different EoL routes at the product life cycle. Germany was chosen as the geographic location for the EoL processing. Further assumptions such as different disassembly depths for the different routes are explained in Sect. 3.2.

Product life cycles with recycling should be assessed in a way that takes into account all inputs and outputs of all connected systems, since this avoids to subjectively choose a method to allocate the environmental benefits of recycling between the life cycle in which the waste is generated and the life cycle in which the recovered material is used [5]. Therefore, the substitution approach according to Geyer et al. was used as methodology for the EoL assessment [4]. Other methods of allocation, such as the circular footprint formula (CFF), can provide false incentives, as e.g., the CFF does not account for the waste disposal avoided through recycling [6].

After research on possible EoL pathways was done, the possible routes were discussed with regards to technical and economic feasibility within the consortium of the project in which this study was carried out. The disassembly depth is one of the main influencing factors with regard to the possible EoL pathways and is discussed in detail in Sect. 3. Within Sect. 3, the detailed life cycle inventory is also provided. Primary data for the component composition and materials were used, as well as literature regarding the EoL pathways and secondary data from the ecoinvent v3.9.1 database [7]. For the Life Cycle Assessment Modelling the python-based software Activity Browser was used [8]. Within the Life Cycle Impact Assessment, the impact categories (IC) acc. to the Product Environmental Footprint 3.1 methodology were used with the corresponding impact indicator and impact assessment methods, excluding toxicity related impact categories.

3   Theory and Calculations.

Within this chapter the theoretical background for the definition of the analysed recycling pathways will be provided and the chosen pathways will be described. Furthermore, an overview about the Life Cycle Inventory (LCI) for these pathways will be provided.

3.1   End-Of-Life Processing of Automotive Components and Recycling of GFRP

The state-of-the-art processes for material separation and sorting of automotive components are a shredding process followed by different process steps for material sorting, such as magnetic and density separators. Typically, only spare parts, comparatively high value components or large plastic components will be detached from the end-of-life Vehicle (ELV) [9]. With upcoming regulations such as the revised ELV directive [3], this might change in the future, leading to a higher disassembly depth in ELV processing. In order to analyse the EoL of automotive components, the disassembly depth of the EoL processes plays a major role, since purer material streams are possible when the input into the shredding process is purer and more predictable [10], thus enabling other recycling pathways for the materials.

Due to the novelty of the application of hybrid steel-FRP components, no recycling pathway is yet established. However, Heibeck et al. have shown, that the material separation of metal and FRP material is possible with a shredding process [10]. Since separation by shredding is already state-of-the-art (SoA) and material sorting technologies enable to form separate material streams of plastics and metal e.g. through magnetic separation, the next step of recycling or treatment is defined by the material streams of the hybrid components, thus steel and GFRP. Since recycling of steel is established, the EoL of steel needs no further investigation, as recycling can be assumed as default pathway.

For GFRP however, multiple EoL routes are possible. These processes can be classified into mechanical, thermal and chemical processes [11]. SoA end-of-life options for GFRP are incineration and the use in the cement industry, both of which can be categorised as thermal processes [11, 12]. During incineration the polymer matrix is used as energy source, whilst the inert glass fibre remains and will be landfilled. When using GFRP for the cement production, the polymer matrix will also be used for energy generation whilst the fibre is used as material for the clinker composition [11]. Multiple recycling processes with a higher degree of material valorisation are technologically possible or under current investigation. Bernatas et al. gave an overview about common recycling processes within their review [11] and also included the use of GFRP in the cement production as one recycling option. These recycling processes result in different uses of the polymer matrix and the glass fibre for GFRP. Within the thermal processes, pyrolysis, fluidised bed technology and microwave pyrolysis all utilse the matrix for energy generation, thus releasing the fibre to be used as reinforcement material. The chemical processes, which are all using the mechanism of solvolysis, separate fibre and matrix to use the fibre for reinforcement, whilst the matrix is turned into monomers or oligomers to be used as organic intermediate material. Within mechanical recycling, GFRP is shredded into small parts and sorted into different sizes resulting in a matrix rich fraction that is used as filler material and a fibre rich fraction that can be used as reinforcement material. From these technologies the pyrolysis, the mechanical recycling and the use in the cement plants are the most mature technologies [11, 12]. Additional pathways for thermoplastics with novel technologies have also been described in [11] with the main aim to keep an optimum of the mechanical properties through different technologies of aligning the reclaimed and often cut fibres. One of the presented technologies also enable the use of fibre and matrix by forming a new thermoplastic GFRP panel through shredding, remelting and forming. Within the review on FRP recycling by Yang et al., the approach of remelting and remoulding thermoplastic FRP has been described as feasible way for recycling, if a reduction of the properties is acceptable [13]. The feasibility of this approach was shown for post-industrial PA66GF35 by Bernasconi et al., who showed that the reduction of the mechanical properties through recycling corelated with the reduction in fibre size and produced and automotive clutch pedal from the recycled material [14]. Similar results were obtained by Kuram et al. with PA6GF30, where post-industrial material was recycled multiple times and the mechanical properties were reduced with each recycling step mainly due to reduction in fibre length [15]. They also concluded that the mechanical properties were still acceptable until the third recycling loop. Otheguy et al. shredded glass fibre reinforced polypropylene (PP) from EoL GFRP laminates (post-consumer) and mixed these with primary PP to produce granules for injection moulding and showed, that properties were achievable, that would be acceptable for non-appearance automotive applications [16]. In conclusion the approach to remelt and remould thermoplastic FRP appears to be a feasible recycling pathway both for post-industrial and post-consumer material if the material streams are pure enough and the reduction of properties is acceptable.

Even though multiple pathways are technologically possible, the economic feasibility is also highly relevant for EoL treatment options. Liu et al. analysed the economic feasibility of recycling pathways for fibre reinforced plastics [17] from wind turbine blades, where relatively predictable and pure material streams can be expected. Their analysis has shown, that due to the relatively low value of glass fibre, only the mechanical recycling and chemical recycling (solvolysis) are economically feasible recycling options. This analysis was done based of thermoset GFRP, thus the results can not directly be transferred to thermoplastic GFRP materials, as the previously described pathway of remelting and remoulding for thermoplastic GFRP has not been included. It is however expected, that this pathway would be even more economically feasible than the mechanical recycling pathway of thermoset GFRP, since the matrix material would also be kept as high value material and not as a filler.

3.2   Definition of Recycling Pathways for Hybrid Components

Based on the described EoL options for ELV components and fibre reinforced plastics, three different EoL pathways have been defined as exemplary scenarios. These scenarios should illustrate differences with regards to the disassembly depth and the valorisation of the GFRP material, yet be considered to be realistic. An overview about the chosen pathways is given in Fig. 1.

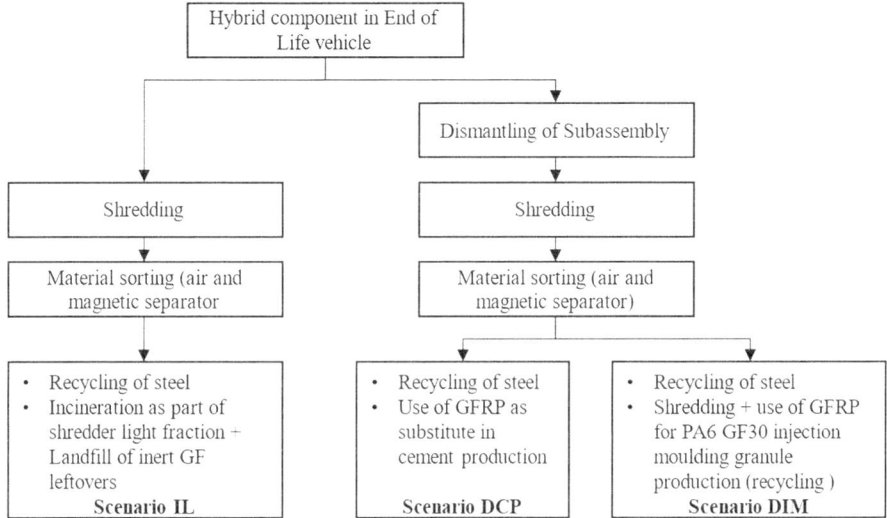

Fig. 1. Analysed end-of-life pathways of the hybrid component

With regards to the disassembly only two options were considered to be realistic:

1) no dismantling of the subassembly, so the bumper crossbeam stays attached to the car body, that will be shredded whole
2) dismantling of the subassembly, so the crash management system (CMS) will be detached from the car body, which includes the bumper crossbeam.

A further dismantling of the bumper crossbeam from the crash management system is unnecessary, as the rest of the CMS is made of metal, so a good material separation is already possible. Also, further detachment of the GFRP organic sheet from the sheet metal part of the bumper cross beam was assumed be unrealistic since the strong bonding between GFRP and steel would require a high effort to separate the materials without crushing the GFRP material, which is assumed to be uneconomic since GFRP is a comparatively low value material. For both options (dismantling or no dismantling of subassembly) the next steps were assumed to shredding and material sorting, since this is the SoA technology and the technology has proven to be viable for hybrid components.

After the material sorting, the steel from the component is assumed to go into the existing recycling stream for all three scenarios. For the scenario without dismantling, it is assumed that no sufficient separation from the other shredder light fraction (SLF) is possible for the GFRP material, so it will be utilised thermally and the inert glass fibre will have to be landfilled (scenario IL). After disassembly two scenarios were considered: use of GFRP as substitute in the cement production (scenario DCP), since this is the SoA for comparatively pure material streams of GFRP; and shredding to smaller size and use for the production of PA6 GF30 injection moulding granule by remelting and remoulding whilst adding primary PA6 to achieve the correct mass ratios (scenario DIM). The scenario DIM is chosen as best-case scenario because of the low grade of material decomposition and under the assumption that the material has similar or only slightly lower mechanical properties compared to primary PA6 GF30 injection moulding granule. This assumption appears reasonable since primary PA6 is added and a fibre laminate with endless GF was shredded, which enables the production of a granule with a similar fibre size as in primary PA6 GF30.

3.3   Life Cycle Inventory

In Table 1 the component properties of the hybrid component are given as well as calorific values that are relevant for the calculation of substituted fuels. The Life Cycle Inventory (LCI) for the LCA of the different EoL scenarios is given in Table 2, where negative values correspond to credits for avoided unit processes. For the DCP and DIM scenarios a share of 12% material loss of GFRP was assumed similar to the mechanical recycling data from [18]. Furthermore, the additional disassembly of the CMS was assumed to be conducted with an electrical impact wrench and the electricity was calculated acc. to an average power for an impact wrench. Only electricity was considered for this disassembly process. The use of GFRP in the cement industry was modelled acc. to [19], however within this study no differences in transport have been considered. Here the mass of hard coal that is avoided was calculated according to the calorific values that are provided in Table 1, however the differences in emission of the incineration for the cement production was not included, as only material credits were given. For the waste incineration process of PA6 credits for electricity and heat were given according to the used ecoinvent dataset [7]. For the production of the injection moulding granule the proxy "extrusion of plastic film" was used.

Table 1. Component properties and relevant material properties for LCA of EoL pathways

Property	Value	Unit
Total component mass	4.000	kg
Steel mass per component	2.800	kg
GFRP mass per component	1.200	kg
Share steel in component	0.700	
Share GFRP in component	0.300	
Share PA6 in GFRP	0.336	
Share glassfibre in GFRP	0.664	
Calorific value hard coal	27.910	MJ / kg
Calorific value PA6	30.400	MJ / kg

Table 2. Life Cycle Inventory of different EoL pathways

Scenario	Process	Value	Unit
All	Shredding and material sorting	1	kg / kg EoL comp
	Steel production EAF route (incl. Sorting + pressing)	1	kg / kg EoL steel
	Steel production: low-alloyed, converter (avoided)	−1	kg / kg EoL steel
IL	PA 6 incinerated	0.336	kg / kg EoL GFRP
	Glassfibre landfilled	0.664	kg / kg EoL GFRP
	Electricity production (avoided)	−0.3248	kWh / kg EoL GFRP
	Heat production (avoided)	−2.3621	MJ / kg EoL GFRP
DCP	Disassembly with impact wrench	0.0093	kWh / kg EoL GFRP
	Material loss GFRP shredding + sorting (SLF)	0.120	kg / kg EoL GFRP
	PA 6 incinerated (SLF)	0.040	kg / kg EoL GFRP
	Glassfibre landfilled (SLF)	0.080	kg / kg EoL GFRP
	Electricity production (avoided)	−0.039	kWh / kg EoL GFRP
	Heat production (avoided)	−0.283	MJ / kg EoL GFRP
	Hard coal (avoided)	−0.322	kg / kg EoL GFRP
	SiO_2 (avoided)	−0.321	kg / kg EoL GFRP
	Al_2O_3 (avoided)	−0.058	kg / kg EoL GFRP
DIM	Disassembly with impact wrench	0.0093	kWh / kg EoL GFRP
	Material loss GFRP shredding + sorting (SLF)	0.120	kg / kg EoL GFRP
	PA 6 incinerated (SLF)	0.040	kg / kg EoL GFRP
	Glassfibre landfilled (SLF)	0.080	kg / kg EoL GFRP
	Electricity production (avoided)	−0.039	kWh / kg EoL GFRP
	Heat production (avoided)	−0.283	MJ / kg EoL GFRP
	Size shredding	0.880	kg / kg EoL GFRP
	Additional PA 6	1.068	kg / kg EoL GFRP
	Extrusion	1.948	kg / kg EoL GFRP
	Substituted PA6 GF30 (avoided)	−1.948	kg / kg EoL GFRP

4   Results and Interpretation.

The relative results of the LCA of the different EoL scenarios are shown in Fig. 2 with the corresponding absolute values in Table 3, where the lowest values mean the highest credits for the EoL treatment through substitution of material and / or energy.

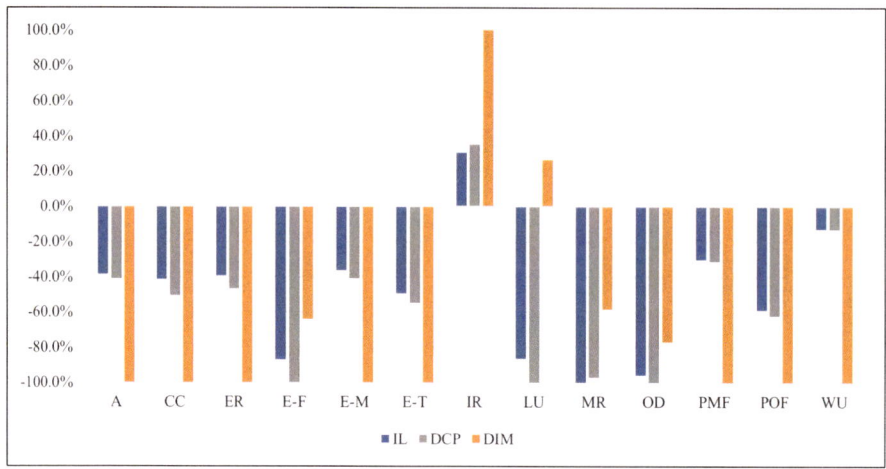

Fig. 2. Relative LCA results of different EoL pathways for multiple impact categories: Acidification (A); climate change (CC); energy resources: non-renewable (ER); eutrophication: freshwater, marine, terrestrial (E-F, E-M, E-T); ionising radiation (IR); land use (LU); material resources: metals/minerals (MR); ozone depletion (OD); particulate matter formation (PMF); photochemical oxidant formation (POF); water use (WU)

All scenarios show a total negative impact in the EoL treatment in all impact categories except in ionising radiation. This stems from the share of nuclear energy in the European electricity mix, which is used within the steel recycling process or for the scenario DIM for the extrusion process. Furthermore, the scenario DIM, in which the EoL GFRP is used to produce injection moulding granule does not have a negative impact value in the impact category land use due to the use of wood-based packaging materials for the extrusion process. Overall, the scenario DIM shows the lowest values in 8 out of 13 impact categories. All higher impact scores in the scenario DIM (E-F, IR, LU, MR, OD) are connected to the production of PA6 or to the extrusion process of the PA6 GF30 granule. In Fig. 3 a heatmap shows the environmental hotspots and subprocesses with the highest credits for the respective scenarios.

Within this figure the efforts and credits for the steel recycling are aggregated since the steel recycling is the same for all processes. Recycling of steel also has the highest credit in most impact categories for the scenarios IL and DCP, thus contributing most to the overall negative impact, whereas the material credits of GFRP gives the highest credits for the scenario DIM. Furthermore, additional efforts of DIM are also environmental hotspots mainly due to the additional primary PA 6 that has to be fed into the process and the electricity for production of new injection moulding granule. However, as can be seen in Fig. 2 they are outweighed by the material credits in many impact categories. For the scenario DCP shredding and sorting has the highest impact in most IC, as relatively low additional effort has to be undertaken (only dismantling) but overall, 4 kg of component have to be shredded and sorted. The overall picture of the heatmap shows a fairly even distribution amongst most impact categories with respect to the subprocesses. The outlier in IR for steel recycling is described above. In the impact category climate change the incineration and landfill subprocess has the highest impact for the scenarios IL and DCP because of the incineration of the polymer.

Table 3. Absolute LCA results of end-of-life scenarios

Scenario	A [mol H + -Eq.]	CC [kg CO2-Eq.]	ER [MJ]	E-F [kg P-Eq.]	E-M [kg N-Eq.]	E-T [mol N-Eq.]	IR [kBq U235-Eq.]
IL	−1.8E-02	−3.6E + 00	−4.2E + 01	−2.1E-03	−3.9E-03	−4.3E-02	1.8E-01
DCP	−1.9E-02	−4.4E + 00	−5.0E + 01	−2.4E-03	−4.4E-03	−4.7E-02	2.0E-01
DIM	−4.6E-02	−8.6E + 00	−1.1E + 02	−1.6E-03	−1.1E-02	−8.6E-02	5.8E-01

Scenario	LU dimension-less	MR [kg Sb-Eq.]	OD [kg CFC-11-Eq.]	PMF [disease incid.]	POF [kg NMVOC-Eq]	WU [m³ world eq. Deprived]
IL	−1.2E + 01	−4.3E-05	−9.3E-08	−2.9E-07	−2.3E-02	−8.1E-01
DCP	−1.4E + 01	−4.2E-05	−9.7E-08	−3.0E-07	−2.4E-02	−8.2E-01
DIM	3.6E + 00	−2.5E-05	−7.5E-08	−9.5E-07	−3.9E-02	−6.3E + 00

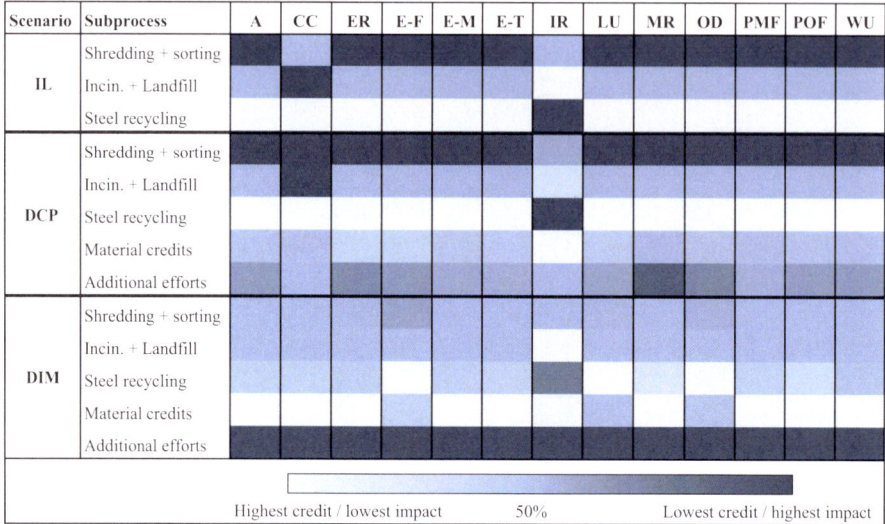

Scenario	Subprocess	A	CC	ER	E-F	E-M	E-T	IR	LU	MR	OD	PMF	POF	WU
IL	Shredding + sorting													
	Incin. + Landfill													
	Steel recycling													
DCP	Shredding + sorting													
	Incin. + Landfill													
	Steel recycling													
	Material credits													
	Additional efforts													
DIM	Shredding + sorting													
	Incin. + Landfill													
	Steel recycling													
	Material credits													
	Additional efforts													

Highest credit / lowest impact   50%   Lowest credit / highest impact

Fig. 3. Heatmaps of LCA results for subprocesses of different scenarios

In order to put some of the results into perspective, the results for the climate change impact values are shown together with the LCA results of the life cycle of the hybrid component in Fig. 4. The use stage is modelled acc. to fuel consumption calculated by using the method of fuel reduction values [20] and have been calculated with a simulation simulated within the project consortium. However, the LCA of the component shall not

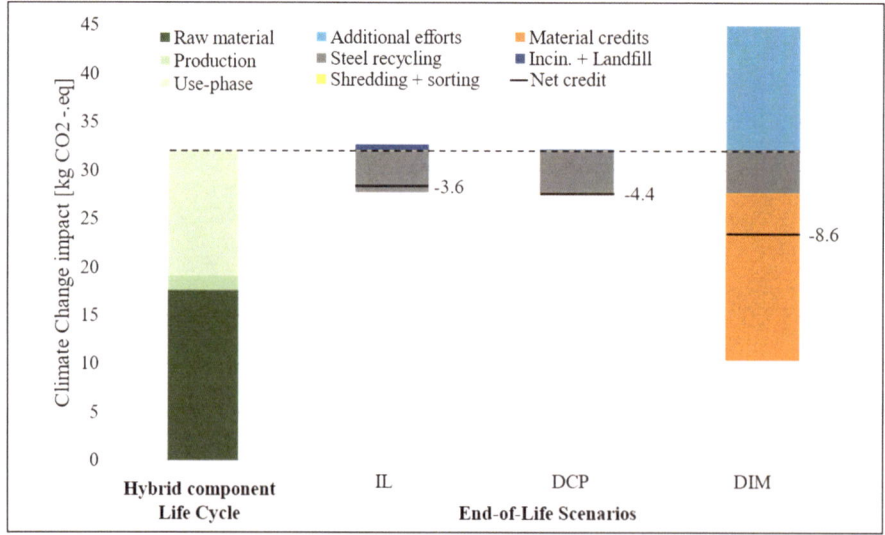

Fig. 4. Climate change impact of the different EoL scenarios in comparison to the remaining life cycle phases of the hybrid component

be described closer here, as the focus this paper is the difference between the EoL scenarios.

Here it can be seen that the overall credit for the scenario DIM is almost twice as high as the credits of the other two scenarios, where the scenario DCP only has a 22% higher credit than the scenario IL. This low difference stems from the relatively high share of the steel recycling credit. Overall, the EoL scenario DIM could reduce the climate change impact of the hybrid component by 26%.

In total the scenario DIM is favourable in 8 of 13 impact categories due to the high material credits for the PA6 GF 30 granule, but also has the highest impacts in the remaining 5 impact categories because of the large additional efforts. The scenario DCP is favourable in all impact categories to the scenario IL. When put into perspective to the environmental impacts of the overall life cycle, high reductions of the environmental impact can be achieved due to a feasible EoL strategy, which would be the scenario DIM for the climate change impact category.

5   Conclusions and Outlook

A LCA of three possible EoL scenarios of an automotive hybrid component made of steel and GFRP has been conducted. For the definition of these pathways different EoL processing technologies have been analysed and three possible EoL scenarios have been defined. It has been shown, that by using thermoplastic matrices instead of thermoset for FRP components, additional EoL pathways can be enabled.

The LCA results have shown, that all three scenarios result in credits in most of the analysed impact categories. The major shares of these credits stem from the recycling of the steel component, if the GFRP material is not recycled. Overall, the recycling of the GFRP material into an injection moulding granule is favourable in 8 of 13 impact categories. Especially when put into perspective to the overall life cycle of the hybrid component, the lever to reduce the environmental impacts through recycling can be quite high. Therefore, it is important to already consider the EoL when designing a component, to enable the most favourable EoL pathway, also with respect to the disassembly of the component group. Ideally, the application for the second life of the recycled material would already be considered in this design phase.

Within the definition of the best-case scenario of recycling of the GFRP material to produce PA6 GF30 injection moulding granule, assumptions were taken regarding the technological feasibility and the processing technology. These assumptions would have to be undermined by additional research. However, since the difference between environmental benefits are quite high compared to the other scenarios in many impact categories, it is likely that even with further additional efforts the recycling into injection moulding granule is to be preferred over the other pathways of incineration and landfill or the use in the cement kiln. Other recycling pathways such as pyrolysis and solvolysis have not been analysed and should also be analysed within future investigations to give a complete picture of the environmental impacts or benefits of all possible recycling pathways.

Acknowledgements. This research and development project is funded by the German Federal Ministry of Education and Research (BMBF) within the funding initiative "Research Campus – Public-Private Partnership for Innovation" (funding number: 02P18Q700) and managed by the Project Management Agency Karlsruhe (PTKA). The author is responsible for the content of this publication.

References

1. C. Herrmann, W. Dewulf, M. Hauschild, A. Kaluza, S. Kara, and S. Skerlos, "Life cycle engineering of lightweight structures," *CIRP Annals*, vol. 67, no. 2, pp. 651–672, 2018, https://doi.org/10.1016/j.cirp.2018.05.008.
2. V. K. Soo, P. Compston, and M. Doolan, "Interaction between New Car Design and Recycling Impact on Life Cycle Assessment," *Procedia CIRP*, vol. 29, pp. 426–431, 2015, https://doi.org/10.1016/j.procir.2015.02.055.
3. European Commission, "End-of-Life Vehicles—European Commission," End-of-Life Vehicles. Accessed: Feb. 20, 2024. [Online]. Available: https://environment.ec.europa.eu/topics/waste-and-recycling/end-life-vehicles_en
4. R. Geyer, B. Kuczenski, T. Zink, and A. Henderson, "Common Misconceptions about Recycling," *J of Industrial Ecology*, vol. 20, no. 5, pp. 1010–1017, Oct. 2016, https://doi.org/10.1111/jiec.12355.
5. T. Ekvall, M. Gottfridsson, M. Nellström, J. Nilsson, M. Rydberg, and T. Rydberg, "Modelling incineration for more accurate comparisons to recycling in PEF and LCA," *Waste Management*, vol. 136, pp. 153–161, Dec. 2021, https://doi.org/10.1016/j.wasman.2021.09.036.
6. R. Pant and L. Zampori, *Suggestions for updating the organisation environmental footprint (OEF) method*. Luxembourg: Publications Office of the European Union, 2019. [Online]. Available: https://doi.org/10.2760/424613
7. G. Wernet, C. Bauer, B. Steubing, J. Reinhard, E. Moreno-Ruiz, and B. Weidema, "The ecoinvent database version 3 (part I): overview and methodology," *Int J Life Cycle Assess*, vol. 21, no. 9, pp. 1218–1230, Sep. 2016, https://doi.org/10.1007/s11367-016-1087-8.
8. B. Steubing, D. De Koning, A. Haas, and C. L. Mutel, "The Activity Browser — An open source LCA software building on top of the brightway framework," *Software Impacts*, vol. 3, p. 100012, Feb. 2020, https://doi.org/10.1016/j.simpa.2019.100012.
9. H. Martens and D. Goldmann, *Recyclingtechnik: Fachbuch für Lehre und Praxis*. Wiesbaden: Springer Fachmedien Wiesbaden, 2016. https://doi.org/10.1007/978-3-658-02786-5.
10. M. Heibeck, M. Rudolph, N. Modler, M. Reuter, and A. Filippatos, "Characterizing material liberation of multi-material lightweight structures from shredding experiments and finite element simulations," *Minerals Engineering*, vol. 172, p. 107142, Oct. 2021, https://doi.org/10.1016/j.mineng.2021.107142.
11. R. Bernatas, S. Dagreou, A. Despax-Ferreres, and A. Barasinski, "Recycling of fiber reinforced composites with a focus on thermoplastic composites," *Cleaner Engineering and Technology*, vol. 5, p. 100272, Dec. 2021, https://doi.org/10.1016/j.clet.2021.100272.
12. P. Weißhaupt, "Faserverbundwerkstoffe: Zukunftsmaterial mit offener Entsorgung," Umweltbundesamt. Accessed: Feb. 01, 2024. [Online]. Available: https://www.umweltbundesamt.de/faserverbundwerkstoffe-zukunftsmaterial-offener
13. Y. Yang, R. Boom, B. Irion, D.-J. Van Heerden, P. Kuiper, and H. De Wit, "Recycling of composite materials," *Chemical Engineering and Processing: Process Intensification*, vol. 51, pp. 53–68, Jan. 2012, https://doi.org/10.1016/j.cep.2011.09.007.

14. A. Bernasconi, P. Davoli, and C. Armanni, "Fatigue strength of a clutch pedal made of reprocessed short glass fibre reinforced polyamide," *International Journal of Fatigue*, vol. 32, no. 1, pp. 100–107, Jan. 2010, https://doi.org/10.1016/j.ijfatigue.2009.02.001.

15. E. Kuram, "Investigating the effects of recycling number and injection parameters on the mechanical properties of glass-fibre reinforced nylon 6 using Taguchi method," *Materials and Design*, 2013.

16. M. E. Otheguy, A. G. Gibson, E. Findon, R. M. Cripps, A. O. Mendoza, and M. T. A. Castro, "Recycling of end-of-life thermoplastic composite boats," *Plastics, Rubber and Composites*, vol. 38, no. 9–10, pp. 406–411, Dec. 2009, https://doi.org/10.1179/146580109X12540995 045642.

17. P. Liu, F. Meng, and C. Y. Barlow, "Wind turbine blade end-of-life options: An economic comparison," *Resources, Conservation and Recycling*, vol. 180, p. 106202, May 2022, https://doi.org/10.1016/j.resconrec.2022.106202.

18. F. Perugini, M. L. Mastellone, and U. Arena, "A life cycle assessment of mechanical and feedstock recycling options for management of plastic packaging wastes," *Environ. Prog.*, vol. 24, no. 2, pp. 137–154, Jul. 2005, https://doi.org/10.1002/ep.10078.

19. A. J. Nagle, E. L. Delaney, L. C. Bank, and P. G. Leahy, "A Comparative Life Cycle Assessment between landfilling and Co-Processing of waste from decommissioned Irish wind turbine blades," *Journal of Cleaner Production*, vol. 277, p. 123321, Dec. 2020, https://doi.org/10.1016/j.jclepro.2020.123321.

20. J. Liebl, M. Lederer, and K. Rohde-Brandenburger, *Energiemanagement im Kraftfahrzeug: Optimierung von CO2-Emissionen und verbrauch konventioneller und elektrifizierter automobile.* in ATZ/MTZ-Fachbuch. Wiesbaden: Springer Vieweg, 2014. https://doi.org/10.1007/978-3-658-04451-0.

Author Index